IET MATERIALS, CIRCUITS AND DEVICES SERIES 38

System Design with Memristor Technologies

Other volumes in this series:

Volume 2	**Analogue IC Design: The current-mode approach** C. Toumazou, F.J. Lidgey and D.G. Haigh (Editors)
Volume 3	**Analogue–Digital ASICs: Circuit techniques, design tools and applications** R.S. Soin, F. Maloberti and J. France (Editors)
Volume 4	**Algorithmic and Knowledge-Based CAD for VLSI** G.E. Taylor and G. Russell (Editors)
Volume 5	**Switched Currents: An analogue technique for digital technology** C. Toumazou, J.B.C. Hughes and N.C. Battersby (Editors)
Volume 6	**High-Frequency Circuit Engineering** F. Nibler *et al.*
Volume 8	**Low-Power High-Frequency Microelectronics: A unified approach** G. Machado (Editor)
Volume 9	**VLSI Testing: Digital and mixed analogue/digital techniques** S.L. Hurst
Volume 10	**Distributed Feedback Semiconductor Lasers** J.E. Carroll, J.E.A. Whiteaway and R.G.S. Plumb
Volume 11	**Selected Topics in Advanced Solid State and Fibre Optic Sensors** S.M. Vaezi-Nejad (Editor)
Volume 12	**Strained Silicon Heterostructures: Materials and devices** C.K. Maiti, N.B. Chakrabarti and S.K. Ray
Volume 13	**RFIC and MMIC Design and Technology** I.D. Robertson and S. Lucyzyn (Editors)
Volume 14	**Design of High Frequency Integrated Analogue Filters** Y. Sun (Editor)
Volume 15	**Foundations of Digital Signal Processing: Theory, algorithms and hardware design** P. Gaydecki
Volume 16	**Wireless Communications Circuits and Systems** Y. Sun (Editor)
Volume 17	**The Switching Function: Analysis of power electronic circuits** C. Marouchos
Volume 18	**System on Chip: Next generation electronics** B. Al-Hashimi (Editor)
Volume 19	**Test and Diagnosis of Analogue, Mixed-Signal and RF Integrated Circuits: The system on chip approach** Y. Sun (Editor)
Volume 20	**Low Power and Low Voltage Circuit Design with the FGMOS Transistor** E. Rodriguez-Villegas
Volume 21	**Technology Computer Aided Design for Si, SiGe and GaAs Integrated Circuits** C.K. Maiti and G.A. Armstrong
Volume 22	**Nanotechnologies** M. Wautelet *et al.*
Volume 23	**Understandable Electric Circuits** M. Wang
Volume 24	**Fundamentals of Electromagnetic Levitation: Engineering sustainability through efficiency** A.J. Sangster
Volume 25	**Optical MEMS for Chemical Analysis and Biomedicine** H. Jiang (Editor)
Volume 26	**High Speed Data Converters** A.M.A. Ali
Volume 27	**Nano-Scaled Semiconductor Devices** E.A. Gutiérrez-D (Editor)
Volume 29	**Nano-CMOS and Post-CMOS Electronics: Devices and modelling** S.P. Mohanty and A. Srivastava
Volume 30	**Nano-CMOS and Post-CMOS Electronics: Circuits and design** S.P. Mohanty and A. Srivastava
Volume 32	**Oscillator Circuits: Frontiers in design, analysis and applications** Y. Nishio (Editor)
Volume 33	**High Frequency MOSFET Gate Drivers** Z. Zhang and Y. Liu

System Design with Memristor Technologies

Lauren Guckert and Earl E. Swartzlander, Jr.

The Institution of Engineering and Technology

Published by The Institution of Engineering and Technology, London, United Kingdom

The Institution of Engineering and Technology is registered as a Charity in England & Wales (no. 211014) and Scotland (no. SC038698).

First published 2018

The Institution of Engineering and Technology
Michael Faraday House
Six Hills Way, Stevenage
Herts, SG1 2AY, United Kingdom

www.theiet.org

British Library Cataloguing in Publication Data
A catalogue record for this product is available from the British Library

ISBN 978-1-78561-561-0 (hardback)
ISBN 978-1-78561-562-7 (PDF)

Typeset in India by MPS Limited

"To teachers and mentors who share not just their knowledge, but their passion."

Contents

Chapter 1

Introduction

Memristors were first hypothesized by Chua [1] in 1971. Memristors, or memristive devices, are nonvolatile circuit components that have a nonlinear relationship between electric charge and magnetic flux. These devices were hypothesized as the missing fourth circuit element, joining the long-standing resistor, inductor, and capacitor.

Since their original introduction, research has focused on developing and formulating memristive devices, which exhibit this nonlinear relationship. Beyond fundamental fabrication, research has pushed to create memristors with characteristics competitive with traditional complementary metal oxide semiconductor (CMOS) circuit components. These memristive devices have progressed to possess smaller form factor and faster switching times than current CMOS equivalents [2,3].

Orthogonally, research has explored various models and their applications for memristive devices [4–9]. Memristive devices encompass a wide range of different physical devices, each exhibiting distinct physical and behavioral characteristics. Thus, numerous models are needed to represent this range of devices and their unique properties. Since memristive devices are still a new concept, it is left to research to determine how exactly to represent these devices with models, and which devices are unique and useful enough to warrant a model. The research field surrounding device modeling for memristors enables many others in the field of memristor research to explore the applications and analysis of memristors and memristor-based designs.

Research into memristor applications attempts to answer the question, "how will memristive devices be used" rather than "how will memristive devices be made?" The answer to this question will solidify the longevity and practicality of memristors in the computer architecture field. Without this answer, it is unclear if memristors will be widely accepted as a front-runner of future technologies. Currently, the applications for memristors fall into two categories, memory and logic.

An important benefit of memristors is the ability to hold data with low area and high density as compared to alternative memory solutions. Thus, the majority of research in memristors has focused on their application to memory [3,10–13], specifically in a crossbar array. However, there are no inherent limitations on memristors that restrict their use as storage elements. Memristors have other characteristics such as dynamic programmability that makes them viable candidates for use in other aspects of system design. For example, recently HP Labs showed that it is possible to implement all Boolean logic functions on memristors. This study used the material implication, or IMPLY, operation [14] in a serialized manner to execute a logically complete set of Boolean functions.

The idea that memristors can perform logic drastically increased the design space for memristors. Nearly every component and logic block in a system is now a potential candidate for a memristor-based design.

Since this finding, researchers have begun to explore arithmetic applications for memristors [13,15–32] in stand-alone logic blocks. Some of these works have focused on the IMPLY operation [13,15–21] and have implemented designs from Boolean logic to adders. Recently, some of these IMPLY-based implementations have come under scrutiny for their complex circuitry and multistep serialized operation [30,31]. Efficiency is becoming ever more critical in future system technologies and some researchers are concerned that the IMPLY approach has not yet caught up to these needs. This has led to an exploration of other research paths to establish alternative memristor implementations for Boolean operations. These include hybrid approaches [22–24], MAGIC gates [12], threshold gates [24,26,27], and others [25,28,29].

This area of research is still in its infancy. Much, if not all, of the preliminary work that has been done has focused on individual logic gates and small circuits such as adders. These works serve an important role as a proof-on-concept for the strengths of memristor-based logic design. However, additional strides are needed to produce larger arithmetic units that are useful in modern, complex system.

This text is focused on that next step, exploring the design of practical, complex, real-world arithmetic units using memristor-based logic. In this work, four distinct approaches to memristor-based arithmetic unit design are proposed and described that extend and optimize upon the current state-of-the-art designs. The four approaches are IMPLY, hybrid-CMOS, threshold gates, and MAD (memristors-as-drivers) gates. These four approaches were selected because each has a unique set of properties in terms of cost and benefit.

As a set, these four approaches represent a wide spectrum of trade-offs in system design, from performance- to area-critical. For a given application, one or more of these approaches may prove to be the most beneficial, while others may have shortcomings. One goal of this text is to highlight the insight behind when memristor-based logic is appropriate and what approaches should be used in different system scenarios.

Implementations of numerous arithmetic units, including adders, multipliers, and dividers, are presented for each of these approaches. Specifically, the adders studied are a ripple-carry adder, a carry-lookahead adder, a carry-select adder, and a conditional-sum adder. The multipliers studied are a shift-and-add multiplier, a Booth multiplier, an array multiplier, and a Dadda multiplier. The dividers studied are a binary restoring divider, an SRT divider, and a Goldschmidt divider. These designs were selected on the basis of their proclivity to memristor optimizations as well as an encapsulation of a wide range of design trade-offs. It is also shown how the MAD gates can be incorporated into a crossbar structure, similar to the IMPLY circuit.

For each arithmetic unit, optimizations are performed on each of the four distinct implementations. In many cases, the traditional CMOS circuitry for the arithmetic unit has been transformed to accommodate memristors and their properties. The ideal or optimized circuitry for a CMOS design may not be the ideal, optimized circuitry for a memristor-based design. These differences are emphasized in this text.

Additionally, some memristor-based designs can be pipelined with minimal to no changes to the circuitry. In these cases, the circuitry has been modified as necessary to allow for pipelining, a key benefit of some memristor-based logic approaches.

Each design is analyzed in terms of delay and complexity and compared against the other memristor-based approaches. The memristor-based designs are also compared against the equivalent CMOS design. These comparisons provide insight into the strengths and consequences of both CMOS and memristor-based design and motivate the designer to use a critical eye when deciding the approach device and approach for a given design. Some of the implementations discussed have prior state-of-the-art in the literature. For these cases, the implementations of the previous literature are compared with the designs presented in this text. The comparison is done in terms of performance, measured in terms of both delay and component costs.

This text presents prior work in this field, the proposed arithmetic designs, and results and suggestions for future work. Chapter 2 describes prior work on memristor-based arithmetic implementations via IMPLY operations, hybrid-CMOS gates, threshold gates, and others. Chapter 3 introduces various adder arithmetic units, including ripple-carry adders and carry-lookahead adders. These units are described at a high level, abstracted away from the underlying CMOS or memristor technologies. Chapter 4 introduces multiple popular multiplier arithmetic units, describing the designs in a similar manner as Chapter 3. Chapter 5 completes the set by discussing common divider arithmetic units. Chapters 6–8 are dedicated to memristor-based arithmetic designs. Chapter 6 focuses on adder arithmetic units, implementing each adder given in Chapter 3 for the arithmetic approaches explained in Chapter 2. Chapters 7 and 8 follow a similar pattern, implementing the multiplier and divider arithmetic units using the various arithmetic approaches outlined in Chapter 2. Chapter 9 serves to motivate the reader for possible future directions for exploration, research, and applications in the field of memristors. Finally, Chapter 10 concludes the text.

1.1 Memristor fundamentals

Memristors were suggested by Chua [1] in 1971 as a two-terminal device with properties that complement the other fundamental electronic components: the capacitor, inductor, and resistor.

The capacitor relates charge to voltage ($dq = Cdv$), the resistor relates voltage and current ($dv = Rdi$), and the inductor relates current and magnetic flux

($d\varphi = Ldi$). The memristor is the final element, which relates magnetic flux and charge. The relationship between magnetic flux and charge is

$$M(q(t)) = \frac{d\varphi(q)}{dq}$$

where $M(q(t))$ is the resistance of the device as it relates to current. The voltage across the memristive device then becomes

$$v(t) = M(q(t)) * i(t)$$

The value, or resistance, of a memristive device (or resistive RAM) at a given moment depends not only on the current passing through it, but also the past. The formulas for a current-controlled memristor can be written as [33]

$$v(t) = r(x, i) * i(t)$$
$$\frac{dx}{dt} = f(x, i)$$

where t is the time, $r(x,i)$ is the device resistance (or memristance), $i(t)$ is the current, and $v(t)$ is the voltage. The variable x can be thought of as the width of the memristive device's conduction channel, which will be discussed in detail later in this section. The power consumption of the device is similar to the traditional equation involving resistance:

$$P(t) = i(t)^2 M(q(t))$$

When the voltage in the resistance is not changing, this simplifies to the traditional I^2R calculation.

Memristors have many properties that make them attractive as storage elements. First, they are nonvolatile devices with high density for easy application to the crossbar context. In addition, memristors are very reliable [3], and the write time is similar to conventional DRAM. Recent advances in the research of fabrication of these devices have produced memristors with smaller form factor and faster switching times [4,5] and memristors that are compatible with CMOS. All these benefits have led to an increase in the exploration of memristors as a future technology [6–9].

In the traditional Von Neumann architecture model, the compute cores and memory are physically separated. As the number of cores and memory controllers grows and a system becomes more complex, this separation increases. Thus, for many of today's big-data applications, a major portion of computational energy and time is spent moving data between the units. To add to the overhead, modern memories run at lower speeds than the cores. The result is a "memory wall," and the major performance bottleneck is moving the data between memory and computation. This has encouraged research in the area of logic-in-memory rather than the traditional Von Neumann structure [3,10–13,34–40].

Logic-in-memory architectures are as their name implies: rather than moving the data between memory and the computation, the computation is performed directly in memory. This drastically reduces the energy and time spent in moving data and improves the computational efficiency of systems. However, logic-in-memory architectures have challenges themselves. For example, it is more difficult to perform complex logic operations and optimizations seen in the state-of-the-art Von Neumann systems. Luckily, memristors can alleviate these challenges.

Memristors offer a promising candidate in the realm of logic-in-memory due to the inherent characteristics, CMOS compatibility, and natural implementation into a crossbar. Memristors consume no static power, have higher endurance than flash, read and write times on the order of 10–30 ns (comparable to DRAM) and high density. They also enable stacking, further improving density, and minimizing data movement overhead. A summary of these qualities in comparison with other conventional memory technologies is shown in Table 1.1.

PCM and STT-RAM are additional emerging technologies for memory that are being explored in parallel to memristors. PCM, or phase-change memory, is another emerging nonvolatile memory. It is based on the notion of manipulating temperatures to change state. STT-RAM uses the principles of spin-transfer torque to implement memory. STT-RAM is an example of magnetic memory, using magnetic fields to change the value of a memory cell.

Memristors have a lower area per bit than PCM or STT-RAM and are similar to DRAM and flash memristors. The energy per bit is lower than all of the other technologies. The read-and-write times are also among the fastest across the technologies. One concern with memristors is their endurance, or lifetime, in a system because it is still not as strong as DRAM or STT-RAM. 3D capability refers to the ability to stack the technology in a memory structure, and it is one of the strongest benefits of memristors.

In addition to these characteristics of memristors, one key characteristic is their ability to both remember and change their resistance value. This enables memristors to behave as not only storage elements, but logic elements too, allowing for logic-in-memory computations. This capability is unique to memristive devices and has led to a surge in memristor popularity in arithmetic applications.

Table 1.1 Memristor technology comparison

	Memristor	**PCM**	**STT-RAM**	**DRAM**	**Flash**	**Hard drive**
Chip area per bit (F^2)	4	8–16	14–64	4–8	4–8	n/a
Energy per bit (pJ)^2	0.1–3	2–100	0.101	2–4	10–10^4	10^6–10^7
Read time (ns)	<10	20–70	10–30	10–50	25,000	5–8*10^6
Write time (ns)	20–30	50–500	13–95	10–50	20,000	5–8*10^6
Retention	>10 years	<10 years	Weeks	<second	~10 years	~10 years
Endurance (cycles)	~10^{12}	10^7–10^8	10^{15}	>10^{17}	10^5–10^6	10^{15}
3D capability	Yes	No	No	No	Yes	n/a

1.1.1 Types of memristors

Since the introduction of memristive characteristics, there has been a breadth of research into creating memristive devices, including unipolar memristors [41] and memristors with SiO_x- and TaN-based gate materials [2,34,35]. This section describes just a few of the types of memristors that have been successfully described in the literature.

1.1.1.1 Unipolar vs. bipolar memristors

Memristive devices are classified as either unipolar or bipolar, depending on their switching characteristics. If opposite-polarity voltages/currents are required for the set and reset operations, then the device is a bipolar device. If same-polarity voltages/currents are required for the set (moving from off to on) and reset (moving from on to off) operations, then the device is a unipolar device. These *IV* characteristics can be visualized in Figure 1.1 [42].

Figure 1.2(a)–(d) shows bipolar devices (a and b) and unipolar (or nonpolar) devices (c and d). The light gray represents the conduction channel, and the dark gray represents the surrounding gate materials. The arrows represent how the conduction channel changes when currents are applied. For example, for the bipolar memristor in Figure 1.2(a), the conduction channel grows and shrinks vertically as currents are applied. In contrast, the unipolar memristors grow/shrink horizontally.

These memristive devices can switch due to both electric field and joule heating and the electron transport mechanisms are shown in Figure 1.2(e)–(i). Usually, when electric-field effects cause the switching behavior, the memristor is bipolar and when joule heating causes the switching behavior, the memristor is unipolar. In unipolar devices, a change in temperature can lead to the formation or destruction of conducting channels in the otherwise insulating oxide layer.

1.1.1.2 SDC memristors

The first patented memristor was a chalcogenide ion-conducting memristor, which later became a self-directed channel (SDC) memristor. This memristor is

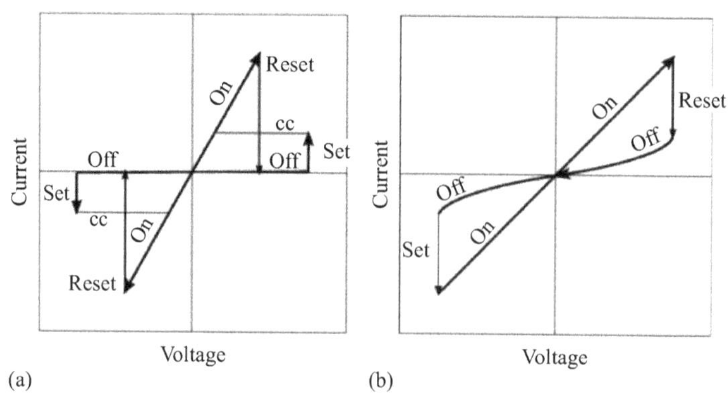

Figure 1.1 IV Characteristics for (a) typical unipolar ReRAM and (b) typical bipolar ReRAM memristors [42]

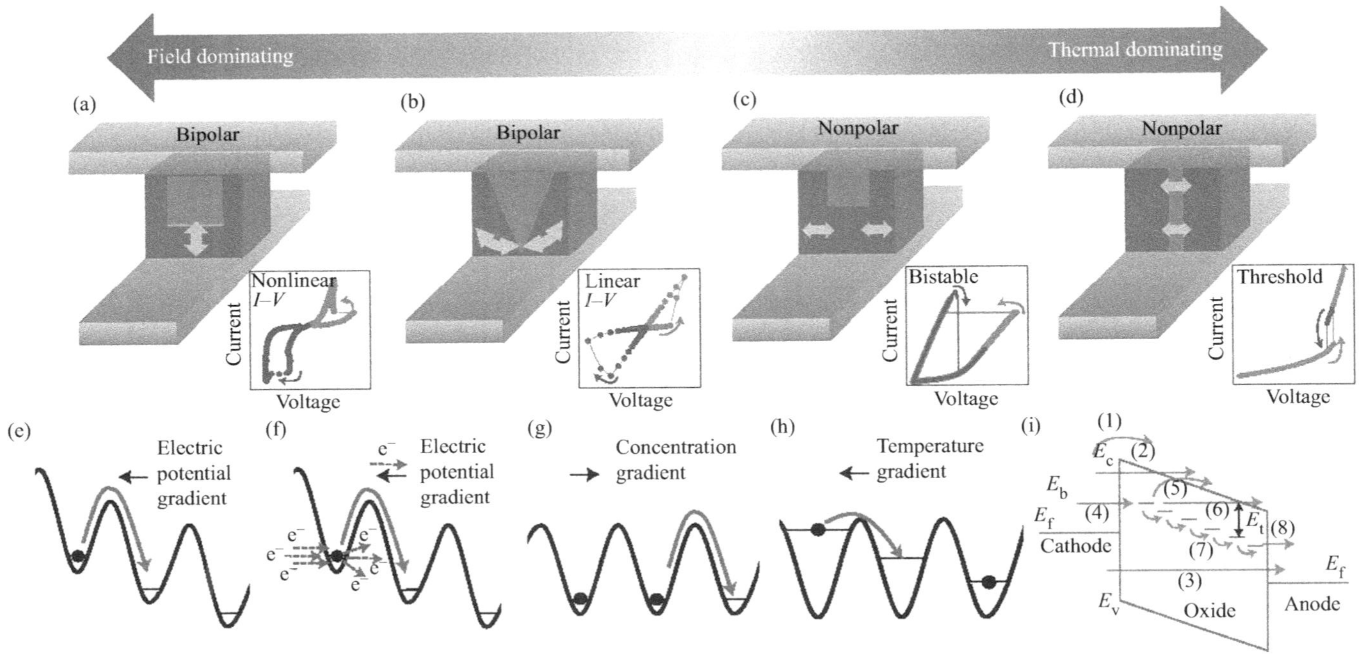

Figure 1.2 Switching in anion-based memristive devices [43]

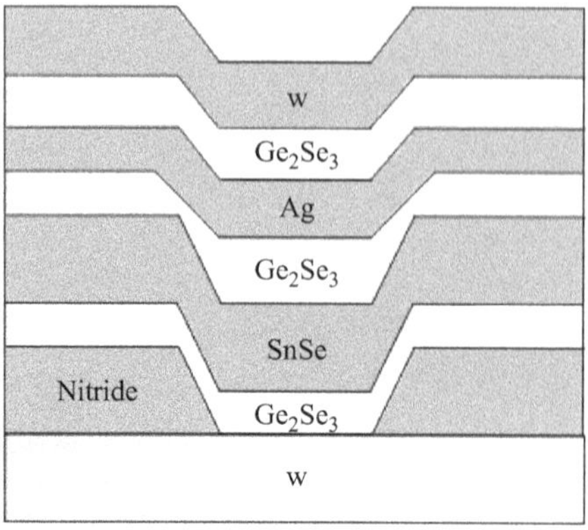

Figure 1.3 Self-directed channel (SDC) memristor

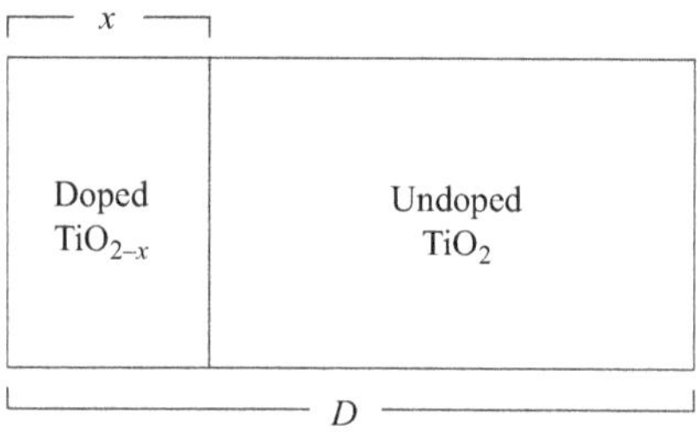

Figure 1.4 Hewlett-Packard memristor device showing the doped TiO_{2-x} and undoped TiO_2 regions

capable of operating at very high temperatures (~250 °C). The memristor consists of Ge_2Se_3 [44].

Figure 1.3 shows a diagram of the layers in an SDC memristor.

1.1.1.3 Titanium dioxide memristors

In 2008, memristive properties were identified by HP in nanoscale titanium dioxide memristors [14]. A visual representation of the TiO_2 memristor implemented by HP is shown in Figure 1.4.

The memristor has two regions, an undoped region and a doped region, sandwiched between two electrodes. The undoped region (TiO_2) is insulating while the doped region (TiO_{2-x}), which has positive oxygen vacancies, is conducting. In Figure 1.4, x represents the width of the doped region and D represents the full width of both regions. In today's memristors, $D = {\sim}3$ nm. The resistance of a memristor is the sum of the resistances of the two regions. The resistance can be adjusted by changing the current through the memristor, known as the compliance current, which in turn affects these regions in the conduction channel in the memristor.

If a positive voltage is applied, the oxygen vacancies in the TiO_{2-x} region are repelled into the TiO_2 region. This changes the TiO_2 region into TiO_{2-x}, increasing the conduction channel (width x) and decreasing resistance. As the doped TiO_{2-x} region grows, the internal memristance approaches R_{on}, or low resistance, turning the device on. If a negative voltage is applied, the undoped TiO_2 region grows. As the undoped region grows, the internal memristance approaches R_{off}, or high resistance, turning the device off. This is shown in Figure 1.5.

In these memristive devices, the memristance at a given time is

$$M(q(t)) = R_{off}\left(1 - \frac{\mu_v R_{on}}{D^2} q(t)\right)$$

where R_{off} is the high resistance of the memristor, R_{on} is the low resistance of the memristor, D is the width of the doped region in the device, and μ_v is the mobility of the dopants in the memristor. The memristor behavior produces the *IV* characteristics seen in Figure 1.6.

The hysteresis curve shows how the memristor changes states from high resistance to low resistance and vice versa at higher magnitude voltages. Notice that the change in resistance has a nonlinear response. When the voltage magnitude is large enough, $x = 0$ and $M = R_{off}$, or $x = D$ and $M = R_{on}$. When this occurs, the resistance remains constant until a voltage of opposite magnitude is applied across the memristor.

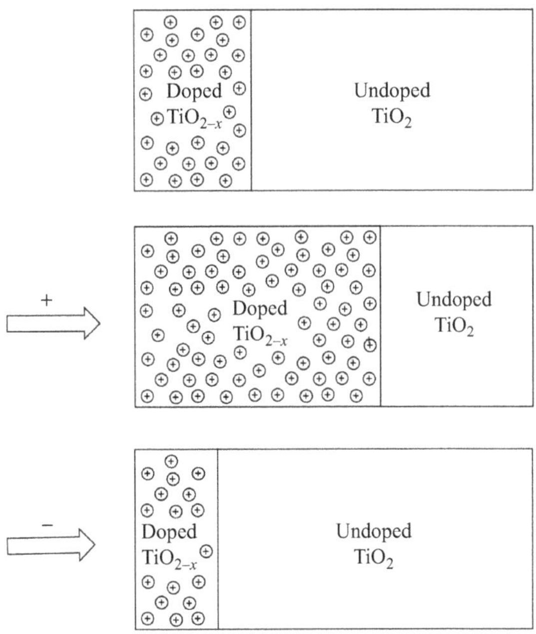

Figure 1.5 The impact of positive and negative current on the doped (TiO_{2-x}) and undoped (TiO_{2-x}) regions of a memristor

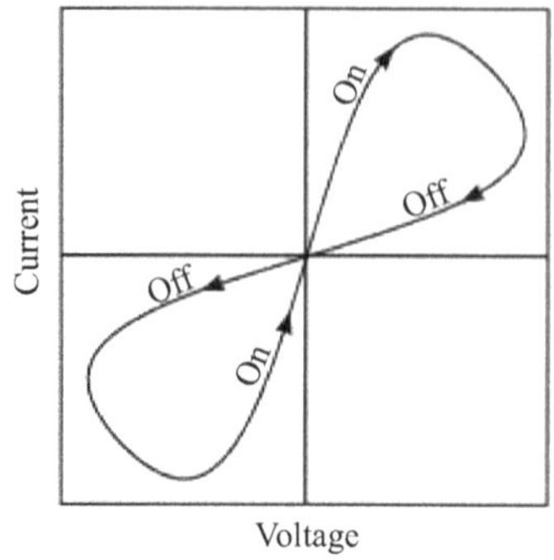

Figure 1.6 Memristor IV characteristics [42]

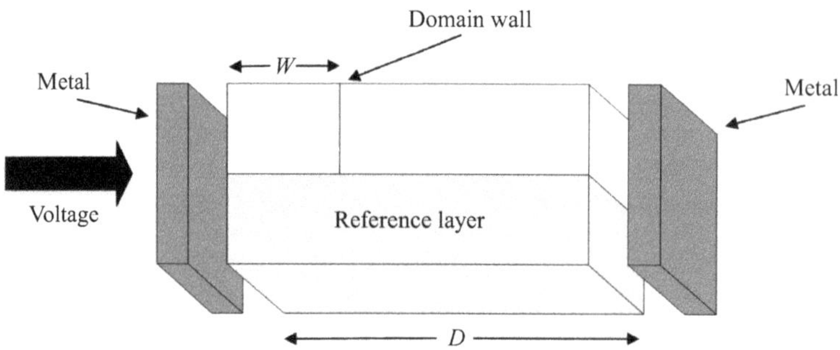

Figure 1.7 Spintronic memristor diagram

1.1.1.4 Spintronic memristors

In 2009, a group of researchers at Seagate Technology introduced various approaches to magnetic memristors using spintronics [45]. Spintronics relates the spin of electrons to the magnetic behavior and charge in solid-state devices. A spintronic memristor has layers placed between two metal plates. The first layer is a reference layer that remains unchanged. The second layer is the free layer that consists of two regions. In the first region, all of the electrons have a spin in a common direction. In the second region, all of the electrons spin in the opposite direction. In this way, a "domain wall" is created between the two regions. As current is applied to either terminal of the device, the electrons being injected have a spin, which in turn affects the size of each of these regions and thus, the resistance [46]. A spintronic memristor can be visualized as shown in Figure 1.7.

1.1.2 Memristor logic

The interesting resistive properties of memristors have led to research into their use in logic. Since the resistance can be changed, it is natural to use this resistance to represent a desired logical value in the context of logic operations. For example, each possible resistance of the memristor could represent a unique stored value. Consider a memristor with a resistance range from 1 to 100 kΩ. Let high resistance $R_{off} = 100$ kΩ and low resistance $R_{on} = 1$ kΩ. The full resistance range is 100 k – 1 k = 99k resistance values. Thus, in an ideal model, this single memristor would be able to represent 99 k unique values with its resistance. However, current technology is not capable of realizing this vision. Instead, the current state-of-the-art research limits the granularity of a memristor to two or four distinct values, dividing the resistance range into quadrants. If this memristor were used to represent four values, then the ranges would be divided as follows:

Value	Resistance values
00	$3(R_{off} - R_{on})/4\ \Omega < R_{off}$
01	$(R_{off} - R_{on})/2$ to $3(R_{off} - R_{on})/4\ \Omega$
10	$(R_{off} - R_{on})/4$ to $(R_{off} - R_{on})/2\ \Omega$
11	R_{on} to $(R_{off} - R_{on})/4\ \Omega$

Notice that the largest resistance values represent 00, and the smallest resistance values represent 11. As the number of resistance regions increases, the regions shrink and the difficulty in differentiating the regions increases. The sensitivity and reliability of the memristors can begin to be an issue if the regions are too small to distinguish. Thus, it is preferential to select a large initial range between R_{on} and R_{off}. As a rule, R_{off}/R_{on} can be around 100.

For the purposes of this text, memristors will have two resistance regions to represent a binary value, "0" or "1." R_{on} will be 1 kΩ and R_{off} will be 100 kΩ. The threshold resistance R_{th} will be selected as $(R_{on} + R_{off})/2$ such that $R_{mem} < R_{th}$ is low resistance, and $R_{mem} > R_{th}$ is high resistance. Applying current through the memristor can either set (to "1") or reset (to "0") the value. The *IV* characteristics for a memristor under this operation are shown in Figure 1.8.

The voltage, V_{reset}, puts the memristor into a high resistance (R_{off}), or open, state corresponding to "0." This is represented by the less steeply sloped line in the figure, corresponding to a higher R value from $R = V/I$. The memristor value then remains unchanged unless a voltage V_{set} puts the memristor into a low resistance (R_{on}), or closed, state corresponding to "1." Once this occurs, the memristor then falls on the more steeply sloped line in the figure until again V_{reset} is applied. V_{cond} is a voltage that is necessarily less than V_{set} and selected such that its magnitude is not large enough to change the internal resistance of the memristor. This voltage is used as the "read" or "sense" voltage, to obtain the current resistance value of the memristor without changing it.

The magnitude of all these voltages will affect the switching frequency of the memristors, with a larger voltage causing a faster switching frequency. However, it will also increase the power consumption. Example values for these voltages are $V_{cond} = -1.6$ V, $V_{set} = -2.5$ V, and $V_{reset} = 5$ V.

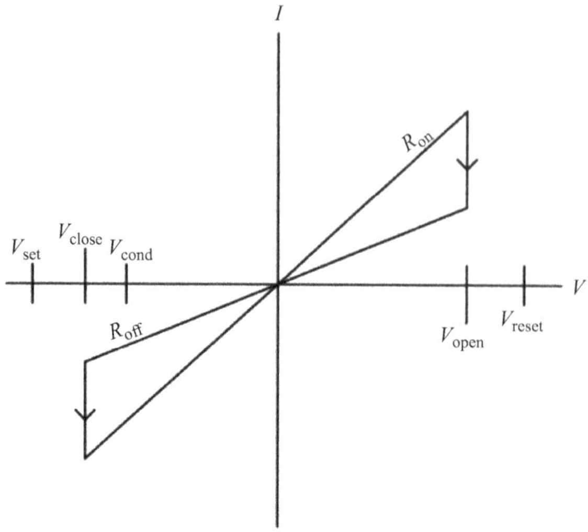

Figure 1.8 Idealized IV characteristics for a bipolar memristor

Applying these voltages will switch the value of a memristor between a logical "0" and "1" and vice versa. However, this only provides the power to set a value into a memristor and retrieve it later. It does not enable any computation on the values within this memristor. By applying combinations of these voltages simultaneously to multiple memristors, logical operations can be operated and stored into memristor results. For example, using these three voltages, an IMPLY operation can be performed.

1.2 Note on simulation methodology

For this research, popular tools and methodologies were employed to allow for reproducibility. The Cadence Virtuoso tool, version 6.1.6-64b, was used for all simulations. The most current version available at the time of the research was chosen. Virtuoso satisfied the basic requirements for a basic memristor simulation—a CMOS component library, the ability to measure power, delay and component counts, use of external stimulation for the driver signals in the designs, and a memristor model. The 45-nm NCSU FreePDK suite was chosen for the CMOS component libraries due to its availability and wide use across prior literature.

There has been substantial prior work in CMOS arithmetic designs and in memristor technologies, but very minimal research in the intersection of these two fields. The majority of the prior work in memristor-based logic has focused on the memory sphere, looking at ways to leverage the small form factor and high density of memristors to perform logic-in-memory. In most cases, the memristors are used in a crossbar structure in which each memristor is used for a memory cell, uniquely identified by its row and column. Outside of the context of memory, there has been minimal research in memristor-based logic. The prior work on arithmetic units has focused on conceptual proofs and simple circuit examples, often optimized for use in memory. Examples of these are IMPLY, hybrid-CMOS, and threshold gates. Adding to the difficulty, memristor models are still in their infancy, with many different models existing for various applications and a lack of support for their use in simulation.

This chapter introduces the memristor and describes its characteristics, behavior, strengths, and weaknesses. It also focuses on the various models that are used to represent memristive devices and the commonalities and differences between them. Finally, the chapter discusses how memristors can be used in a crossbar and to perform logic.

1.3 Memristor models

Numerous memristor models have been proposed and used, each with distinct properties [4–9,47–49]. Some of the existing models are the linear ion drift model, the nonlinear ion drift model, the Simmons Tunnel Barrier model, and the TEaM (ThrEshold Adaptive Memristor) model [7].

The linear drift model [5] represents memristive devices with a doped TiO_{2-x} region (width w) and an undoped TiO_2 region. The total width of both regions is D. This model assumes that there is average ion mobility and represents the doped and undoped regions as two resistors. The nonlinear drift model is similar in its representation but tries to represent fabricated memristive devices more closely. It is asymmetric, representing both a tunneling phenomenon as well as an ideal diode equation. In the Simmons Tunnel Barrier model rather than using two resistors as the underlying constructs, there is a resistor in series with an electron tunnel barrier. This model is widely accepted as the most accurate model for current memristive devices.

The TEaM model [7] is capable of representing each of these models as well as others. It accepts any current voltage relationship and uses simplified expressions without detrimental loss in accuracy. Other models have more complex expressions, but they are less general and are often compute intensive. For larger circuits and simulations, this can become a practical challenge, requiring large amounts of memory and simulation time with multiple memristors. The TEaM model [7] simplifies the derivative of the memristor's conduction channel width x as

$$\frac{dx(t)}{dt} = \begin{cases} k_{\text{off}}{}^{*}\left(\dfrac{i(t)}{i_{\text{off}}} - 1\right)^{a_{\text{off}}} * f_{\text{off}}(x), & 0 < i_{\text{off}} < i \\ 0, & i_{\text{on}} < i < i_{\text{off}} \\ k_{\text{on}}{}^{*}\left(\dfrac{i(t)}{i_{\text{on}}} - 1\right)^{a_{\text{on}}} * f_{\text{on}}(x), & i < i_{\text{on}} < 0 \end{cases}$$

k_{off}, k_{on}, a_{off}, and a_{on} are all constants set as parameters by the designer to create the desired nonlinear dependence on current. i_{off} and i_{on} are current thresholds: no change is made to the internal channel width when the current magnitude is below these thresholds. f_{off} and f_{on} are functions that represent the dependence on the width of the conduction channel itself. The model represents the change in memristance with an exponential dependence on x. Let the bounds of the width of the conduction channel be called x_{off} and x_{on}. Now, the voltage and current relationship can be simplified and written as

$$v(t) = R_{on}e^{(\lambda/(x_{off}-x_{on}))(x-x_{on})} * i(t)$$

where R_{off} is the high resistance of the memristor, R_{on} is low resistance, and λ is defined as [7]

$$e^{\lambda} = \frac{R_{off}}{R_{on}}$$

By exposing these constants and parameters to the user, the TEaM model can be easily fitted to other popular memristive models. Another key benefit of the TEaM model is its compatibility with CMOS. This is necessary for simulations and larger, more complex circuits and hybrid systems. Also, the TEaM model is one of the few openly available models that offers a Verilog description for schematics. This is also the model used in many research works, allowing for direct comparisons to be made. Because of these benefits, the TEaM model is used for the simulations in this text.

The TEaM model has a nondestructive read operation, a characteristic crucial for logic and memory applications. This allows the value of the memristor to be read (or sensed) without changing its value. Its internal state changes significantly when a high magnitude voltage is applied, but does not change or changes minimally when a lower magnitude voltage is applied. The model also has fast read-and-write times, a characteristic that becomes increasingly important as the number of accesses to the memristor increases.

In order to use the TEaM model, various memristor parameters must be chosen carefully. Two such parameters are the resistances R_{high} and R_{low}, where $R_{high} \gg R_{low}$ and a memristor with R_{high} resistance is considered in the 0 state, and a memristor with R_{low} resistance is in the 1 state. The specific values of R_{low} and R_{high} depend on other properties of the memristors such as doping width and internal thresholds. A full list of the memristor parameters is given in Table 1.2.

The parameters given in Table 1.2 are consistent for all designs discussed in this work and were chosen to match prior work.

All schematics were designed with the Virtuoso Schematic Editor L. All memristors and any other relevant values in the designs are initialized to 0, or high resistance at $t = 0$ ns. All of the driver signals (time and voltage) are generated using external files written in Microsoft Excel. All simulations were performed with transient analysis in the Virtuoso ADE L and the Spectre simulator. Specific values selected for voltages and other parameters are discussed for each design approach in the corresponding future sections.

When implementing memristor-based designs, it is not practical to directly compare to the traditional CMOS design and sometimes against other memristor designs. This is for multiple reasons. First, the concept of a "gate delay" does not

Table 1.2 List of selected parameters for the TEaM memristor model

Parameter	Value
R_{high}	100 kΩ
R_{low}	1 kΩ
D	3e−9
U_v	1e−15
w Multiplied	1e9
p Coefficient	2
J	1.5
Window noise	1e−18
Threshold voltage	0
C_{off}	3.5e−6
C_{on}	4e−5
I_{off}	20e−6
I_{on}	−20e−6
X_c	107e−11
B	0
A_{on}	1.8e−9
A_{off}	1.2e−9
K_{on}	−1
K_{off}	10
Alpha_{On}	1
$\text{Alpha}_{\text{Off}}$	1
V_{on}	−1.78
V_{off}	0.0115
IV relation	1
X_{on}	3e−9
X_{off}	0

exist in an IMPLY or MAD design since they are based on the applications of voltages to perform operations. Thus, the step delay in these designs is not equivalent to the time delay of a gate in CMOS or other memristor-based designs. Second, memristors themselves are still largely not understood in terms of their interference and other practical limitations in a circuit. Thus, the actual dimensional area requirements of the memristor model are not available, which makes area comparisons between designs impractical. Third, the components being used are often not the same and it is challenging to find a fair, impartial means for comparison. Last, the support circuitry, which should be included in the component area, counts for the designs are not always clear. For example, in the logic-in-memory context, there is disagreement on whether driver logic should be included and to what extent. Beyond driver logic, there is discussion on whether the logic to implement and control the driver logic should be measured. In this text, the data path driver logic is reported and included in component counts, but the control logic for the drivers is not.

1.4 Crossbar technology

A crossbar consists of a matrix of rows and columns that together make up a memory structure. Each juncture between a row *r* and a column *c*, called a crosspoint, can be seen as a switch. Thus, for a crossbar with *R* rows and *C* columns, there are a total of *R**C* switches. When a voltage is applied to a row *r* and a column *c*, the switch at location (*r*,*c*) is activated.

Early crossbars consisted of metal wires with various mechanisms for creating contact between the row and column. However, crossbars have recently gained attention for their proclivity toward use with memristors. A memristor can be used at each juncture to create a connection between the row wire and column wire and each crosspoint. Then, when a voltage is applied to the column *c* and ground is applied to row *r*, current will flow through the memristor at (*r*,*c*). This can be visualized in Figure 1.9.

Each memristor can be used as a memory cell by representing either the value "0" or "1" when voltages are applied to its row and column. Now, a crossbar matrix with *R* rows and *C* columns has a total or *RC* memristors, or *RC* memory cells. Further details on memristor crossbar operation are given in the following chapter.

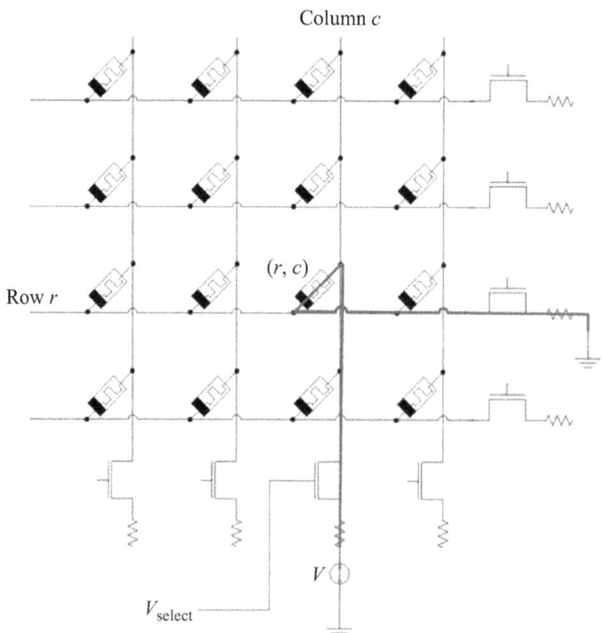

Figure 1.9 Single bitcell in a memristor crossbar

Chapter 2
Memristor logic gates

There are numerous approaches to implementing logic with memristors. The most popularized approach is the IMPLY operation [14], most commonly known from HP Labs' groundbreaking study. However, since this study, numerous other approaches have been proposed, which offer differing properties and trade-offs. Some of these approaches include memristors-as-drivers (MAD) gates [29], hybrid-CMOS gates [22,23], MAGIC gates [12], threshold gates [24–28], Zhang [3,10], and others. This chapter will explore two of these approaches in detail: the IMPLY and MAD operations because they have the widest applicability and the most research in their use. The remaining approaches will be discussed briefly for completion.

2.1 IMPLY gates

The IMPLY operation is the most popular approach to memristor-based logic in modern literature. The IMPLY operation was first demonstrated to successfully perform Boolean operations using the circuitry shown in Figure 2.1.

This circuitry is similar to that of a crossbar memory, with each memristor representing a separate cell. The circuitry requires two memristors, P and Q, to hold the input operands. These memristors are connected to a pull-down resistor, R_g.

The value of R_g is selected based on the memristor parameters, driver voltages, and other factors. The methodology is discussed thoroughly in prior work [20]. It is important that $R_{on}<R_g<R_{off}$ so R_g will be selected as 2 kΩ for this work.

As discussed, memristors have a threshold T such that if the voltage drop across the memristor is greater than T, its internal resistance will change. If the voltage across the memristor is less than T, the internal resistance will remain unchanged. To successfully perform an IMPLY operation P IMP Q, the value of P must remain unchanged and the value of Q must change to hold the result. Thus, R_g must be selected such that the voltage drop across P is less than T and the voltage across Q is greater than T when the operation executed and R_g lies between R_{on} and R_{off}.

To execute the IMPLY operation, voltages are applied via the driver circuitry. The operation P IMP Q is performed by applying a voltage V_{cond} to memristor P and a voltage V_{set} to memristor Q. The result ends up in the Q memristor, overwriting the original input operand. In other words, the memristance value of

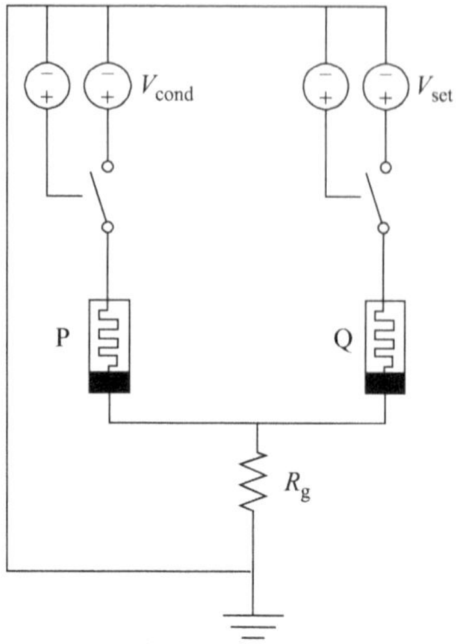

Figure 2.1 Circuitry for the IMPLY operation P IMP Q

Table 2.1 Truth table for the IMPLY operation

P	Q	P IMP Q
0	0	1
0	1	1
1	0	0
1	1	1

memristor Q will represent the result of the operation. A "0" will be represented with a resistance of R_{off}, and a "1" will be represented with a resistance of R_{on}. The corresponding truth table for the IMPLY operation is given in Table 2.1.

Consider each input combination in turn. If P is "0," or high resistance, when V_{cond} is applied then the voltage across the Q memristor approaches V_{set}, regardless of the internal resistance of Q. As defined earlier, V_{set} is a large enough voltage to change the internal resistance of the Q memristor to "1." Thus, for input combinations P = 0, Q = 0 and P = 0, Q = 1, the output of Q will be "1." If P is "1," low resistance when V_{cond} is applied, then the voltage across the Q memristor approaches $V_{set} - V_{cond}$ which by definition is not large enough to change the internal resistance of Q. Thus, Q remains unchanged as shown in the bottom two rows. Thus, for input combination P = 1, Q = 0, the output of Q is 0 and for input combination P = 1, Q = 1, the output of Q is 1. This circuitry effectively satisfies the IMPLY operation truth table.

Since the circuitry required for the IMPLY operation is two memristors and a pull-down resistor, the operation can be mapped directly onto a traditional memristor crossbar. A memristor crossbar is shown in Figure 2.2.

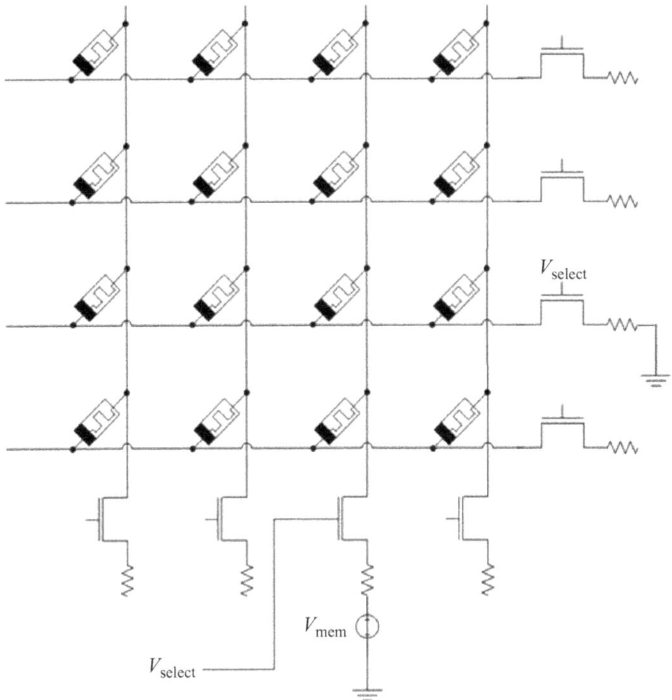

Figure 2.2 Traditional memristor crossbar

Figure 2.2 shows a 4 × 4 crossbar, with a total of 16 storage cells, each comprising a single memristor. There are many alternatives to the exact structure of a memristor crossbar, but this design is used for argument sake. To access cell (3, 3), the select lines for row 3 and column 3 are driven high, while row 3 is connected to ground and column 3 is connected to a voltage V_{mem}. This voltage either sets or reads the cell value. If $V_{mem} = V_{set}$, the memristor in (3, 3) is put into low-resistance state and the value "1" is written into the memory cell. This is shown with the path highlighted in Figure 2.3.

The pull-down resistor on each row behaves the same as the pull-down resistor R_g described for the IMPLY operation. Let $R_g = 2$ kΩ, $R_{off} = 100$ kΩ, and $R_{on} = 1$ kΩ. If $V_{mem} = V_{cond}$, the voltage across the memristor in (3, 3) is not enough to change its value and instead the value is read. To read the value of the memristor, the row line is probed. If the memristor in (3, 3) is a "0," or high resistance, then the voltage on row 3 when V_{cond} is applied is $V_{cond}/52$ or 0.03 V when $V_{cond} = 1.6$ V. The circuit is essentially a voltage division circuit for the two pull-down resistors (2 kΩ) and the memristor (100 kΩ). If the memristor in (3, 3) is a "1," or low resistance, then the voltage on row 3 is $2V_{cond}/5$ or 0.64 V based on the voltage divider circuitry explained earlier. By identifying these two voltages as 0 and 1, the memristor value can be read with the V_{cond} voltage. To obtain a greater differential in these voltages, R_g and V_{cond} can be changed accordingly.

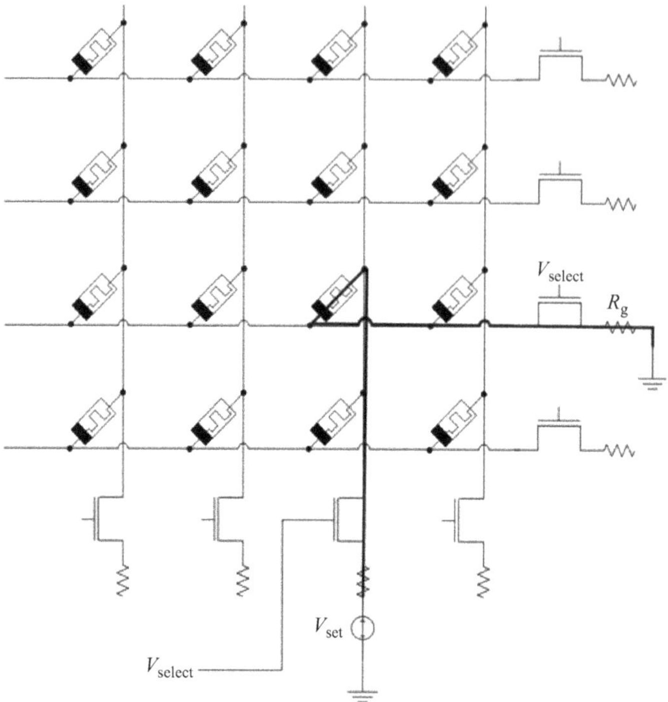

Figure 2.3 V_{set} *operation in a crossbar*

To perform an IMPLY operation within the crossbar, two memristor cells are used as input operands P and Q, and both V_{set} and V_{cond} are driven onto the memristors, respectively, as shown in Figure 2.4.

V_{cond} is driven on the P memristor, and V_{set} is applied to Q to perform P IMP Q while the select lines for the two column lines and common row line are driven high. The result ends up in Q. Note, there is great flexibility with the selection of operands for the IMPLY operation. The only limitation is that the memristor operands P and Q must exist on either the same row or the same column. There is no requirement that they be adjacent cells.

2.1.1 IMPLY Boolean logic

Notice from the IMPLY truth table that it is possible to perform a Boolean NOT operation simply by performing an IMPLY operation in which the value of Q begins as a "0." This takes a single step. Unfortunately, the remaining Boolean operations require numerous IMPLY steps and more than two memristors to obtain the final result. The execution of Boolean operations with the IMPLY operation and their costs are given in Table 2.2.

First, consider the IMPLY NOT operation. This operation is the simplest IMPLY operation and only requires one step. This circuit for this operation looks

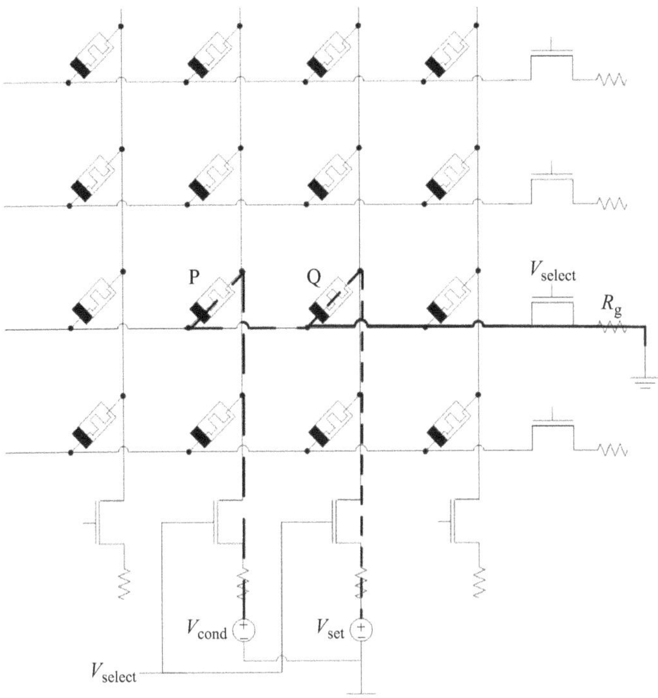

Figure 2.4 An IMPLY operation in a traditional memristor crossbar memory

Table 2.2 IMPLY implementations of Boolean operations [14]

Operation	Implementations	No. of steps	No. of memristors
NOT P	P IMP 0	1	2
P NAND Q	P IMP (Q IMP 0)	2	3
P AND Q	(P IMP (Q IMP 0)) IMP 0	3	4
P NOR Q	((P IMP 0) IMP Q) IMP 0	5	6
P OR Q	(P IMP 0) IMP Q	4	5
P XOR Q	(P IMP Q) IMP ((Q IMP P) IMP 0)	8	7

identical to the basic IMPLY circuitry in Figure 2.1. Let the input operand lie in the P memristor in Figure 2.1, and let Q be a reset memristor holding logical "0." Now, a single IMPLY operation is performed as described previously. The read voltage V_{cond} is applied to the P memristor, and the set voltage V_{set} is applied to the Q memristor. The IMPLY truth table shows that the result in the Q memristor at the end of the operation will be logical "0" if P = 1 and logical "1" if P = 0. This performs the NOT operation with the result in the Q memristor. It takes a single step and two memristors, one for each operand.

Now consider the NAND operation. An example circuit of extending the IMPLY circuitry to implement a NAND operation is shown in Figure 2.5.

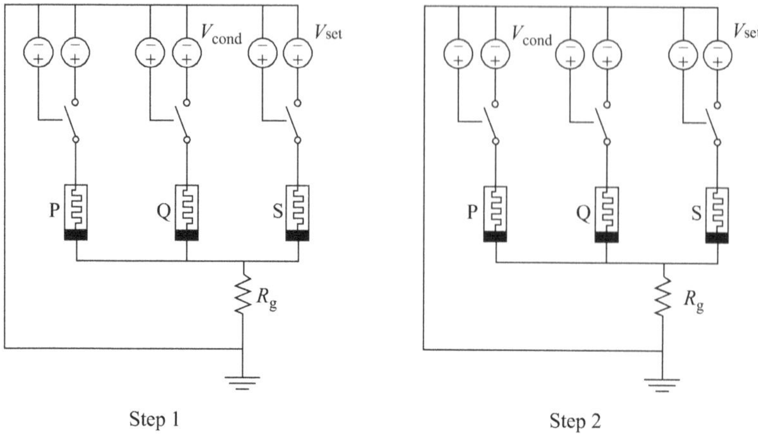

Figure 2.5 Circuitry for the IMPLY NAND operation

A third memristor, S, is added to the original IMPLY circuit, S is cleared, or initialized to 0, or high resistance. In the first step, Q IMP 0 is performed by applying the read voltage V_{cond} to the Q memristor and the write voltage V_{set} to the S memristor. No voltage is applied to the P memristor. At the end of this step, the result of Q IMP 0 lies in the S memristor. Effectively, a NOT operation has been performed on Q, and the result has been placed in S. In the second step, the IMPLY operation is performed on the P memristor and the result from step 1. P IMP (Q IMP 0) is performed by applying V_{cond} to the P memristor and V_{set} to the S memristor (where the result from step 1 is). Again, the result of step 2 lies in the S memristor. P NAND Q has been completed.

Once the result of P NAND Q is obtained, it is possible to obtain P AND Q by simply taking the inverse of the result of P NAND Q, performing NOT (P NAND Q) == P AND Q. A NOT operation is performed by applying V_{cond} to the initial operand and V_{set} to a cleared result memristor. This is shown with the introduction of memristor *T* in Figure 2.6.

Now consider the OR operation. First, V_{cond} is applied to P and V_{set} is applied to a cleared memristor, call it M_v holding the value 0, or high resistance. This executes P IMP 0 and the result lies in M_v. Second, V_{cond} is applied to M_v and V_{set} is applied to the Q memristor to perform (P IMP 0) IMP Q. The result lies in Q. This requires three memristors and two steps. However, Table 2.2 shows that the OR operation requires five memristors and four steps. This is because the value of Q is overwritten in the OR operation. Assuming this value is needed in future operations, the value of Q must be copied, or saved, before Q is overwritten. This is one key downside of the IMPLY operation.

To perform a COPY, two consecutive NOTs are performed. First, Q IMP 0 is performed onto a cleared memristor M_w such that M_w now holds (NOT Q). Then, M_w IMP 0 is performed onto another cleared memristor M_x such that M_x holds a copy of Q. Now, the OR operation can be performed. In this way, rather than

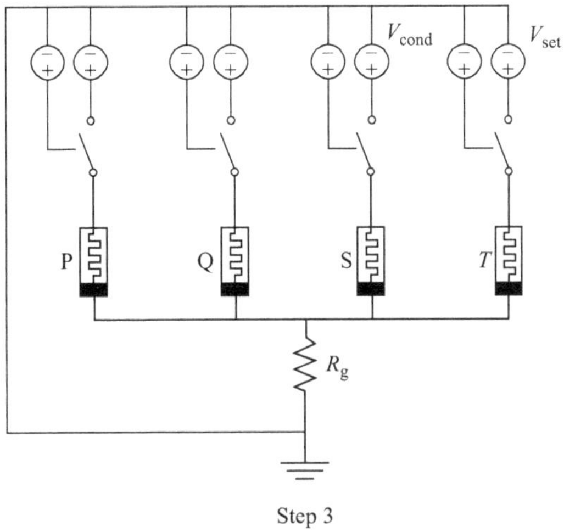

Figure 2.6 Circuitry for the IMPLY AND operation

executing (P IMP 0) IMP Q and overwriting Q, (P IMP 0) IMP M_e is performed, overwriting its value, but maintaining Q. Since this copy operation requires two memristors and two IMPLY steps, the total cost of an OR operation is five memristors and four steps.

The NOR operation is performed by performing the IMPLY OR operation, followed by the IMPLY NOT operation. Both these operations have been discussed already, so it is simply a matter of serializing the two operations. All of the IMPLY steps are performed as described for the OR operation. At the end of this operation, the result lies in the Q memristor. A sixth memristor, S, is introduced to perform the last NOT operation. This memristor will be reset to logical "0." A fifth step performs (Q IMP S). Since S is initialized to 0, Q IMP S is equivalent to Q IMP 0. Thus, this final step essentially NOTs the result of the OR operation and stores the result into S, completing the NOR operation.

A major hindrance to both delay and latency is the high cost of the XOR operation in IMPLY because the XOR operation requires copying both operands before execution.

The XOR operation is calculated as (P IMP Q) IMP ((Q IMP P) IMP 0). The first step is either (P IMP Q), which would overwrite the input operand Q, or (Q IMP P), which would overwrite the input operand P. Performing either of these steps (or a future operation that requires either operand) will make the other step impossible since the original operand has been overwritten. Thus, both must be copied to secondary memristors before beginning the XOR operation. The breakdown of the full steps for an XOR operation is the following:

1. P IMP 0
2. (P IMP 0) IMP 0 == COPY of P
3. Q IMP 0

4. (Q IMP 0) IMP 0 == COPY of Q
5. P IMP Q
6. Q IMP P
7. (Q IMP P) IMP 0
8. (P IMP Q) IMP ((Q IMP P) IMP 0)

An improved IMPLY implementation for the XOR operation [18] is given by

$$\text{P XOR Q} = (\text{P IMP Q})\text{IMP}\left(\left(\overline{\text{P}}\ \text{IMP}\ \overline{\text{Q}}\right)\text{IMP}\ 0\right)$$

By rewriting the XOR operation with these IMPLY steps, the operand Q is no longer overwritten. The updated steps are shown in Table 2.3.

This reduces the number of steps to seven and the memristor count to six. This is an example of the fact that there are numerous ways to compute Boolean operations using IMPLY and memristors. Often times, it is beneficial to consider the specific context of the operation before determining the appropriate steps. For example, consider whether the inputs will be needed again or can be overwritten without a COPY operation. This is often true in the case that one or more of the input operands is an intermediate result and has a fan-out of one. This will become clearer as the units become more complex.

2.1.2 Real implementation

As the memristor becomes more feasible and popular, research has begun into fabrication of memristors to perform logic applications. For example, one research group demonstrated the correct execution of an IMPLY operation on unipolar SiO_x-based memristors [50]. The memristors were constructed in a 4×4 1D1R crossbar such that each cell contained one diode and one resistor.

A full adder was also implemented on the 4 × 4 crossbar to calculate the carry-out and sum results in 43 steps. In this implementation, low resistance for the memristors was 93 kΩ, and high resistance was 260 MΩ. The pull-down resistors in the crossbar were selected to be 5.7 kΩ. The steps for a full adder in the context of this 4×4 crossbar are shown in Table 2.4 for $X + Y + C_{in}$.

Each row denotes a step in the addition. The memristors are each labeled beginning with an “M,” and their number corresponds to their location in the

Table 2.3 Steps for improved IMPLY XOR operation

Step	Operation	Memristors used
1	P IMP 0	2
2	Q IMP 0	2
3	(Q IMP 0) IMP 0 == COPY of Q	1
4	(P IMP 0) IMP (Q IMP 0) == ($\overline{\text{P}}$ IMP $\overline{\text{Q}}$)	0
5	($\overline{\text{P}}$ IMP $\overline{\text{Q}}$) IMP 0	1
6	P IMP (Q IMP 0 IMP 0) == (P IMP Q)	0
7	(P IMP Q) IMP (($\overline{\text{P}}$ IMP $\overline{\text{Q}}$) IMP 0)	0

Table 2.4 Full adder steps in 4 × 4 SiO_x-based memristor crossbar

Step no.	Operation	M11	M12	M13	M14	M21 (Count)	M22	M23	M24
0	INIT all	0	0	0	0	0	0	0	0
1	INIT M11 with X	X							
2	INIT M22 with Y						Y		
3	INIT M44 with C_{in}								
4	M11 IMP M14				X IMP 0				
5	M14 IMP M12		X						
6	M14 IMP M13			X					
7	M22 IMP M23							Y IMP 0	
8	M23 IMP M21					Y			
9	M44 IMP M42								
10	M42 IMP M43								
11	INIT M14 with 0				0				
12	INIT M23 with 0							0	
13	M21 IMP M11	Y IMP X							
14	M11 IMP M14				(Y IMP X) IMP 0				
15	M12 IMP M22						X IMP Y		
16	M22 IMP M23							(X IMP Y) IMP 0	
17	M23 IMP M24								X IMP Y
18	M24 IMP M14				X XOR Y				
19	INIT M23 with 0							0	
20	M21 IMP M23							Y IMP 0	
21	M13 IMP M23							X IMP (Y IMP 0)	
22	INIT M21 with 0					0			
23	M23 IMP M21					X AND Y			
24	INIT M24 with 0								0
25	M14 IMP M24								(X XOR Y)

(Continues)

Table 2.4 (Continued)

Step no.	Operation	M11	M12	M13	M14	M21 (Count)	M22	M23	M24
									IMP 0
26	M24 IMP M34								
27	M44 IMP M34								
28	M34 IMP M31								
29	M14 IMP M44								
30	INIT M42 with 0								
31	M44 IMP M42								
32	M42 IMP M41								
33	M41 IMP M31								
34	INIT M13 with 0			0					
35	M14 IMP M13			(X XOR Y) IMP 0					
36	M43 IMP M13			C_{in} IMP ((X XOR Y) IMP 0)					
37	M13 IMP M33								
38	M33 IMP M32								
39	INIT M22 with 0						0		
40	INIT M42 with 0								
41	M32 IMP M42								
42	M42 IMP M22						(C_{in} AND (X XOR Y)) IMP 0		
43	M22 IMP M21					(C_{in} AND (X XOR Y)) OR (X AND Y)			

Step no.	Operation	M31 (S)	M32	M33	M34	M41	M42	M43	M44
0	INIT all	0	0	0	0	0	0	0	0
1	INIT M11 with X								
2	INIT M22 with Y								
3	INIT M44 with C_{in}								C_{in}
4	M11 IMP M14								
5	M14 IMP M12								
6	M14 IMP M13								
7	M22 IMP M23								
8	M23 IMP M21								
9	M44 IMP M42						C_{in} IMP 0		
10	M42 IMP M43							C_{in}	
11	INIT M14 with 0								
12	INIT M23 with 0								
13	M21 IMP M11								
14	M11 IMP M14								
15	M12 IMP M22								
16	M22 IMP M23								
17	M23 IMP M24								
18	M24 IMP M14								
19	INIT M23 with 0								
20	M21 IMP M23								
21	M13 IMP M23								
22	INIT M21 with 0								
23	M23 IMP M21								
24	INIT M24 with 0								
25	M14 IMP M24								
26	M24 IMP M34				X XOR Y				
27	M44 IMP M34				C_{in} IMP (X OR Y)				
28	M34 IMP M31	C_{in} IMP (X XOR Y) IMP 0							

(Continues)

Table 2.4 (Continued)

Step no.	Operation	M31 (S)	M32	M33	M34	M41	M42	M43	M44
29	M14 IMP M44								(X XOR Y) IMP C_{in}
30	INIT M42 with 0						0		
31	M44 IMP M42						((X XOR Y) IMP C_{in}) IMP 0		
32	M42 IMP M41					(X XOR Y) IMP C_{in}			
33	M41 IMP M31	((X XOR Y) IMP C_{in}) IMP (C_{in} IMP (X XOR Y) IMP 0)							
34	INIT M13 with 0								
35	M14 IMP M13								
36	M43 IMP M13								
37	M13 IMP M33			C_{in} AND (X XOR Y)					
38	M33 IMP M32	(C_{in} AND (X XOR Y)) IMP 0							
39	INIT M22 with 0								
40	INIT M42 with 0						0		
41	M32 IMP M42						(C_{in} AND (X XOR Y)) IMP 0) IMP 0		
42	M42 IMP M22								
43	M22 IMP M21								

crossbar. For example, memristor M14 is the memristor located in the crossbar cell in row 1 and column 4.

The IMPLY operations have the limitation that they must be performed on two memristor operands that reside in the same row of the crossbar. The result is stored in the second operand, so it is also on the same row. IMPLY operations cannot be performed across rows. In order to use two operands on differing rows in an IMPLY operation, one operand must first be moved, or copied, to the other's row. Each time an IMPLY operation in the full adder execution completes, it must be copied to the same row as the next operand (if it is not already) it needs before the next execution step can proceed. This increases the number of steps required to execute the full adder due to the limited dimensions of the crossbar memory.

This implementation is the first known implementation of its kind. There was no previous work that designed and tested a full adder on fabricated memristors in a crossbar.

2.2 MAD gates

MAD gates are a recent development in memristor logic. MAD gates, or memristors-as-drivers gates, were introduced to overcome the long delays of the IMPLY operations and the fan-out, signal degradation and buffering issues in hybrid-CMOS operations (discussed below). MAD gates take a similar form to IMPLY gates, but they have a slightly more complex cell structure and require fewer steps to execute Boolean operations. MAD gates are built on the concept of voltage division. Voltage is leveraged to infer and procure information about the internal resistances of the operand memristors in order to compute a Boolean operation.

The introduction to MAD gates begins with simple gate structures in order to understand the underlying concepts of the gate functionality. These gates can be used in stand-alone logic and arithmetic units in modern ALUs, GPUs, hardware-specialized architectures, and the like. They can also be transformed to function effectively in a standard crossbar circuitry, similar to the way IMPLY functions enable logic-in-memory.

Consider a simple memristor circuit consisting of a single voltage source with magnitude V_{cond} and two memristors A and B connected in series with two 10-kΩ resistors as shown in Figure 2.7.

The magnitude of V_{cond} is less than the threshold of the memristor and is not high enough to change their internal resistance. The memristors in this context are essentially resistors that are either in $R_{off} = 100$ kΩ or $R_{on} = 1$ kΩ. Consider the node V_{lo} as the circuit output. This is a circuit reminiscent of a voltage division circuit. The voltage at V_{lo} for each combination of A and B can be seen in Table 2.5.

Similarly, the voltages at V_{hi} are in Table 2.6.

Notice that these nodes show opposite trends in the output node's voltage as the input changes from 00 to 11. Also, in both cases, the output values for 00 and 11 are different from the output voltage for 01 or 10. Thus, it is possible to know the

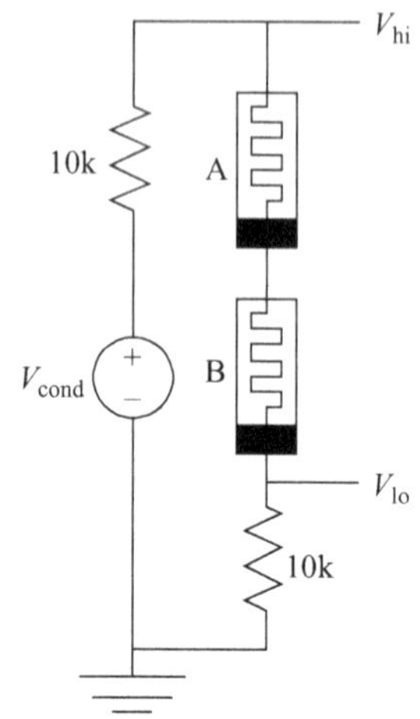

Figure 2.7 Simple memristive voltage division circuit

Table 2.5 Possible voltage outputs at V_{lo}

AB	V_{lo}
00	$(10/220)*V_{cond}$
01	$(10/121)*V_{cond}$
10	$(10/121)*V_{cond}$
11	$(10/22)*V_{cond}$

Table 2.6 Possible voltage outputs at V_{hi}

AB	V_{hi}
00	$(210/220)*V_{cond}$
01	$(111/121)*V_{cond}$
10	$(111/121)*V_{cond}$
11	$(12/22)*V_{cond}$

internal resistances of A and B from sensing these voltages. For example, if V_{lo} is greater than $(10/121)*V_{cond}$, then the inputs A and B must both be 1. This knowledge can be used to execute Boolean operations.

A MAD gate implementation of the AND operation is given in Figure 2.8.

The basic structure is identical to the voltage division circuit introduced earlier. There is a third memristor, labeled AND in this example, to hold the result of the operation that is connected to a voltage source V_{set} and a pull-down resistor $R_g = 10\ k\Omega$. This is a key divergence from a traditional IMPLY circuitry: The result

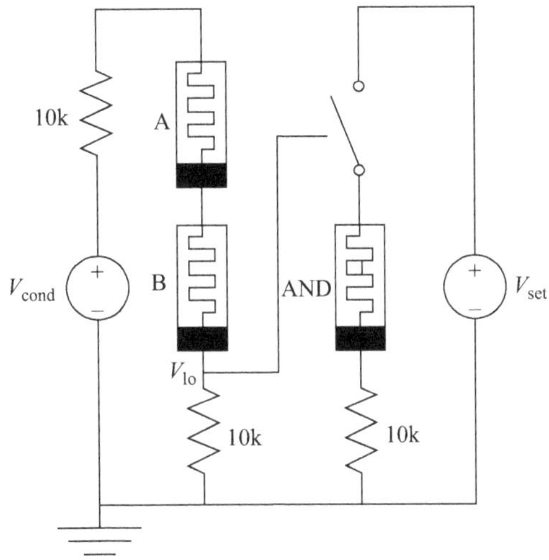

Figure 2.8 MAD AND gate

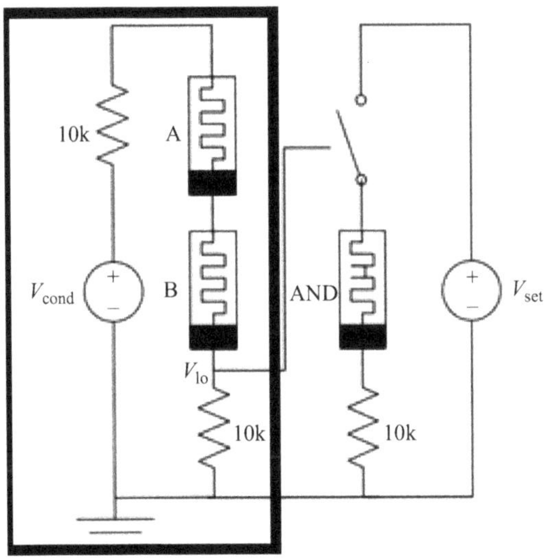

Figure 2.9 Voltage division operation in the MAD AND gate

does not overwrite either of the initial operands. V_{cond} and V_{set} are selected by the same methods so $V_{cond} = 1.6$ V and $V_{set} = 2.5$ V.

To perform a MAD operation, the circuit essentially performs two parallel operations at once: a voltage division and a set operation. Together, these two operations complete the Boolean operation.

The voltage division is performed with the part of the circuit indicated in the bold box in Figure 2.9.

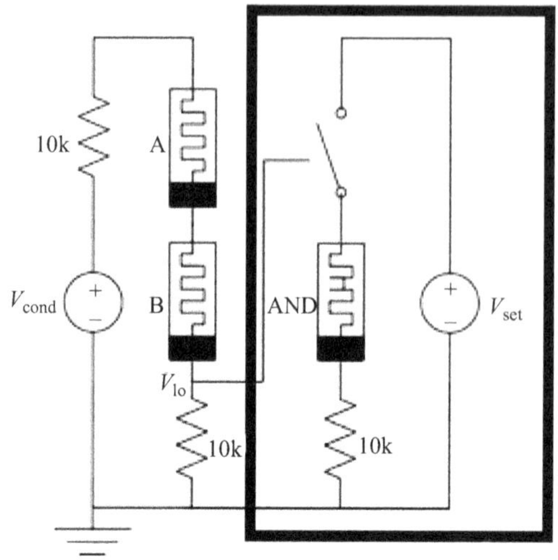

Figure 2.10 Set operation in the MAD AND gate

The read voltage V_{cond} is applied to the two resistors and input memristors in series. The voltage at node V_{lo} is defined as was discussed in the voltage division circuit. When both input memristors are logical "1," the voltage at node V_{lo} is $(10/22)*V_{cond}$. For the input combinations AB = "01" and AB = "10," V_{lo} is $(10/121)*V_{cond}$. For the input combination AB = "00," V_{lo} is $(10/220)*V_{cond}$.

In parallel, the set operation is performed with the part of the circuit indicated in the bold box in Figure 2.10. Clearly, the result memristor should be set when V_{lo} is $10/22*V_{cond}$ and only then.

The set voltage V_{set} is applied to the n-terminal of the result memristor through a switch. This switch is driven by the voltage at V_{lo}. Thus, the voltage V_{lo} determines if the switch will close or not and whether the memristor will be set to "1" or not. If the voltage sensed at node V_{lo} is greater than the threshold of the switch, the switch will close and the V_{set} voltage will set the result memristor to a logical "1," else the memristor will remain a logical "0." In the design of an AND gate, the switch should close if and only if the values of the input memristors A and B are both logical "1." As described previously, when both A and B are a "1," the voltage at V_{lo} is $(10/22)*V_{cond}$. Thus, the threshold of the switch, called V_{th}, must be less than $(10/22)*V_{cond}$. However, the switch should not close if either of the input memristors is a "0." When one input is a "0," the voltage at V_{lo} is $(10/121)*V_{cond.}$ When both inputs are "0," the voltage at V_{lo} is $(10/220)*V_{cond}$. Thus, the threshold of the switch must be greater than $(10/121)*V_{cond}$. In this way, the switch will only close during the set operation if both of the input memristors are a "1." Given that $V_{cond} = 1.6$ V, the threshold of the switch must satisfy

$$10/121 * 1.6\text{ V} < V_{th} < 10/22 * 1.6\text{ V}$$
$$\text{or } 16/121\text{ V} < V_{th} < 16/22\text{ V}$$
$$\text{or } 1.32 < V_{th} < 0.727$$

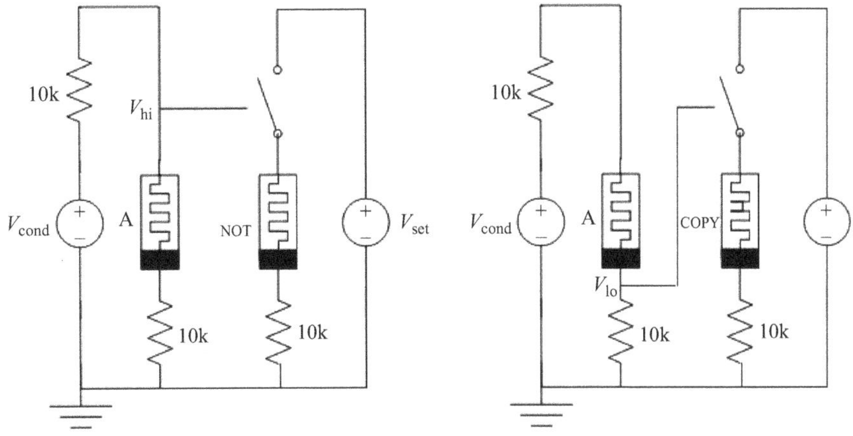

Figure 2.11 MAD NOT and COPY gates

Here, V_{th} is chosen to be 0.5 V. The same circuit can be modified to accomplish the remaining Boolean operations by varying the value of V_{th}. For example, for the OR operation, operation of the gate is identical, performing both a voltage division and a set operation. The voltage division portion of the gate is identical because it is only sensing the inputs, independent of the Boolean operation. The set operation still drives a set voltage to the output memristor, gated by a single switch. This switch still closes to set the output to a "1" and remains open to set the output to a "0." The only difference between the OR gate and the AND gate is that the scenarios that should close the switch have changed. Now, input combinations AB = 01, AB = 10, and AB = 11 should all close the switch. Thus, based on Tables 2.5 and 2.6, V_{th} should be selected such that $16/220 < V_{th} < 16/121$ V. This achieves an OR gate.

A similar circuit can also be used to accomplish COPY and NOT operations. The NOT and COPY operations only require a single input memristor since they are unary operators but take the same form otherwise. The respective circuits for the NOT and COPY operations are shown in Figure 2.11.

The circuitry for the set operation remains unchanged. The circuitry for the voltage division now only has a single-input memristor rather than two. However, the fundamental behavior of the gate is identical.

For the NOT operation, V_{cond} is applied to the input memristor A. If A is 0, or high resistance, then the voltage at node V_{hi} will be 176/120 V. If A is 1, or low resistance, then the voltage at node V_{hi} will be 17.6/21 V. These are the only two possibilities at V_{hi} when the voltage division occurs. To perform a NOT, the voltage when A is a "0," 176/120 V, needs to be large enough to close the gate on the output memristor and set it to a 1. The voltage when A is a "1," 17.6/21 V, should not be large enough to close the switch on the output memristor. Thus, the threshold of the switch V_{th} must satisfy $17.6/21\ \text{V} < V_{th} < 176/120$ V. By setting V_{th} to, for example 1.2 V, V_{set} will be gated to the output memristor and set it to logical "1"

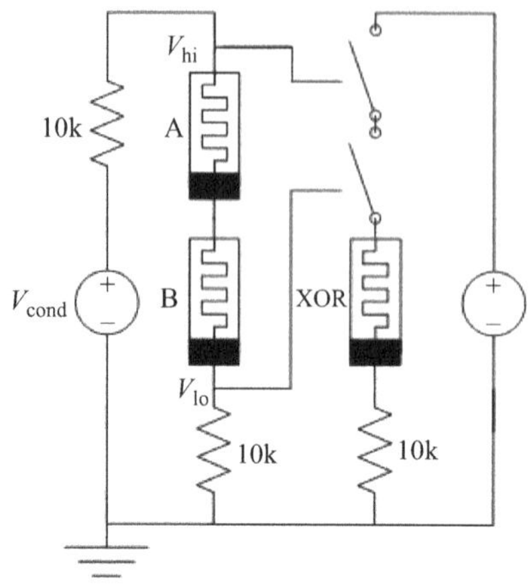

Figure 2.12 MAD XOR gate

only when A is 0. Else, when A is 1, the switch remains open and the result memristor remains 0. Thus, the NOT operation is achieved.

The copy operation uses nearly identical circuitry but uses the voltage at V_{lo} rather than V_{hi} to drive the switch on the output memristor. When the input memristor A = "0," the voltage at V_{lo} is 16/120 V. When the input memristor A = "1," the voltage at V_{lo} is 16/21 V. For a copy operation, the output memristor should be set to a "1" if the input memristor A = "1" which corresponds to $V_{lo} = 16/21$ V. The output memristor should remain a "0" if the input memristor A = "0" which corresponds to $V_{lo} = 16/120$ V. Thus, the threshold of the switch on the output memristor should satisfy 16/120 V $< V_{th} <$ 16/21 V. This correctly performs the copy operation of A into the result memristor.

Finally, the XOR MAD gate shown in Figure 2.12 is described.

The XOR operation is the most complex MAD gate variation. This gate has the same voltage division circuitry as the AND and OR MAD gates. However, the circuitry on the result memristor changes. Now, there are two switches gating the set voltage V_{set} to the output memristor. Each has a distinct V_{th} voltage and is driven by distinct nodes. The top switch is driven by the voltage at V_{hi} and the bottom switch is driven by the voltage at V_{lo}. Consider the four possible values of input memristors A and B and again let $V_{cond} = 1.6$ V. If A = B = "0," the voltage at $V_{hi} = 336/220$ V and $V_{lo} = 16/220$ V. If A = B = "1," the voltage at $V_{hi} = 19.2/22$ V and $V_{lo} = 16/22$ V. If A = 0, B = 1 or A = 1, B = 0, then the voltage at $V_{hi} = 177.6/121$ V and $V_{lo} = 16/121$ V. These voltages are summarized in Table 2.7.

For an XOR operation, the output memristor is set to 1 if and only if AB = 01 or AB = 10. This corresponds to the second and third row in the table. Note that it is not possible to use a single switch with a single threshold voltage to perform an XOR.

Table 2.7 *Voltages at V_{lo} and V_{hi} in XOR MAD gate voltage division circuit*

Inputs AB	V_{lo} (V)	V_{hi} (V)
00	16/220	336/220
01	16/121	177.6/121
10	16/121	177.6/121
11	16/22	19.2/22

There is no threshold voltage that would produce the correct result. To use V_{lo} to drive a single switch on the output memristor, the threshold voltage of the switch would need to satisfy

$$\frac{16}{220}\text{V} < V_{th} < \frac{16}{121}\text{V}$$

Thus, input combination 00 does not have a high enough magnitude to close the switch, but input combinations 01 and 10 do. However, the threshold voltage also needs to satisfy

$$\frac{16}{22}\text{V} < V_{th}$$

In this case, input combination 11 does not have a high enough magnitude to close the switch. There is no solution that satisfies both equations.

Similarly, if V_{hi} is used to drive a single switch, the threshold voltage of the switch needs to satisfy

$$\frac{19.2}{22}\text{V} < V_{th} < \frac{177.6}{121}\text{V}$$

Input combination 11 does not have a high enough magnitude to close the switch, but input combinations 01 and 10 do. However, the threshold voltage also needs to satisfy

$$\frac{336}{220}\text{V} < V_{th}$$

Input combination 00 does not have a high enough magnitude to close the switch. Again, there is no solution that satisfies both equations.

Instead, two switches can be used with the voltages at both nodes V_{lo} and V_{hi}, to perform an XOR operation. One switch can be used to eliminate input combination 00, and the other can be used to eliminate input combination 11. By placing the switches in series with the output memristor, both input combinations will be eliminated, and only input combinations 01 and 10 will successfully set the output memristor to 1.

Notice that the voltage at V_{lo} increases when going from input combination 00 to 11 in Table 2.7, and the voltage at V_{hi} increases when going in the opposite direction, from input combination 11 to 00 in the table. Thus, the first switch will be driven by V_{lo} and will have a threshold voltage that satisfies the first equation:

$$\frac{16}{220}\text{V} < V_{th} < \frac{16}{121}\text{V}$$

Now, if the input memristors A and B are "00," the threshold of the first switch will not be met and the output memristor will not be set, regardless of the second switch. The remaining input combinations 01, 10, and 11 will all have a large enough voltage magnitude to close the first switch.

The second switch will be driven by V_{hi} and will have a threshold voltage that satisfies the previous equation:

$$\frac{19.2}{22}\text{V} < V_{th} < \frac{177.6}{121}\text{V}$$

Now, if the input memristors A and B are "11," the threshold of the second switch will not be met and the output memristor will not be set, regardless of the first switch. The remaining combinations 00, 01, and 10 will be large enough to close the second switch.

With these two switches configured as such in the XOR circuitry, only input combinations 01 and 10 will satisfy the threshold voltages of both switches as required to set the output memristor. Thus, the XOR operation is achieved.

2.2.1 MAD metrics, pros, and cons

The main benefit of MAD gates is their ability to perform all Boolean operations in just a single step. This is a stark contrast to the serialization issues in the IMPLY operation. A summary of the number of steps required for each approach for Boolean operations is shown in Table 2.8.

For both IMPLY and MAD operations, a step is executed each time voltage(s) is(are) applied to the circuit. MAD gates are able to achieve these step counts that maintain the same number of memristors as the equivalent IMPLY operation and in many cases reducing the number of memristors. Although the IMPLY operation does not require a separate result memristor, it can require multiple intermediate memristors for the results of intermediate steps. The MAD gates only require one

Table 2.8 Comparison of execution steps for IMPLY vs. MAD Boolean ops

Operation	IMPLY steps	MAD steps
P NAND Q	2	1
P AND Q	3	1
P NOR Q	5	1
P OR Q	4	1
P XOR Q	7	1
NOT P	1	1
COPY P	2	1

memristor per operand. A summary of the number of memristors required for each approach is shown in Table 2.9.

However, for practical purposes, it is also important to consider the overhead of the driver logic and any other circuit components besides the memristors themselves required for each of the operations. A complete breakdown of the component comparisons for the various Boolean operations is given in Table 2.10.

In general, an IMPLY operation is going to require one voltage driver for each memristor because each memristor will need to be driven by V_{cond} or V_{set} individually. A MAD operation will require two drivers: one for the input operands

Table 2.9 Comparison of memristor counts for IMPLY vs. MAD Boolean ops

Operation	IMPLY memristors	MAD memristors
P NAND Q	3	3
P AND Q	4	3
P NOR Q	6	3
P OR Q	5	3
P XOR Q	6	3
NOT P	2	2
COPY P	3	2

Table 2.10 Component comparisons for IMPLY vs. MAD gates

Op	IMPLY	MAD
NAND	3 memristors 3 drivers 3 transistors	3 memristors 2 drivers 2 transistors
AND	4 memristors 4 drivers 4 transistors	3 memristors 2 drivers 2 transistors
NOR	6 memristors 6 drivers 6 transistors	3 memristors 2 drivers 2 transistors
OR	5 memristors 5 drivers 5 transistors	3 memristors 2 drivers 2 transistors
XOR	6 memristors 6 drivers 6 transistors	3 memristors 2 drivers 3 transistors
NOT	2 memristors 2 drivers 2 transistors	2 memristors 2 drivers 2 transistors
COPY	3 memristors 3 drivers 3 transistors	2 memristors 2 drivers 2 transistors

together and one for the output memristor. For both IMPLY and MAD, in order to gate the voltages to each memristor, every driver will require a transistor. For the MAD XOR and XNOR operations, there is an additional switch since two nodes are sensed and used to drive switches on the output memristor.

IMPLY and MAD operations require three different voltages in total: V_{cond}, V_{set}, and V_{reset}. One complication that MAD gates have is that the threshold voltages on the transistor switches vary depending on the operation. The exact values for these threshold voltages depend on the various parameters of the memristors and circuitry; however, it adds a layer of complexity that the IMPLY operation does not have. The IMPLY operation can use a standard transistor switch with a common threshold voltage across all.

One benefit of MAD gates is their ability to offer a uniform standardized cell for all Boolean operations. The threshold of the switch can be changed to perform the desired operation, without a change to the general structure or circuitry. MAD gates have the additional benefit of preserving both of the initial operands and storing the result in a separate memristor. This removes the need for intermediate COPY operations, which can increase the number of memristors and steps significantly. It also allows for higher fan-out. For example, the same two operands can be used repeatedly for a series of Boolean operands, and the cost of the operations can be amortized. Consider performing A OR B followed by A AND B in the IMPLY and MAD contexts. In the IMPLY context, you would perform the OR operations that require five memristors and four steps. Then, you would perform an AND operation, requiring two more memristors and three steps for a total of seven memristors and six steps. Now consider the MAD context. In both operations, the same two memristors A and B are driven with the V_{cond} voltage, and the voltage at node V_{lo} is used to drive the switch for the result memristor. Thus, the operations can perform in parallel in a single step. V_{cond} is applied to the input memristors A and B in series and V_{set} is applied to both output memristors. V_{lo} drives the switches on each output memristor. One switch has the threshold voltage corresponding to the OR operation and the other has the threshold voltage for the AND operation, respectively. In this way, both the OR and AND operation require just two steps and four memristors. Extending to N Boolean operation, the MAD gates would only require $N + 2$ memristors: two for the inputs and N for the N Boolean outputs. This is an example of how the MAD context has savings both within operations and across them. These improvements in area and performance can lead to significant energy savings.

In addition to low latency and low complexity, MAD gates also offer increased flexibility over previous approaches. In the IMPLY circuitry, the result memristor inherently must lie on the same row as the input memristors since one of the inputs is used for the output. In a MAD operation in a crossbar, the result can lie on any row line and bit line. This is important for full-system designs if logic-in-memory is going to be harnessed for performance gains. If data are frequently being moved around in the memory in order to read, write, and execute logic, the advantages of performing the logic in the memory rather than the CPU are diminished or even eliminated. MAD gates do not suffer from this overhead.

2.2.2 Implementation in a crossbar

MAD gates can be transposed into a crossbar structure with a few minor adaptations to the circuitry. First, in the standard MAD Boolean gate, the input memristors are connected in series by connecting their opposite-polarity terminals, i.e., one input memristor's p-terminal is connected to the second input memristor's n-terminal. This is required in order to be able to maintain the values of the input memristors when V_{cond} is applied. Consider the voltage division circuit's functionality if the common-polarity terminals are connected in series as shown in Figure 2.13.

Recall the description of the memristor *IV* characteristics and the effect current has on the conduction channels in the memristors. When a negative voltage is applied, the resistance increases and when a positive voltage is applied, the resistance decreases. Now let the memristors in the voltage division circuit both be "1," or low resistance, and apply V_{cond}. Due to the polarity orientation of the memristors, this is a positive voltage for memristor A and a negative voltage for memristor B. Thus, memristor A behaves in the normal manner for V_{cond}, with it behaving as a read voltage. However, memristor B potentially increases its resistance and risks being reset and put into a high-resistance state. When the memristors are connected in series with their opposite-polarity terminals, both see V_{cond} as a positive voltage and treat it as a read voltage, maintaining their original values as desired.

2.2.3 Gate adaptation

MAD gates must be adapted in order to exist in a traditional crossbar structure. In a crossbar structure, the p-terminals of both input memristors necessarily share a

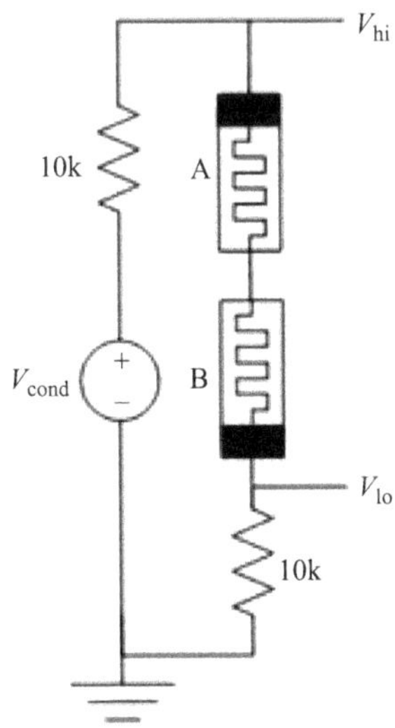

Figure 2.13 Memristor voltage division circuit with common-polarity terminals connected in series

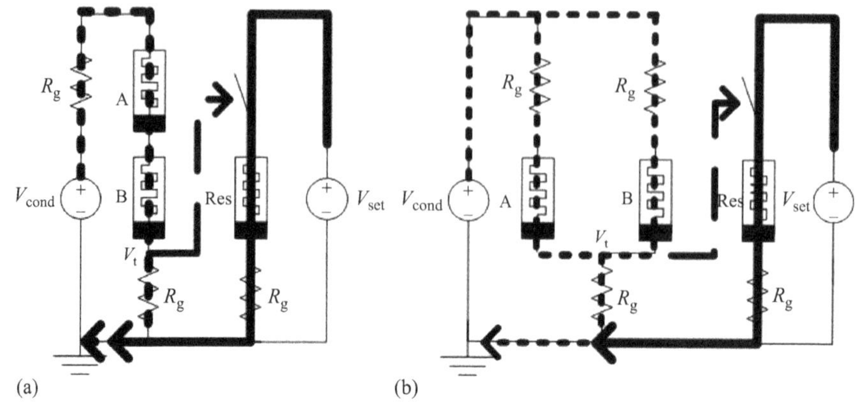

Figure 2.14 Translation of (a) a MAD AND gate into (b) a crossbar form

common row line, connecting their common polarity in series. Because of this, the MAD circuitry must be slightly changed as shown in Figure 2.14.

The finely dotted path follows the V_{cond} signal, the solid path follows the V_{set} signal, and the coarsely dotted highlighted path follows the voltage division threshold signal. The only necessary change is on the voltage division circuitry for the input memristors and the V_{th} of the switch. The other signals remain unchanged. Instead of connecting the input memristors in series and applying V_{cond} and GND to the terminals, V_{cond} is applied to the n-terminal of both input memristors, and their shared row line is connected to GND via the pull-down memristor. The sensed node V_t is now shared by both input memristors. The resistor R_g values are selected as described previously.

Consider the updated functionality of the gate for each input combination. Let $R_g = 10$ kΩ. When

$$\begin{aligned} \text{AB} = 00 &: V_t = (10/65) * V_{cond} \\ \text{AB} = 01 &: V_t = (1/2) * V_{cond} \\ \text{AB} = 10 &: V_t = (1/2) * V_{cond} \\ \text{AB} = 11 &: V_t = (20/31) * V_{cond} \end{aligned}$$

So V_t is still capable of discerning the input combination 11 from the others by requiring V_{th} on the switch to satisfy $(1/2) * V_{cond} < V_{th} < (20/31) * V_{cond}$. Similarly, the OR operation can use the updated crossbar circuitry and change its V_{th} on the switch to satisfy $(10/65) * V_{cond} < V_{th} < (1/2) * V_{cond}$. Similar steps can be done for the remaining Boolean operations as will be shown in the following sections.

Figure 2.15 shows the implementation of a MAD AND gate in a crossbar structure using the new circuitry.

The inputs p and q are two memristors on the same row in the crossbar. Here, the solid line follows the set operation, the finely dotted line follows the read operation, and the coarsely dotted line follows the switch driver signal.

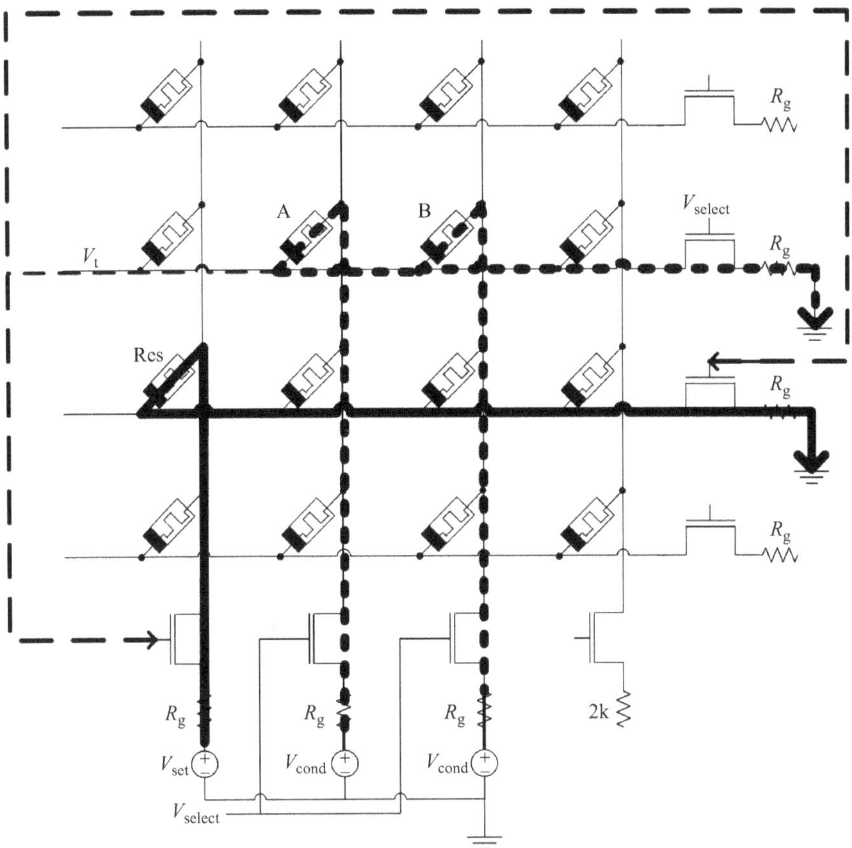

Figure 2.15 MAD AND gate in a crossbar structure

To sense the values of A and B, the row and bit lines of A and B are selected, and V_{cond} is driven on both bit lines while the row is grounded. The row line now represents the sensed node voltage, V_t. In the second step, the sensed value at V_t is used to drive the select lines on the result memristor's row and bit lines. If the value of V_t is greater than the threshold of the select line gates, then V_{set} and GND will be driven to the result memristor Res, and it will be set to 1. Otherwise, the row and bit lines will not be selected, and Res will remain at 0. The same operation exists for an OR operation, except the threshold of the switch is different.

For the XOR and XNOR operation, the same crossbar circuit is used, but now V_a and V_b are sensed rather than V_t. Figure 2.16 shows this for the XNOR operation.

For example purposes, let $R_g = 2$ kΩ. When both input memristors are 0, the voltages at V_a and V_b are 0.975 V. You can think of nodes V_a and V_b as V_{lo} and V_{hi} from the original explanation of a MAD XNOR gate. When input memristor A is a 1 and B is a 0, the voltage at V_a is 0.6 V and the voltage at V_b is 0.982 V. If the values of the inputs are swapped, the voltages at V_a and V_b swap accordingly.

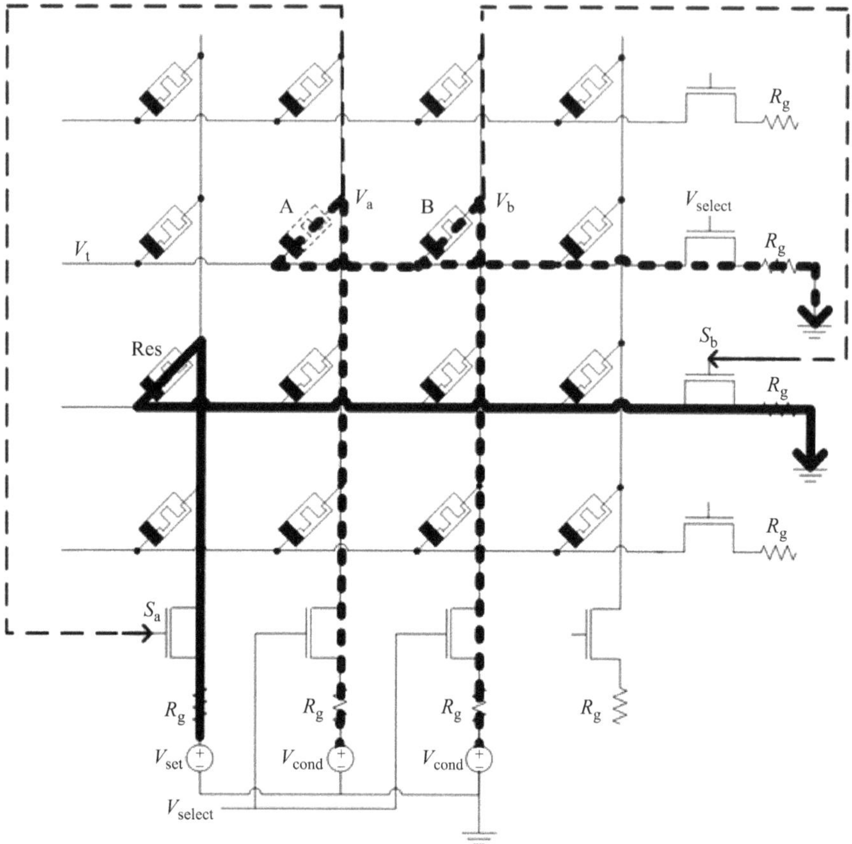

Figure 2.16 MAD XNOR gate in a crossbar structure

If both inputs are a 1, the voltages at V_a and V_b are 5/7 V. Thus, V_a and V_b can be used to differentiate the four input scenarios.

In the original XNOR MAD gate, two separate conditions on two switches must both be true to gate the V_{set} signal to the result memristor. This was implemented by placing two switches in series between the V_{set} signal and the result memristor, one gated by V_a (with threshold V_{thA}) and the other gated by V_b (with threshold V_{thB}). To correctly perform an XNOR, $V_{thA} = V_{thB} = 0.6$ V. The result memristor will only be set to 1 when both of these conditions are true, which only occurs when both inputs are 0 or both inputs are 1.

However, this structure is not conducive to the context of a crossbar since it requires two switches. Instead, a crossbar can achieve the same functionality by placing one switch on the bit line gating V_{set}, S_a, and the other switch on the row line gating GND for the result memristor, S_b. The S_a switch will still be gated by the voltage V_a and S_b will be by the voltage V_b. This effectively achieves the same operation as the original XNOR gate.

Similarly, the XOR operation can use two switches in parallel and vary the driving voltages. One switch will be driven by the voltage at V_a and the other will be driven by V_b. Both switches will have a V_{th} of 0.98 V. If and only if either V_a or V_b is above 0.98 V, corresponding to the input combinations 01 and 10, the output memristor will be set to 1. Practically speaking, the sensitivity of the circuit would not be capable of distinguishing between 0.975 and 0.98 or 0.982 V. In a real implementation, care would be taken in selecting the pull-down memristor as well as the various device parameters and voltages in order to produce feasible ranges.

Recall that the NOT and COPY operations only require a single-input memristor in the MAD gate design. Since these two gates do not use input memristors in series, there are no issues with the way the terminals of the memristors are connected in the original MAD gate design. Thus, their gate designs do not need to be altered at all to translate into the crossbar context. This can be seen in Figure 2.17.

V_{select} is driven high for the row and bit line of the input memristor to create a circuit from V_{cond}, through the input memristor and pull-down resistor, to GND. In the

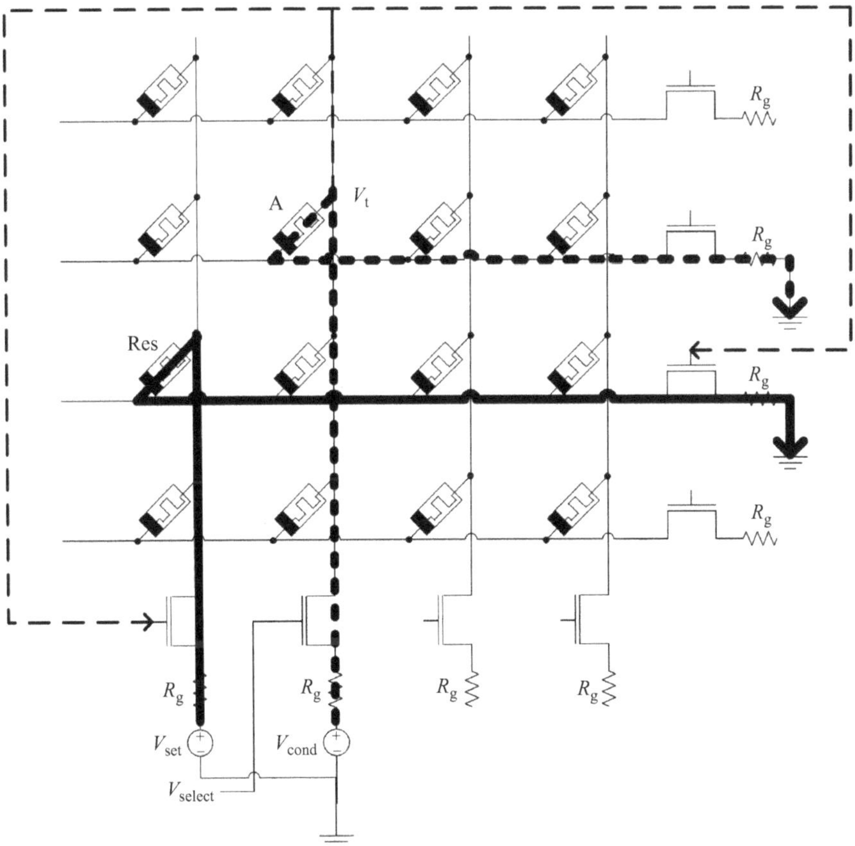

Figure 2.17 MAD NOT gate in a crossbar structure

next step, the voltage sensed at V_t is used to determine if V_{set} should be gated to the result memristor Res. This is identical to the original MAD NOT gate execution.

With the modified MAD gates for the crossbar, some of the complexities inherently change. For example, to prevent interference between the result and inputs, the result memristor must be written in a different cycle than the input operands are read. Thus, the number of steps for each MAD operation increases from one to two. However, the complexity in terms of transistors, drivers and memristors, remains the same.

2.2.4 Crossbar adaptation

Alternatively, the crossbar structure itself can be transformed to accommodate the voltage division circuitry of the MAD gate, without any gate adaptation. Consider the crossbar structure in Figure 2.18.

The only change in the architecture is that the polarity of alternating memory cells has reversed. For example, the memristor in row 0 column 1 now has its n-terminal connected to the column line and p-terminal connected to the row line.

Now, the original MAD gate can be directly mapped into the crossbar because neighboring memristor cells have matching polarity when placed in series. Thus, the original voltage division circuit does not need to change. This can be seen in Figure 2.19.

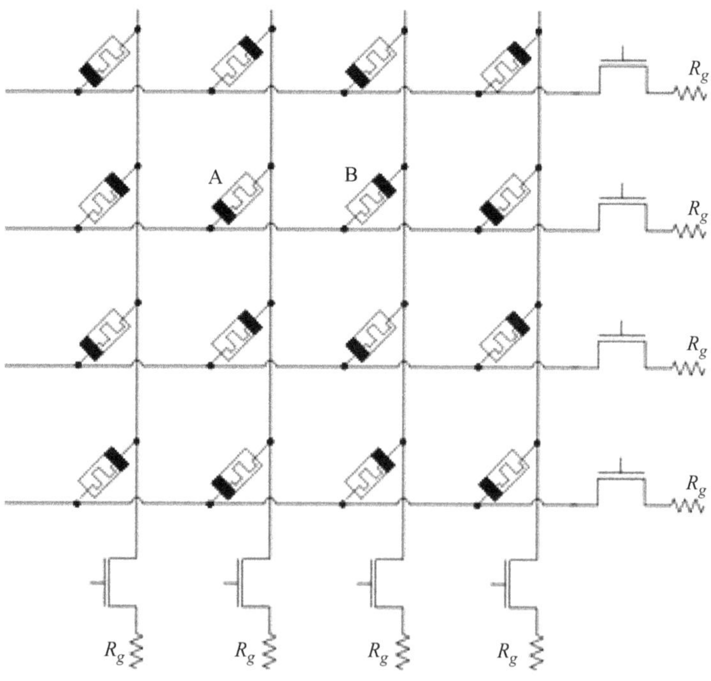

Figure 2.18 Alternative memristor crossbar with alternating memory cell polarity

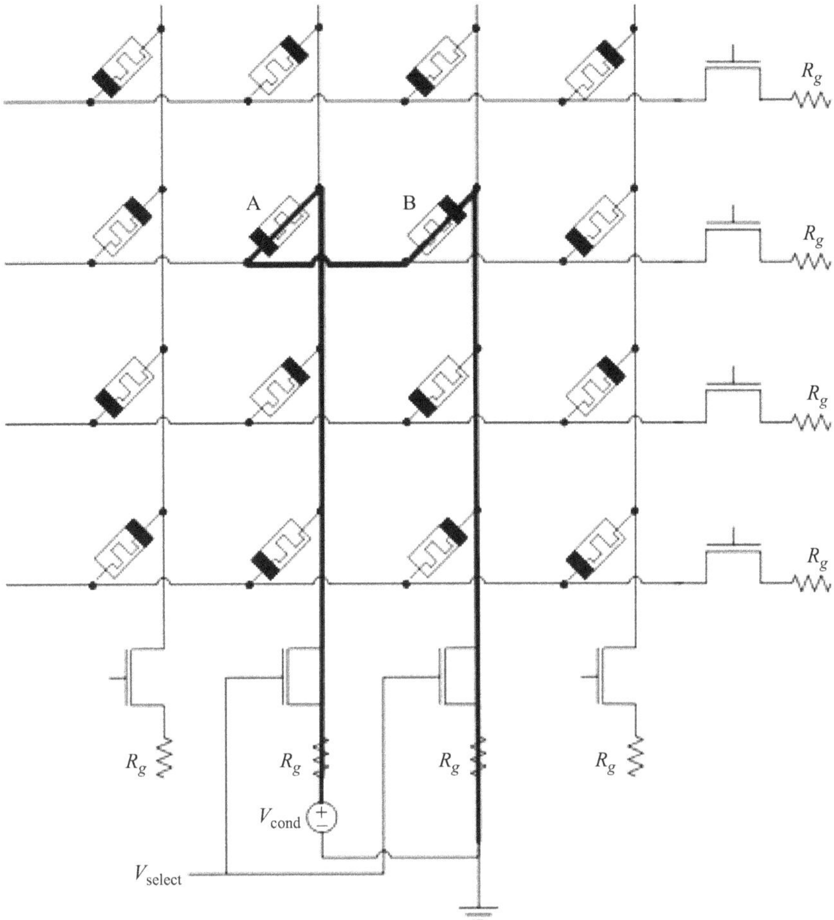

Figure 2.19 Voltage division circuit in adapted crossbar structure

The figure shows the same two memory cells A and B as in the previous examples. V_{cond} is applied to column 1 and GND is applied to column 2. The row line is left unselected. Memristors A and B are now connected by their opposite-polarity terminals, and V_{cond} properly behaves as a read voltage for both. The full MAD gate implementation is shown in Figure 2.20.

V_{lo} is probed on column 1 since it is the p-terminal of the B memristor. It is then used to drive V_{set} and GND on the select lines on the row and column of the result memristor, identical to the set operation in the original crossbar. The same mapping can be done for all Boolean operations since their original MAD gate forms do not need to change. The OR gate takes the same form as the gate in Figure 2.20, but with an adapted threshold on the select lines. The NOT gate remains unchanged since it only uses a single memristor.

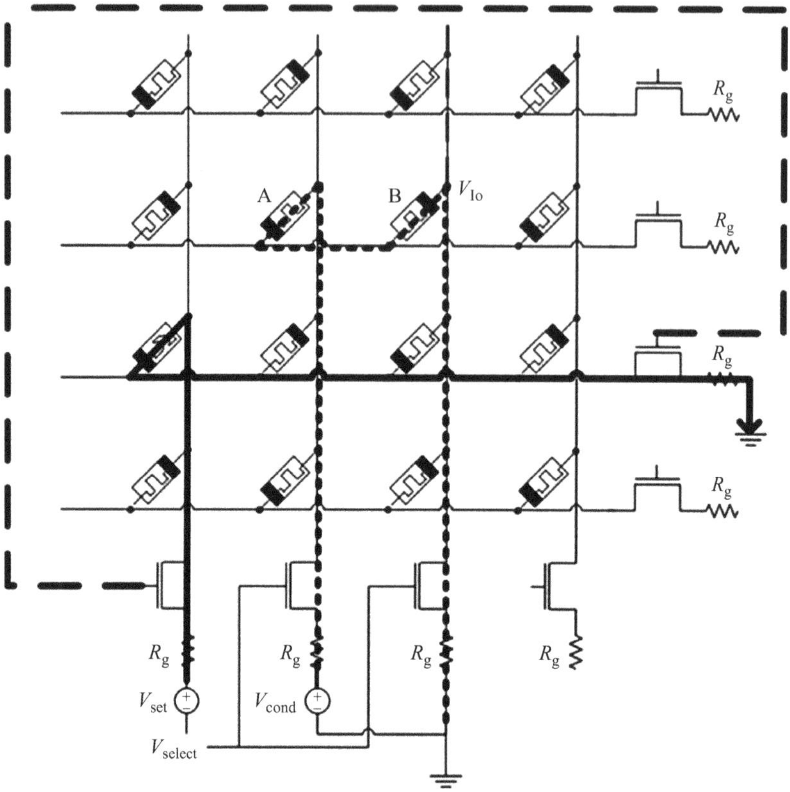

Figure 2.20 MAD gate in the adapted crossbar structure

With the adapted crossbar context, all Boolean MAD operations require two steps per operation but maintain their original complexities of one memristor per operand. Note that there are some consequences to this crossbar structure that need to be satisfied during operation. For example, a MAD operation cannot be performed on two cells that have the same directional polarity, for example, cells (0,0) and (0,2). This can be solved by selecting beneficial result memristors during the set operation. The result memristor has no limitations on its location. Thus, for each MAD operation, a result memristor in a suitable cell with the correct polarity can be selected to enable the next operation in the sequence. The same logic should be taken when selecting a memristor when a value is first loaded, set, or copied into the crossbar. This eliminates the overhead of any necessary data movement in nearly all designs.

Either the gate adaptation or the crossbar adaptation can be performed in order to translate a MAD gate logic unit into a crossbar. It is left to the designer to decide which is preferable for the context of the system. Keep this in mind in future sections when the MAD logic units are described in detail. As an exercise, try to envision adapting the logic units with both gate adaptations and crossbar adaptations.

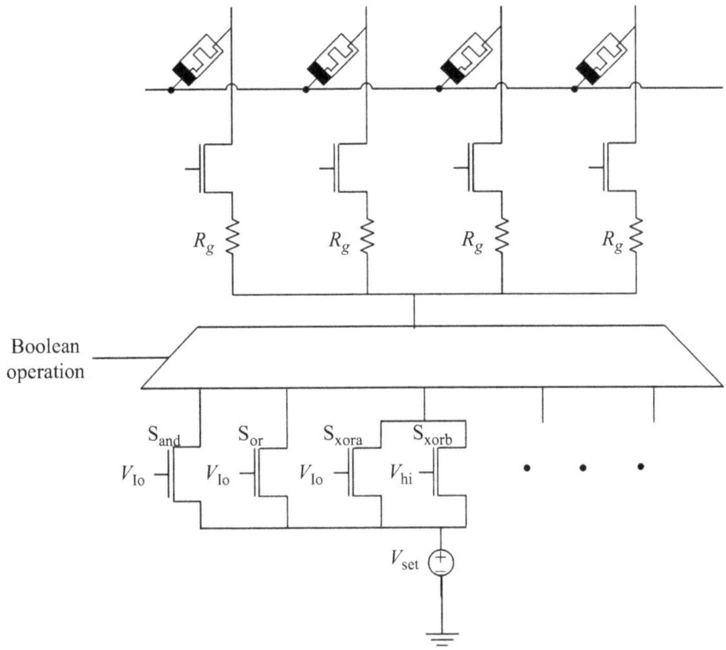

Figure 2.21 Driver logic for selecting MAD crossbar Boolean operations

In many instances, one of the adaptations produces a more optimal design over the other.

2.2.5 Boolean operation and threshold gate selection

Depending on the Boolean operation being performed on a MAD gate or MAD crossbar, the threshold voltage of the switch on the output memristor will vary. However, the hardware cannot be reconfigured dynamically to change the threshold voltage on a given switch. This poses a challenge for crossbars performing MAD operation. Note, this is not an issue for IMPLY. For IMPLY operations, the threshold voltages on the switches are constant throughout execution because every IMPLY operation is identical.

Since the threshold voltage of a switch cannot be reconfigured dynamically, a MAD crossbar needs more than a single switch on each row and column. The rows and columns of the crossbar must be able to be driven by one of many switches, each corresponding to a different Boolean operation. An extra logic block is added to perform this task. One instance of this block is used for the entire crossbar memory.

To perform a MAD Boolean operation, the correct switch will be selected using a multiplexer as shown in Figure 2.21.

Each input to the multiplexer is a switch circuit corresponding to a Boolean operation. The switch is configured with the threshold voltage necessary for that

Boolean operation. The select lines are the encoding of the Boolean operations. Based on the desired Boolean operation being performed, the correct switch with the corresponding threshold will gate the voltage V_{set} to the memristor crossbar.

For example, let the first input to the multiplexer correspond to the AND operation. Thus, a single switch, S_{and}, will lie on this path and have its corresponding V_{th}. When an AND operation is being performed, the select lines are driven with S = "000" in order to select the AND switch input to appropriately gate V_{set} to the output memristor in the crossbar.

Let the second path correspond to the OR operation. A single switch, S_{or}, will lie on this path with its calculated V_{th} value. To perform an OR operation, the select lines are driven with S = "001." The third path will have two parallel switches for the XOR operation, S_{xora} and S_{xorb}, driven by the voltages V_{lo} and V_{hi} as described previously. The select lines are driven with S = "010." The process will continue for each of the remaining operations. Similar logic exists for the row lines.

The select lines on each row and column will now only depend on the address of the cell itself rather than the result of the node voltage during the read operation.

2.3 N–N memristor gates

This section proposes a novel implementation of a memristor-based logic gate that is designed for seamless integration into a crossbar to allow for logic-in-memory applications. This gate is called an N–N memristor gate because it can directly implement the NAND and NOR gates in a single step. This is coupled with the standard IMPLY NOT function to create a logically complete set. The N–N logic gate is also capable of implementing the remaining Boolean logic gates in two steps.

The N–N logic gate has a very similar circuit design to the IMPLY gate but uses a third memristor. This third memristor is used in order to be able to perform the NAND and NOR operation in a single step and store the result in a separate memristor, without overwriting either input memristor. The IMPLY NAND operation requires two steps and three memristors, and the NOR operation requires three steps and four memristors. The proposed gate requires one step and three memristors for each of the two ops. Figure 2.22 shows the logic circuit for the proposed gate.

N–N memristor gates use the same fundamental reasoning and calculations as the IMPLY and MAD gates. P and Q are the input operand memristors, and S is the result memristor. Let the internal resistances of memristors, P, Q, and S, be denoted by R_p, R_q, and R_s, and let the common node between each of the operands be V_{lo}. Initially, S = 0 (R_s is high resistance). The circuit is essentially a voltage divider at V_{lo} as it was in MAD such that V_{lo} is represented as

$$V_{lo} = \frac{V_{cond}{}^{*}\left(R_p + R_q\right) * R_g}{R_pR_q + R_g{}^{*}\left(R_p + R_q\right)}$$

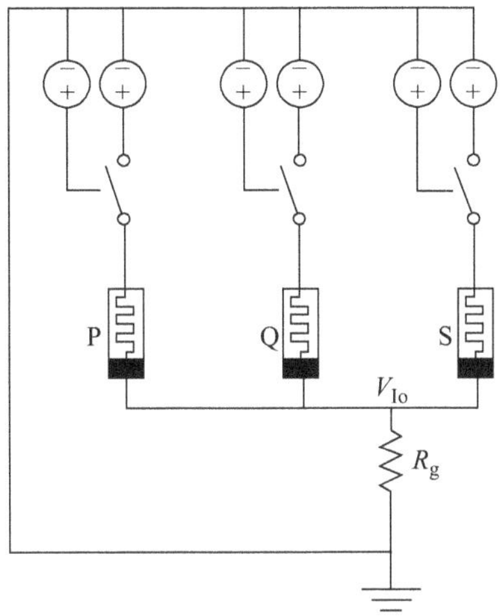

Figure 2.22 Proposed memristor logic circuit

Table 2.11 V_{lo} voltages for proposed logic circuit

Input PQ	V_{lo}
00	0.038 * V_{cond}
01	0.67 * V_{cond}
10	0.67 * V_{cond}
11	0.8 * V_{cond}

For $R_g = 2$ kΩ, $R_{high} = 100$ kΩ, and $R_{low} = 1$ kΩ. For example, for P = 0 ($R_p = 100$ kΩ) and Q = 0 ($R_q = 100$ kΩ).

$$V_{lo} = \frac{V_{cond}{}^{*}200k^{*}2k}{100k^{*}100k + 2k^{*}200k}$$

$$V_{lo} = \frac{V_{cond}{}^{*}400}{10{,}400}$$

$$V_{lo} = \frac{V_{cond}{}^{*}2}{52}$$

$$V_{lo} = 0.038 * V_{cond} \text{ volts}$$

Similarly, for each input combination of P and Q, the voltage at V_{lo} is shown in Table 2.11.

Similar reasoning to that used for the IMPLY operations is used to determine the voltage drop across memristor S. By voltage division, when V_{set} is applied to the S memristor, the voltage across the S memristor is $V_{set} - V_{lo}$. For $V_{cond} = 1.6$ V, the voltage across memristor S is shown for $V_{set} = 2.85$ V and $V_{set} = 2.6$ V in Table 2.12.

Table 2.12 Voltage drop on the result memristor S for $V_{cond} = 1.6$ V

Input PQ	V_{drop} on S, $V_{set} = 2.85$ V (V)	V_{drop} on S, $V_{set} = 2.6$ V (V)
00	2.79	2.54
01	1.78	1.53
10	1.78	1.53
11	1.57	1.32

Table 2.13 NAND and NOR operations in IMPLY vs. N–N gates

NAND	IMPLY			Proposed		
	V_p	V_q	V_s	V_p	V_q	V_s
1		V_{cond}	$V_{set} = 2.5$ V	V_{cond}	V_{cond}	$V_{set} = 2.6$ V
2	V_{cond}		$V_{set} = 2.5$ V			

NOR	IMPLY				Proposed		
	V_p	V_q	V_r	V_s	V_p	V_q	V_s
1	V_{cond}		$V_{set} = 2.5$ V		V_{cond}	V_{cond}	$V_{set} = 2.85$ V
2		$V_{set} = 2.5$ V	V_{cond}				
3		V_{cond}		$V_{set} = 2.5$ V			

Thus, when $V_{set} = 2.85$ V, the voltage across the S memristor has a magnitude large enough to set the memristor into low-resistance state, or logical "1," when the input is 00, 01, or 10, but not 11. This corresponds to a NOR operation on P and Q. When $V_{set} = 2.6$ V, the voltage across the S memristor has a magnitude large enough to set the memristor into low-resistance state, or logical "1," when the input is 00, but not when the input is 01, 10, or 11. This corresponds to a NAND operation on P and Q. The steps for these operations are shown in Table 2.13.

By coupling the NOR and NAND operation with the traditional IMPLY NOT operation, a complete Boolean logic set is established. The standard IMPLY NOT is chosen since it uses the same two voltages and only one memristor for the result. The XOR and XNOR operations are constructed from NOR, NAND, and NOT operations as follows:

P XNOR Q: (P NAND Q) → M0
(P NOR Q) → M1
NOT (M1) → M2
(M0 NAND M2) → M3
P XOR Q: (P NAND Q) → M0
(P NOR Q) → M1
NOT (M1) → M2 = (P OR Q)
(M0 NAND M2) → M3
NOT (M3) → M4 = (P NAND Q) AND (P OR Q) = P XOR Q

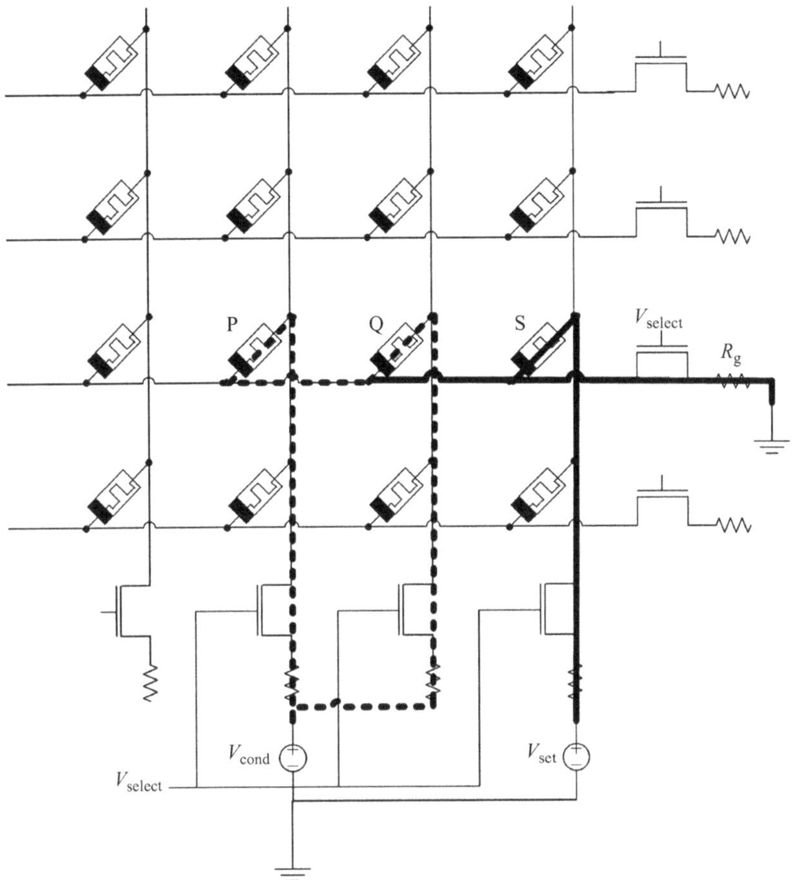

Figure 2.23 N–N gate execution in a memristor crossbar

This logic gate can be incorporated into a crossbar in the exact same manner as an IMPLY gate as shown in Figure 2.23.

The path of the read voltage V_{cond} is shown with the dotted line, and the path of the write voltage V_{set} is shown with the solid line. The only difference is the additional variant V_{set} values to allow each operation to perform. For standard IMPLY operations, the voltages V_{reset}, V_{cond}, and V_{set} are required. For the N–N gates, the additional voltage $V_{set} = 2.85$ V is required. N–N gates reduce the number of steps for NAND and NOR to 1 and AND and OR to 2. A summary of the number of steps for N–N gates in comparison to IMPLY and MAD gates in a crossbar is shown in Table 2.14.

This also reduces power consumptions since voltages are now applied for fewer steps. This logic gate requires fewer steps than the MAD gate crossbar implementation except for the XOR and XNOR operations. The component costs are also competitive with both IMPLY and MAD as shown in Table 2.15.

One benefit that N–N gates have over MAD gates is that they do not require any threshold detection across steps. In the crossbar, MAD gates have a "sense" stage, in which the input memristors are driven with V_{cond}, and one or more nodes are probed in the crossbar. These values are stored and used in the next step to drive the select lines on the result memristor. This incurs a slight overhead that N–N gates do not. Also, N–N gates do not require the additional logic block for selecting which switches and corresponding threshold voltages need to be driven for a given

Table 2.14 Step counts for IMPLY, MAD, and N–N gates

Operation	IMPLY steps	MAD steps	N–N steps
P NAND Q	2	2	1
P AND Q	3	2	2
P NOR Q	5	2	1
P OR Q	4	2	2
P XOR Q	7	2	5
P XNOR Q	8	2	4
NOT P	1	1	1
COPY P	2	2	2

Table 2.15 Component counts for IMPLY, MAD, and N–N gates

Op	IMPLY	MAD	N–N
NAND	3 memristors 3 drivers 3 transistors	3 memristors 2 drivers 3 transistors	3 memristors 2 drivers 3 transistors
AND	4 memristors 4 drivers 4 transistors	3 memristors 2 drivers 3 transistors	4 memristors 3 drivers 4 transistors
NOR	6 memristors 6 drivers 6 transistors	3 memristors 2 drivers 3 transistors	3 memristors 2 drivers 3 transistors
OR	5 memristors 5 drivers 5 transistors	3 memristors 2 drivers 3 transistors	4 memristors 3 drivers 4 transistors
XOR	6 memristors 6 drivers 6 transistors	3 memristors 2 drivers 4 transistors	7 memristors 7 drivers 7 transistors
XNOR	7 memristors 7 drivers 7 transistors	3 memristors 2 drivers 4 transistors	6 memristors 6 drivers 6 transistors
NOT	2 memristors 2 drivers 2 transistors	2 memristors 2 drivers 2 transistors	2 memristors 2 drivers 2 transistors
COPY	3 memristors 3 drivers 3 transistors	2 memristors 2 drivers 2 transistors	3 memristors 3 drivers 3 transistors

Boolean operation. Instead, only a different V_{set} value is applied. This again saves the additional overhead of the complexity for this block in the MAD gates. However, N–N gates have the same limitation as IMPLY gates that the input memristors and output memristor must be on the same row in the crossbar.

This logic gate can also be extended to N-bit operations by simply changing the value of V_{set}. A V_{cond} signal is applied to all of the input memristors P0–P7 in parallel, while the output memristor S is applied a V_{set} voltage, which can be calculated in the same manner. For example, for an 8-bit NOR, the result memristor should only be set to "1" if all of the input memristors are "0," or high resistance of 100 kΩ. When this is true, the voltage at V_{lo} is 0.22 V. When one of the input memristors is a "1," the voltage at V_{lo} is 1.09 V. Thus, the same voltage $V_{set} = 2.6$ will work for an N-bit OR too. This is especially useful for performing logical operations on two N-bit operands in parallel within memory. No other proposed logic gates have shown this functionality and is a key strength of the gate. For applications that perform a large number of vector and array operations, this gate offers high performance and is preferred.

2.4 Hybrid-CMOS gates

One consequence of the IMPLY operation and other logic-in-memory memristor gates is their serial nature. For example, in order to perform the OR operation with IMPLY operations, four steps are required instead of a single CMOS gate delay. Also, many memristors are used to hold intermediate results rather than useful data. This is partially due to the fact that the IMPLY operation overwrites one of the input operands each step.

In order to avoid the series of execution steps and accompanying driver circuitry for the IMPLY operation, prior work has proposed approaches that use both memristors and CMOS structures for logic operations in order to leverage the benefits of each domain. These are collectively known as hybrid-CMOS gates. They are intended for use in stand-alone logic blocks outside of memory for specialized arithmetic, acceleration, or other logic. In hybrid-CMOS, memristors are used to construct AND and OR gates, while traditional CMOS is used for inverters [22,23]. Vias are required to transition from the CMOS domain to the nanoscale memristor domain. This increases the overall area and cost of the design, so the number of transitions between CMOS and memristors should be minimized at design time.

An AND and OR gate constructed from memristors are shown in Figure 2.24.

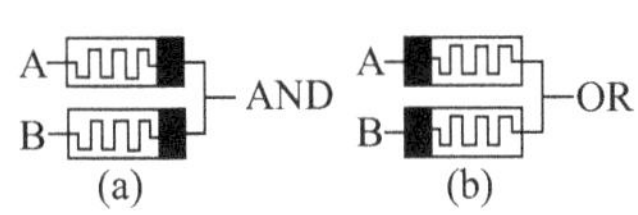

Figure 2.24 Hybrid-CMOS (a) AND gate and (b) OR gate

Voltages are applied to the A and B inputs to represent their values and a result voltage is output, where a low voltage is "0," and a high voltage is "1." Let $V_0 = 0$ V and $V_1 = 3$ V. Note that the only difference between the two gates is the polarity of the memristors. The polarity of the memristors determines which direction of current flow is considered negative vs. positive, which in turn determines how the resistance of the memristors changes as the current changes.

Consider the AND gate. When A and B are both "0," or low voltage, no current flows through the memristors and the voltage at the AND output is a "0." Similarly, when A and B are both "1," no current flows and the voltage at the output is a "1." When A is "0" and B is "1," 3 V flows from the B terminal to the A terminal. This causes the resistance in memristor B to increase and the resistance in memristor A to decrease. Thus, the voltage at the AND output is low. The circuit behaves similarly when A is "1" and B is "0." This is an AND operation. If a CMOS NOT gate is put at the output of this memristor gate, it performs a NAND operation.

For the OR gate, the only difference is the behavior of the memristor resistances in the case where one input is a 0 and the other is a 1. In this case, the resistance in memristor A increases and the resistance in memristor B decreases, causing the opposite effect of the AND gate and the voltage at the output goes high. This is an OR operation. If a CMOS NOT gate is put at the output of this memristor gate, it performs a NOR operation.

Note that the operation of the gates relies heavily on the internal resistance of the memristors when the input voltages are applied. For example, when A = "1" (3 V) and B = "0" (0 V) into an OR gate, the result at the output of the OR gate is not precisely 3 V. These complications become of greater concern when larger logic blocks are constructed from these gates. When this result feeds into another gate as an input, it is no longer a full strength "1." As the chain of gates or fan-out increases in a design, the voltage strength can continue to decrease to a point where the gates can no longer differentiate it as a "0" or "1" properly. However, the remaining downstream components must be able to distinguish a "0" from a "1" at this output to function. The exact behavior depends on many factors including the internal resistance of the memristors, the memristor characteristics, the magnitude of the voltage, the frequency, and the values of R_{on} and R_{off}. This creates a challenge to have highly sensitive, reliable, and consistent components for these gates.

Designs using hybrid-CMOS gates must take steps to ensure that the internal resistances of the memristors do not get too biased during execution. One way to avoid this issue is to periodically perform a "refresh" on the memristor gates, pulsing reset sequences to recalibrate the internal resistances. Alternatively, signal restoration buffers can be placed at the outputs of gates where necessary to restabilize the output to its intended value for the remaining downstream components. A consequence of this option is that it increases both the delay and complexity of the design.

By using the AND and OR gates presented and CMOS NOT gates, a hybrid-CMOS XOR gate can be constructed as shown in Figure 2.25.

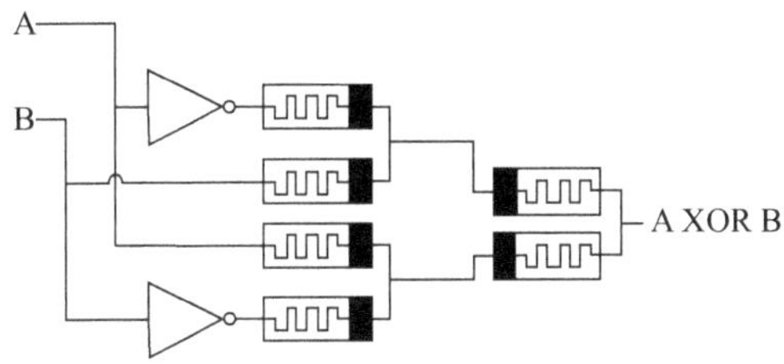

Figure 2.25 Hybrid-CMOS implementation of an XOR gate

This gate requires four MOSFETs for the NOT gates and six memristors for the two AND gates and the OR gate. Voltage flows from left to right as is done in traditional CMOS. As described, it may be necessary to buffer the output depending on the context that the XOR gate is used in.

2.5 Threshold gates

Threshold gates are studied because they lend themselves to the operations in adder computations by simplifying logic and thus reducing gate delays. They also do not exhibit the serialization issues of the IMPLY operation or the signal degradation or CMOS issues of the hybrid-CMOS approach. However, threshold gates are still in their infancy stages of fabrication and are discussed here as an emerging technology in memristor-based logic design.

A threshold gate with threshold T is a gate that behaves as follows: for N inputs, X_0 to X_N, with weights W_0 to W_N,

$$\text{Output} = \left(\sum_{i=1}^{N} W_i * X_i\right) \geq T$$

Essentially, a threshold gate has a threshold voltage T such that if the sum of the weighted input voltages is greater than or equal to T, the output of the threshold gate is "1," else it is "0." The weights for each input voltage are independent of each other and can be selected in order to achieve a certain functionality from the gate.

Threshold gates can be used to implement Boolean gates too. Consider a two-input AND gate. The output of the gate is only "1" if both inputs are "1." Also, each of the inputs has an equal weight since they equally determine the output. Thus, an AND gate can be created using the threshold gate shown in Figure 2.26.

Both inputs have the same weight $W = 1$, and the threshold on the gate is 2. When either of the inputs is a "0," the summation check fails and the output is 0. If both of the inputs are "1," the summation check is 2 and the output is 1. Thus, an AND gate is implemented.

An OR gate has a similar form but has a different threshold of 1. Now, if either of the inputs is a "1," the summation check will be at least 1 and the threshold gate

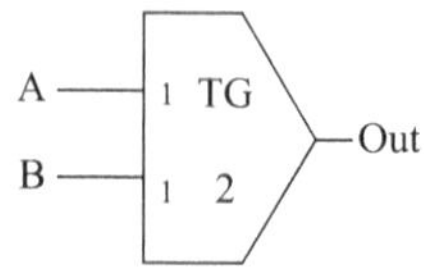

Figure 2.26 AND gate using a threshold gate

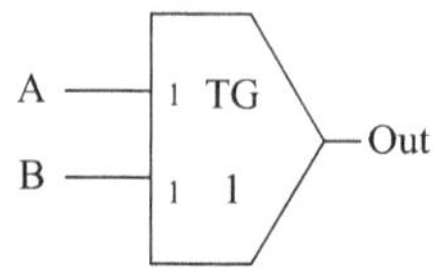

Figure 2.27 OR gate using a threshold gate

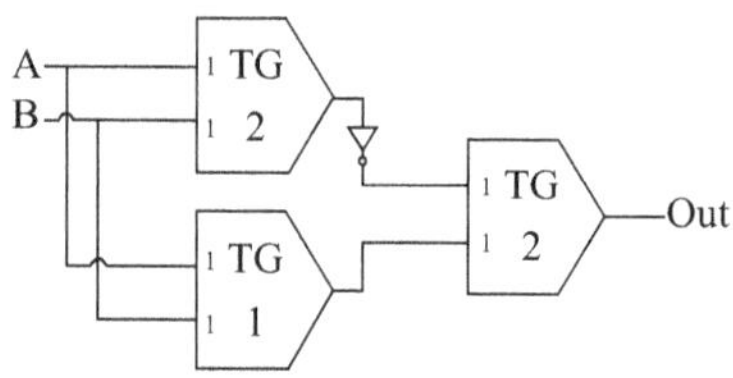

Figure 2.28 XOR operation using threshold gates

will output a 1. If both inputs are "0," the summation check will be 0, which is less than the threshold and the output will be 0. This OR gate is shown in Figure 2.27.

It is not possible to implement an XOR gate with a single threshold gate. This is because it is not possible to distinguish between the cases where inputs AB = 01 and where inputs AB = 10. Any single threshold gate will output the same value for both of these inputs. Thus, more threshold gates are required. Figure 2.28 shows an implementation for an XOR gate using threshold gates.

The XOR operation essentially performs $\text{A XOR B} = \overline{(\text{A AND B})}\ \text{AND}\ (\text{A OR B})$. In all, this operation requires two AND threshold gates, one OR threshold gate and a CMOS NOT gate. Alternative implementations for XOR operations exist, but all require at least this complexity.

The remainder of this section discusses two implementations of threshold gates: GOTO pairs and CSTG gates.

2.5.1 Threshold gates implemented with GOTO pairs

Memristor-based threshold gates can be constructed from a structure known as a GOTO pair [26]. GOTO pairs are a convenient option for memristor-based gates because they require less area and remove the need for internal vias to connect

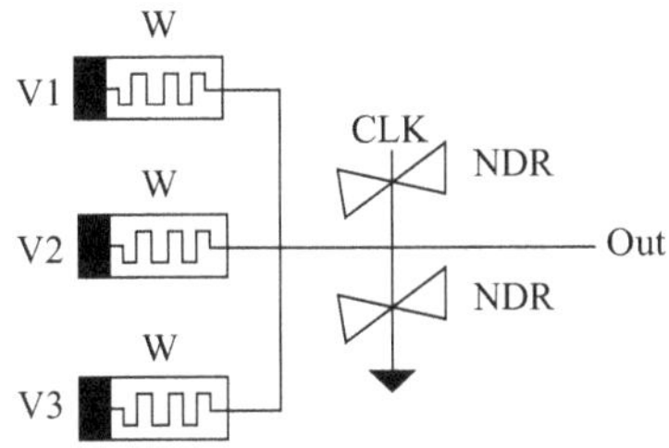

Figure 2.29 Majority gate implemented with memristors and a GOTO pair

to CMOS components. One shortcoming is that they are currently difficult for metal–oxide-based fabrication. A deeper explanation of GOTOs can be found in prior work [26], but an example of a threshold gate is shown in Figure 2.29.

The gate requires one memristor for each input, V1, V2, and V3, each with a memristive weight W_i. The weight for each memristor is selected on the basis of their relative weights in the desired Boolean equation. For example, for a simple majority gate, each weight W_i would be equal to 1.

NDR Load and NDR Drive are two resonant tunneling diodes, which perform the detection. These diodes create two stable operating points, which essentially serve as logical "0" and "1" for the output of the threshold gate. If the sum of the incoming currents, SUM(V_i*W_i), is above the threshold, the output will pull toward the logical "1" point, else it will pull toward the logical "0" point.

NOT logic is not implemented with this scheme. Although preliminary research has shown that it is possible to perform a NOT with memristors and GOTO pairs, it is highly sensitive and exhibits high variability. Thus, for this work, NOT logic is performed with traditional CMOS logic.

2.5.2 Threshold gates implemented with CSTG gates

Due to the difficulties of GOTO pair fabrication, a CMOS buffer is considered as an alternative for the threshold detector. This can be implemented using charge-sharing threshold gates (CSTGs) as shown in prior work [24] such as the one in Figure 2.30.

A single CSTG requires four MOSFETs to perform the threshold detection on the output plus one memristor, one resistor, and two MOSFETs per input as shown for the three inputs on the left of the figure. As M_i approaches high resistance, the node at W_{Ni} approaches V_{Mi}. As M_i approaches low resistance, the node at W_{Ni} approaches 0. Thus, the larger the weight of M_i, the stronger its contribution will be in the overall threshold detection at V_s.

An N-input threshold gate requires N memristors, N resistors, and $2N+4$ MOSFETs. Although greater in complexity, CSTGs generate both the output and its inverse as a by-product of the threshold detection. This can be leveraged to achieve lower area and lower delay in some arithmetic designs as described in later sections.

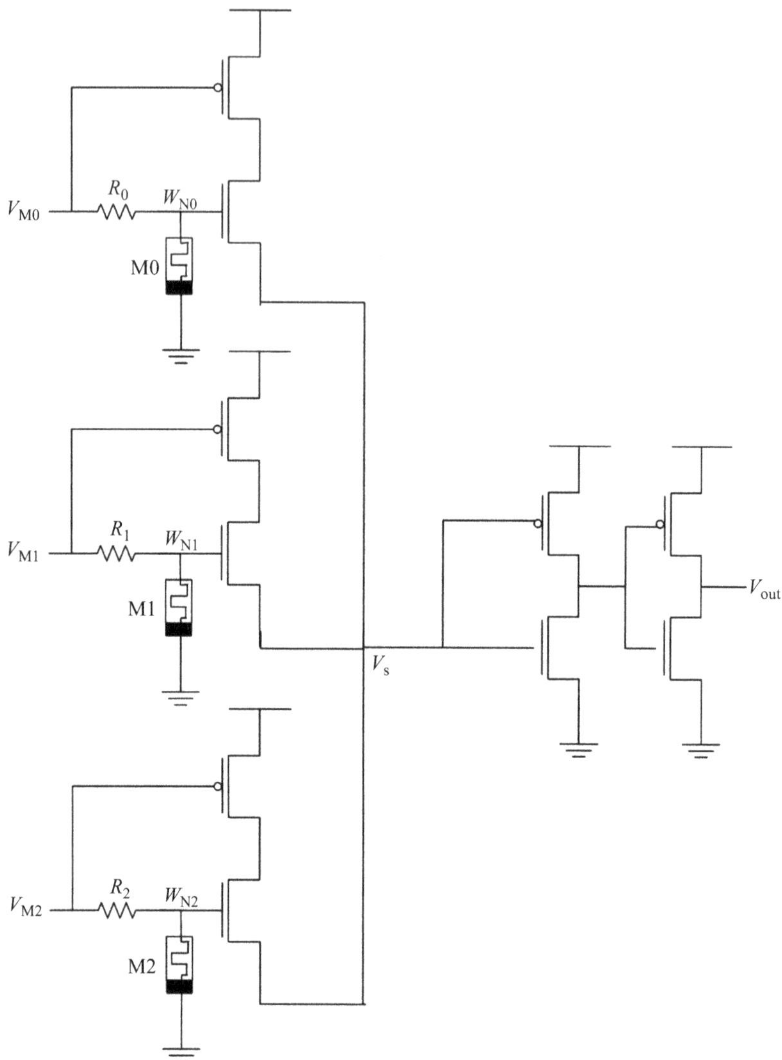

Figure 2.30 Example CSTG for threshold detection

2.6 MAGiC gates

Another type of logic gate for memristors in the MAGiC gate as introduced in previous literature [12]. MAGiC gates are capable of performing all Boolean operations however only the NOR is compatible with a traditional crossbar memory. For this reason, their adoption has not been as rapid, but they do offer competitive step and component counts. MAGiC gates take two input memristors as operands, holding a logical “0” or “1” in the same manner as discussed for other logic gates.

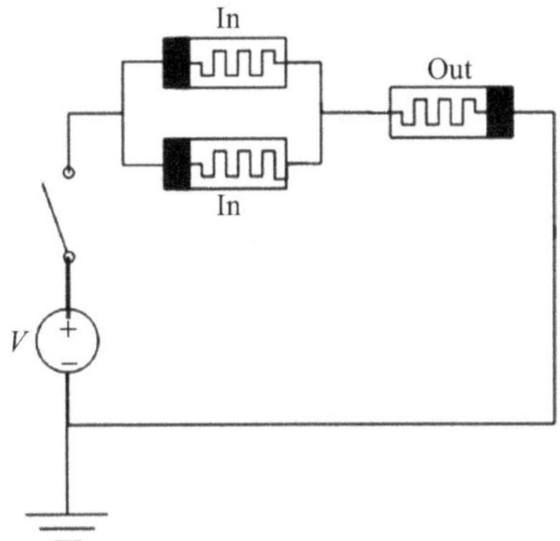

Figure 2.31 MAGiC NOR Gate [12]

The result is stored into a third memristor. The NOR gate behaves as follows. The output memristor is initialized to logical "1," a difference from the other logic gate approaches discussed. Then, a voltage V_0 is applied to the input memristors, while the output memristor is tied to GND as shown in Figure 2.31. Note that all memristors must lie on the same row in the crossbar.

When both inputs are "0," the voltage at the junction between the input and output memristors approaches 0, and the voltage across the output memristor is not enough to change its value. When one or both of the inputs are "1," the voltage is higher and approaches $0.67{*}V_0$ which is large enough to change the internal resistance of the output memristor to 0. Thus, a NOR is performed. The NOR gate requires one cycle after initialization and a single voltage and GND to perform. These are fewer voltages than the IMPLY, MAD, or N–N approaches discussed. However, only the NOR gate can exist in a traditional memristor crossbar. The remaining Boolean operations are not compatible due to the polarity of the memristors in their circuitry.

Chapter 3

Adder arithmetic units

Adders are a universal component of nearly every computing system, whether traditional or modern. Even the most fundamental machine, which performs no arithmetic, requires an adder to calculate address offsets and branch instructions. Adders are also generally less complex than other arithmetic units found in system. This chapter introduces four different adders: a ripple-carry adder, a carry-lookahead adder, a carry-select adder, and a conditional-sum adder. The traditional CMOS circuits for each of the adders are described, and the performance and area of these designs are each analyzed. The adders vary in terms of their complexity. The simplest design is the ripple-carry adder, and the most complex design is the carry-lookahead adder. In general, as the complexity of the adder increases, the speed and performance improve.

3.1 Ripple-carry adders

Ripple-carry adders are considered the baseline adder model and prioritize simplicity over performance. An example of a 4-bit ripple-carry adder can be seen in Figure 3.1 [15].

Each bit has a full adder (FA) with three inputs, A, B, and a carry-in, and two outputs, a sum and carry-out output. The carry-out output of each bit j is used as the carry-in input for bit $j+1$, building a chain. The equations for the sum and carry-out bits are as follows:

$$\begin{aligned} \mathrm{Sum} &= \mathrm{A\ XOR\ B\ XOR\ C_{in}} \\ \mathrm{C_{out}} &= \mathrm{(A\ AND\ B)\ OR\ ((A\ OR\ B)\ AND\ C_{in})} \end{aligned}$$

A traditional N-bit CMOS ripple-carry adder requires nine gates for each FA as shown in Figure 3.2.

At first glance, the critical path to resolve the sum is six cycles from input B and the critical path to resolve the carry-out is five cycles from B. However, this latency can be hidden, or amortized, across adders. Let us consider a 4-bit adder. All of the A and B inputs are available to all of the FAs at $t=0$. Only the carry-in inputs are not known yet. Thus, all the gates and paths, which do not require the carry-in, can resolve. This means that the first four gates in Figure 3.2 can all resolve in cycles $t=1$, 2, and 3. Now, when each FA obtains its carry-in, there is only a two-cycle delay to the carry-out signals and a three-cycle delay to the sum

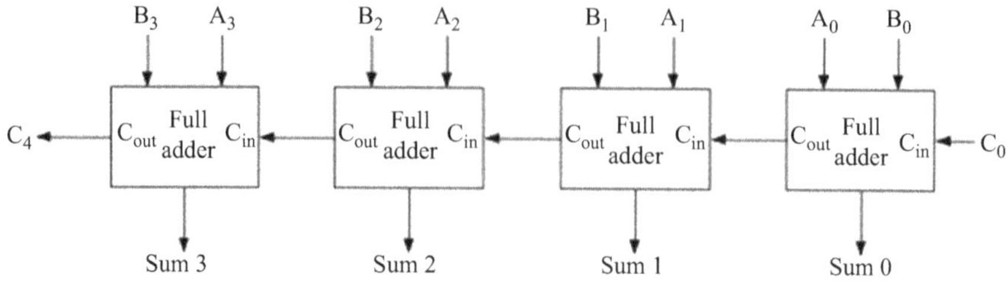

Figure 3.1 Ripple-carry adder diagram

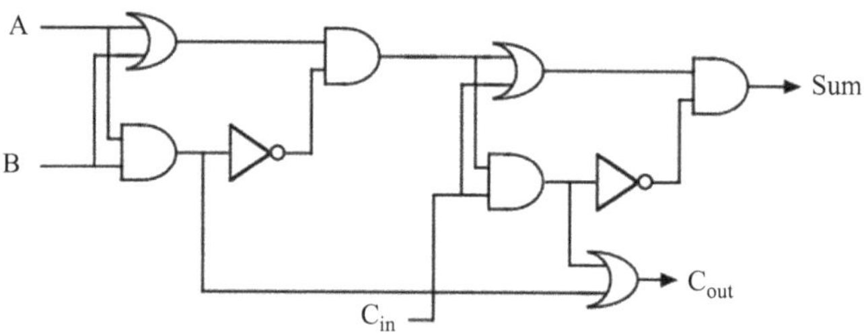

Figure 3.2 Traditional CMOS implementation of a full adder

signal. Let us consider the steps of the execution. This first FA, for bit 0, begins at $t=0$ and produces its sum at $t=6$ and its carry-out at $t=5$. Now the FA of bit 1 can take the carry-in at $t=5$ and produces its sum at $t=8$ and carry-out at $t=7$. The full execution is shown in Table 3.1.

This "ripple" effect through the adders is where the ripple-carry adder gets its name. This N-bit adder has a delay of $2N+4$ gate delays for the final sum and requires $9N$ gates or equivalently $34N$ MOSFETs.

3.2 Carry-lookahead adders

Carry-lookahead adders were first explored as an alternative to ripple-carry adders to eliminate the long delay imposed by the carry chain [51]. However, carry-lookahead adders require significantly more complexity than ripple-carry adders. Of the four adder designs discussed, carry-lookahead adders require the most complexity and the fewest steps. A diagram for a carry-lookahead adder is shown in Figure 3.3.

Carry-lookahead adders accelerate delay from $O(N)$ to $O(\log N)$ by defining propagate, P_i and generate bits, G_i and combining these bits into larger blocks. The propagate bit is "1" if the FA for this bit would propagate an incoming carry-in bit to the carry-out bit. So if carry-out is 1 when carry-in is 1 in an FA i then $P_i=1$. The generate bit is "1" if the FA for this bit will produce a carry-out of "1"

Table 3.1 Execution steps for a 4-bit ripple-carry adder

t	**Full adder bit 0**	**Full adder bit 1**	**Full adder bit 2**	**Full adder bit 3**
1	A_0 AND B_0 = G0 A_0 OR B_0 = G1	A_1 AND B_1 = G0 A_1 OR B_1 = G1	A_2 AND B_2 = G0 A_2 OR B_2 = G1	A_3 AND B_3 = G0 A_3 OR B_3 = G1
2	NOT G0 = G2	NOT G0 = G2	NOT G0 = G2	NOT G0 = G2
3	G1 AND G2 = G3	G1 AND G2 = G3	G1 AND G2 = G3	G1 AND G2 = G3
4	G3 OR C_{in0} = G4 G3 AND C_{in0} = G5			
5	NOT G5 = G6 G5 AND G0 = C_{out0}			
6	G6 AND G4 = S_0	G3 OR C_{in1} = G4 G3 AND C_{in1} = G5		
7		NOT G5 = G6 G5 AND G0 = C_{out1}		
8		G6 AND G4 = S_1	G3 OR C_{in2} = G4 G3 AND C_{in2} = G5	
9			NOT G5 = G6 G5 AND G0 = C_{out2}	
10			G6 AND G4 = S_2	G3 OR C_{in3} = G4 G3 AND C_{in3} = G5
11				NOT G5 = G6 G5 AND G0 = C_{out3}
12				G6 AND G4 = S_3

regardless of the value of carry-in. For each bit *i* in the addition, P_i and G_i have the following equations:

$$P_i = A_i \text{ XOR } B_i$$

$$G_i = A_i \text{ AND } B_i$$

These bits can be produced in a single CMOS gate delay at $t = 1$.

Consider a 16-bit carry-lookahead adder with 4-bit block. A "block" contains four FAs and their corresponding G and P logic. Each bit in the *N*-bit addition computes its P_i and G_i. Then, these individual P and G bits can be used to create the group propagate (GP) and group generate (GG) bits for each *n*-bit block as follows:

$$GP = P0 * P1 * P2 * P3$$
$$GG = G3 + G2 * P3 + G1 * P2 * P3 + G0 * P1 * P2 * P3$$

The GP and GG bits have the same meaning as the propagate and generate bits, but represent the behavior for the carry-in and carry-out of the *n*-bit block as a whole

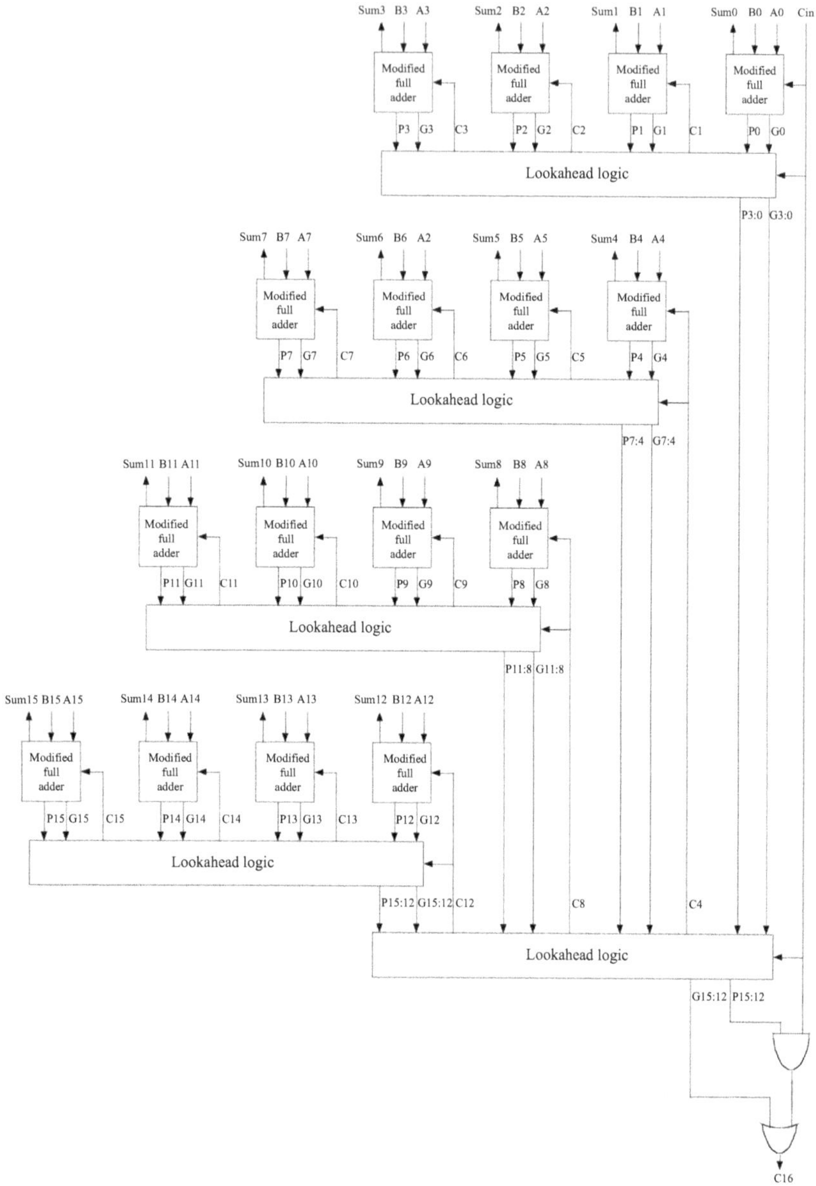

Figure 3.3 16-bit carry-lookahead adder diagram

rather than the individual adders. The GP signals are resolved at $t = 2$, one cycle after the individual P bits, and the GG signals are resolved at $t = 5$.

Then, each intermediate carry-out bit can be represented in terms of these P_i and G_i bits coupled with C_0, the initial carry-in bit. Consider the first 4-bit block for bits 0–3.

$$\begin{aligned}
C_{out0} &= G_0 + P_0C_{in} \\
C_{out1} &= G_1 + P_1C_{in} \\
C_{out1} &= G_1 + P_1[G_0 + P_0C_{in}] \\
C_{out1} &= G_1 + P_1G_0 + P_1P_0C_{in} \\
C_{out2} &= G_2 + P_2C_{in} \\
C_{out2} &= G_2 + P_2[G_1 + P_1G_0 + P_1P_0C_{in}] \\
C_{out2} &= G_2 + P_2G_1 + P_2P_1G_0 + P_2P_1P_0C_{in} \\
C_{out3} &= G_3 + P_3C_{in} \\
C_{out2} &= G_3 + P_3[G_2 + P_2G_1 + P_2P_1G_0 + P_2P_1P_0C_{in}] \\
C_{out2} &= G_3 + P_3G_2 + P_3P_2G_1 + P_3P_2P_1G_0 + P_3P_2P_1P_0C_{in}
\end{aligned}$$

Or alternatively, C_{out2} can be represented using the GG and GP bits since it is the carry-out bit of the block:

$$C_{out2} = GG_0 + GP_0C_{in}$$

With these equations, a 4-bit block can resolve its four carry-out bits in parallel, completing at $t = 3$, $t = 4$, $t = 5$, and $t = 6$, respectively. The resolution of the block's carry-out bit C_{out2} is on the critical path as it is used as the carry-in bit for the second block containing bits 4–7. Once C_{out2} is resolved at $t = 6$, the block for bits 4–7 can compute its carry bits using the same equations as above.

Notice that all of the N-bit blocks compute their GP bit at $t = 2$ and GG bit at $t = 5$. Therefore, for each subsequent block, from block 2 to block $(N/n) - 1$, the delay from carry-in to carry-out for the entire block is only two cycles. The sum bits are also calculated for each bit, but these do not fall on the critical path and can be computed in parallel with other operations. They use the same equations as a standard FA and are resolved three cycles after the carry-in is resolved. Specifically,

$$\begin{aligned}
Sum_0 &= A_0 \text{ XOR } B_0 \text{ XOR } C_{in} \\
Sum_1 &= A_1 \text{ XOR } B_1 \text{ XOR } [G_0 + P_0C_{in}] \\
Sum_2 &= A_2 \text{ XOR } B_2 \text{ XOR } [G_1 + P_1G_0 + P_1P_0C_{in}] \\
Sum_3 &= A_3 \text{ XOR } B_3 \text{ XOR } [G_2 + P_2G_1 + P_2P_1G_0 + P_2P_1P_0C_{in}]
\end{aligned}$$

Thus, the final sum bit for bit $N-1$ is the final signal to resolve and completes at $t = 17$ for a total delay of 17 cycles for a 16-bit carry-lookahead adder. This approach can be extended in logarithmic fashion to larger sizes of N and n.

3.3 Carry-select adders

The carry-select adder attempts to perform more logic in parallel to amortize the cost of the adder as the size N grows. It incorporates greater complexity than a ripple-carry adder in order to decrease delay. In general, the critical path for an adder lies on the carry-in to carry-out path, so the carry-select adder focuses on removing delay here.

The carry-select adder splits the N bits of the input words into groups of n bits and assigns each group to two n-bit blocks for a total of $2(N/n)$ blocks. It then operates on each block in parallel. Each block is an n-bit ripple-carry adder. The difference between the two blocks for a given n-bit group is that one assumes a carry-in input $C_{in0} = 0$ and the other assumes $C_{in0} = 1$. The only exception is for the lowest n-bit block, since the carry-in bit for this block is known at $t = 0$. Now, all of the n-bit ripple-carry adders can begin at $t = 0$ and complete at $t = 2n + 3$ for the carry-out signals and $t = 2n + 4$ for the sum. Block 0 is complete once these two signals are resolved. Block 1 receives its true carry-in from block 0 and uses it to select the appropriate sum output and carry-out. Execution continues in this manner through each block such that the delay through each n-bit block is now the delay of a mux rather than the delay of an n-bit ripple-carry adder.

Figure 3.4 shows an example for a 16-bit carry-select adder with 4-bit block sizes.

The total delay of an N-bit adder with (N/n) n-bit blocks is calculated as follows:

$$\begin{aligned}
&2n + 3 \text{ steps to resolve the carry-out of the first } n\text{-bit block} \\
&+2(N/n - 2) \text{ steps between each block to resolve the next carry bit} \\
&+3 \text{ steps to resolve the final multiplexer for the sum in block } n - 1 \\
&= \mathbf{2(N/n) + 2n + 2 \ \ steps}
\end{aligned}$$

The optimal size of the blocks n is computed by taking the derivative of the delay function with respect to n:

$$\begin{aligned}
&d(2(N/n) + 2n + 2) = 0 \\
&-2Nn^{-2} + 2 = 0 \\
&Nn^{-2} = 1 \\
&N = n^2 \\
&\boldsymbol{n} = \sqrt{N}
\end{aligned}$$

So the delay for the optimal block size is $4\sqrt{N} + 2$. The number of gates required for an N-bit carry-select adder with n-bit block sizes is as follows:

$$\begin{aligned}
&[2(N/n) - 1] * 9n \text{ gates for the ripple-carry adders} \\
&+(N/n - 2) * 2 \text{ gates for the carry propagation} \\
&+(N/n - 1) * 4n \text{ gates for the sum multiplexers} \\
&= 2(N/n) - 4 + 4N - 4n + 18N - 9n \text{ gates} \\
&= \mathbf{22N - 13n + 2(N/n) - 4 \ \ gates}
\end{aligned}$$

For the optimal block size $n = \sqrt{N}$, this becomes $22N - 11\sqrt{N} - 4$ gates.

3.4 Conditional-sum adders

Conditional-sum adders use a hierarchy of multiplexers in order to improve the carry propagation time over ripple-carry adders and other adder designs. The first

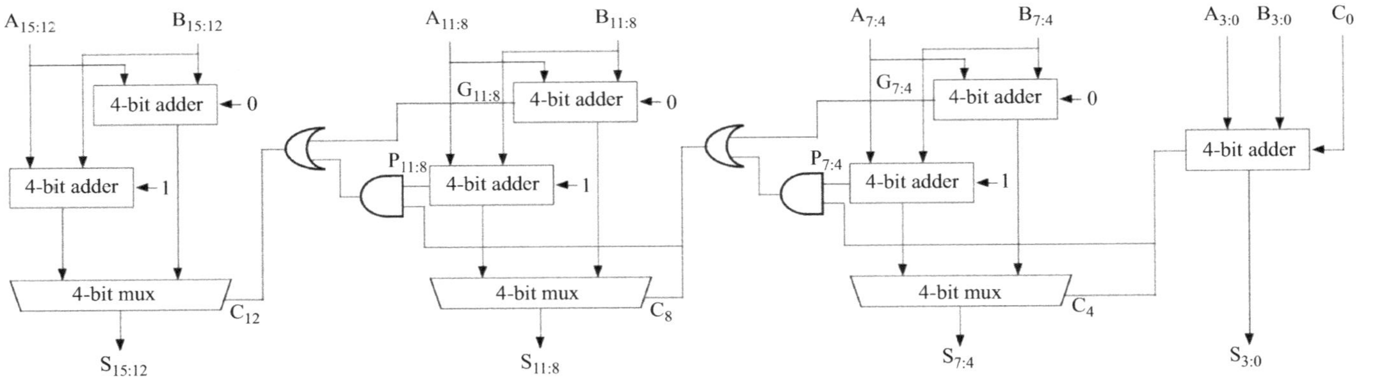

Figure 3.4 16-bit carry-select adder

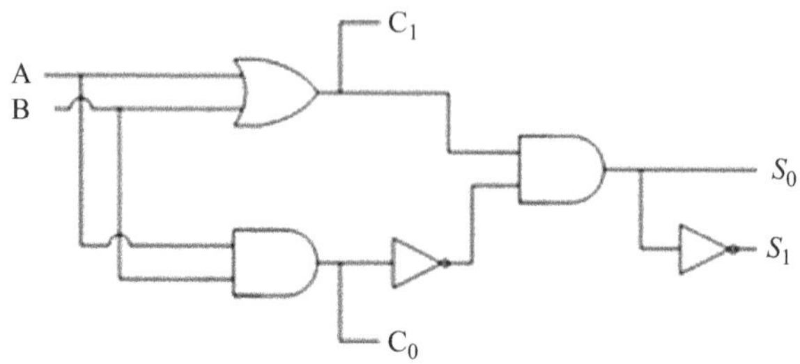

Figure 3.5 Modified half adder circuit

bit uses a standard FA cell since the carry-in bit is known at time 0. The remaining sets of input bits each feed modified half adders (MHAs), which produce four output bits, C_0, C_1, S_0, and S_1. An MHA is shown in Figure 3.5 [16].

C_0 and S_0 correspond to the carry-out and sum bits if the carry-in bit is a 0. C_1 and S_1 correspond to the carry-out and sum bits if the carry-in bit is a 1. Each of these four output bits can be generated as soon as inputs A_i and B_i arrive at $t=0$. C_0 and C_1 are resolved at $t=1$ and S_0 and S_1 are resolved at $t=3$ and $t=4$, respectively. Once the carry-in bit is available, it is used as the select line to multiplexers to select C_1 and S_1 if the carry-in is 1 or C_0 and S_0 if the carry-in is 0. This minimizes the carry propagation delay from one bit to the next to the delay of a multiplexer. An example of an 8-bit conditional-sum adder is given in Figure 3.6.

Conceptually, each of the three tiers of multiplexers "narrows down" the possibilities for the output. The outputs of the MHAs are the sum and carry-out bits if the carry-in bit is a 0 or 1. In the first tier of multiplexers, each pair of MHAs is given a pair of 2-bit multiplexers, one for the scenario in which the carry-out from the first MHA is a 0, and the other for the scenario in which the carry-out bit from the first MHA is a 1. All four outputs serve as inputs to the two 2-bit multiplexers. These multiplexers are essentially "narrowing down" the output possibilities based on the information gained about the carry-out bits from the first MHA in each pair. Now, proceed to the next tier of multiplexers in which more information has been collected. The definitive value of the carry-out value of MHA1 is now known as well as the two possible carry-out values from MHA5. The carry-out value of MHA1 is used to resolve the sum bits of MHA2 and MHA3 and the carry-out of MHA3. In the last tier of multiplexers, the possible sum values of MHA4, MHA5, MHA6, and MHA7 only depend on the carry-in value of MHA3. Thus, in a final multiplexer stage, all remaining output bits are resolved. In essence, execution progresses logarithmically through the stages, with each stage taking the delay of a multiplexer.

Now consider how to calculate the delay of the adder. The FA for bit 0 and the MHAs for bits 1 to $(N-1)$ all begin at $t=0$. At $t=1$, C_0, and C_1 are resolved in all of the MHAs. At $t=3$ and $t=4$, S_0 and S_1 are resolved in all of the MHAs. Now the hierarchy of multiplexers can begin execution. However, the multiplexer for MHA1 must wait on FA0 to finish computing its carry-out bit at $t=5$. Consider the

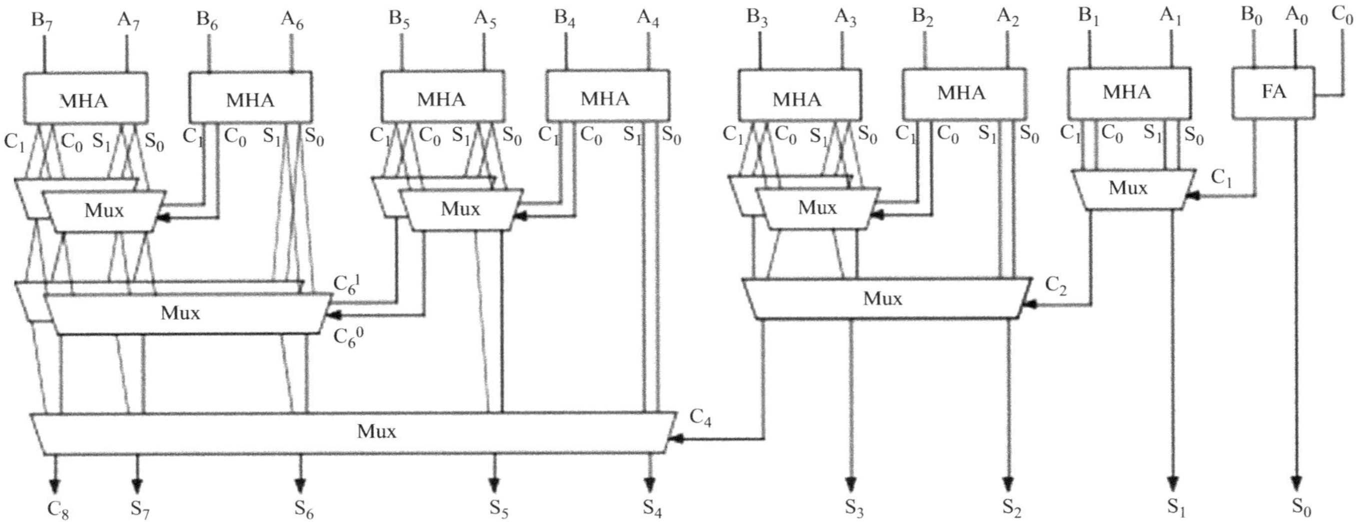

Figure 3.6 8-bit conditional-sum adder

Table 3.2 Carry and sum bit resolution in the conditional-sum adder

t	**Bit 7**	**Bit 6**	**Bit 5**	**Bit 4**	**Bit 3**	**Bit 2**	**Bit 1**	**Bit 0**
1	$C7_0$, $C7_1$	$C6_0$, $C6_1$	$C5_0$, $C5_1$	$C4_0$, $C4_1$	$C3_0$, $C3_1$	$C2_0$, $C2_1$	$C1_0$, $C1_1$	
2								
3								
4	Mux level 1 C_{out}		Mux level 1 C_{out}		Mux level 1 C_{out}			
5								C0
6	Mux level 1 Sum		Mux level 1 Sum		Mux level 1 Sum			S0
7	Mux level 2 C_{out}	Mux level 2 Sum						
8							C1, S1	
9	Mux level 2 Sum							
10								
11					C3, S3	S2		
12								
13								
14	C7, S7	S6	S5	S4				

pair of multiplexers coupled with MHA2 and MHA3. MHA2 produces its carry-out bits $C2_0$ and $C2_1$ at $t = 1$, providing the select lines for the multiplexers. MHA3 also produces its carry-out bits $C3_0$ and $C3_1$ at $t = 1$, so the two multiplexers can resolve their carry outputs at $t = 4$. The sum bits are ready at $t = 3$ and $t = 4$, so the multiplexers can resolve their sum outputs at $t = 6$. This timing is true for all of the pairs of multiplexers for the MHAs. In parallel, FA0 resolves its carry-out bit C0 at $t = 5$ and sum bit S0 at $t = 6$. Thus, the first multiplexer resolves its outputs C1 and S1 at $t = 8$. The resolution of the carry and sum bits now continues in a hierarchical fashion, with each level of multiplexers adding a constant delay. The breakdown of the sum and carry-bits resolution is shown in Table 3.2.

The total delay for the 8-bit conditional-sum adder is 14 steps. Notice that in each stage of multiplexers, twice as many bits are resolved as the previous stage. This is because of the logarithmic nature of the adder. The delay of an N-bit conditional-sum adder is proportional to $\log_2(N)$, increasing by one tier of multiplexers (and a delay of three cycles) each time the adder width is doubled. This is a significant improvement over ripple-carry adders and other designs but comes at the expense of great complexity for the hierarchy of multiplexers.

The total cost of an N-bit conditional-sum adder can be deduced recursively. Consider Figure 3.6 of the conditional-sum adder. The logic for MHA1, MHA2, and MHA3 is replicated twice to make the logic for MHA4, MHA5, MHA6, and MHA7. The only extra hardware is the additional tier's multiplexer. Given this knowledge, the complexity for an N-bit conditional-sum adder is as follows:

$$
\begin{aligned}
\text{4-bit:}\ & 9 + 5n \text{ gates for the MHAs and FA} \\
& + 9^{*}3 + 4 \text{ gates for the nine 1-bit multiplexers and four select lines} \\
& = 31 \text{ gates for multiplexers} + (5n + 9) \text{ gates for FA and MHAs} \\
& = 60 \text{ gates}
\end{aligned}
$$

$$
\begin{aligned}
\text{8-bit:}\ & 9 + 5n \text{ gates for the MHAs and FA} \\
& + 31^{*}3 \text{ gates for the double replication of multiplexers} \\
& + 5^{*}3 + 1 \text{ gates for the five 1-bit multiplexers and 1 select line} \\
& = 109 \text{ gates for multiplexers} + (5n + 9) \text{ gates for FA and MHAs} \\
& = 158 \text{ gates}
\end{aligned}
$$

$$
\begin{aligned}
\text{16-bit:}\ & 9 + 5n \text{ gates for the MHAs and FA} \\
& + 109^{*}3 \text{ gates for the double replication of multiplexers} \\
& + 9^{*}3 + 1 \text{ gates for the nine 1-bit multiplexers and 1 select line} \\
& = 355 \text{ gates for multiplexers} + (5n + 9) \text{ gates for FA and MHAs} \\
& = 247 \text{ gates}
\end{aligned}
$$

And can continue in this manner for any adder width N.

Chapter 4

Multiplier arithmetic units

Multipliers are becoming ubiquitous in modern computing systems. With the drive for big data and machine-learning applications, multipliers have become an underlying construct for nearly all state-of-the-art workloads. Depending on the specific system, design constraints, and expected application, the optimal multiplier implementation can vary. Sometimes, a simpler, slower multiplier is preferred and sometimes a more complex, faster multiplier is. This chapter presents four popular multipliers, which span these two metrics: shift-and-add multipliers, Booth multipliers, array multipliers, and Dadda multipliers. The performance and area of these designs are each analyzed in the CMOS domain.

4.1 Shift-and-add multipliers

Shift-and-add multipliers consist of three main components, an adder (usually a ripple-carry adder), a multiplexer, and shift registers. Both the product and the multiplier reside in a shift register. The multiplicand resides in a standard register. An example diagram for a shift-and-add multiplier is shown in Figure 4.1.

At initialization, the product register is set to 0 and the multiplicand and multiplier are loaded into the A and B registers. For an N-bit multiplier, the product register will be size $2N$ to hold the maximum value. At the beginning of each clock cycle, bit 0 of the multiplier is used to select either the multiplicand or the value "0." This value is then added to the N most significant bits of the running sum in the product register and stored back into the product register into its most significant bits. At the end of the cycle, both the multiplier and the product register are shifted one bit to the right, discarding the bit in the multiplier and shifting the product down by one bit. This process repeats for N cycles. Essentially, the multiplier is adding the multiplicand A to itself B times. By storing the product into the most significant bits and shifting to the right after each addition, the multiplicand is added with the correct weight each time. Table 4.1 shows an example for the multiplication 0101 * 1011 = 00110111.

The total complexity of the multiplier is the N-bit multiplicand register, N-bit multiplier shift register, $2N$-bit product shift register, and an N-bit adder. The total delay is N iterations of the addition + shift operations.

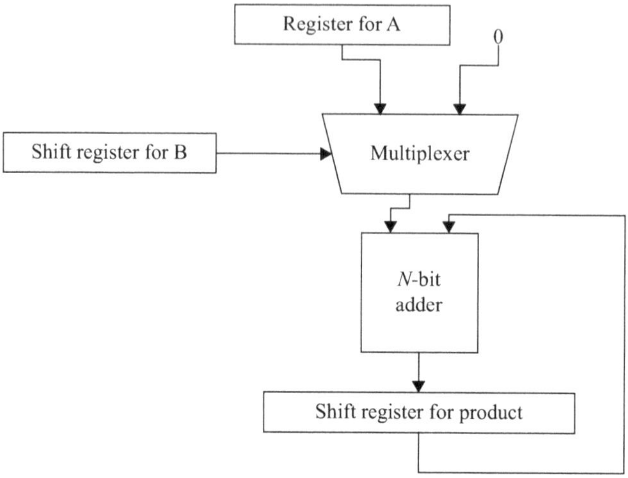

Figure 4.1 Shift-and-add multiplier flow diagram

*Table 4.1 Execution steps of a 4-bit shift-and-add multiplier for 0101 * 1011 = 00110111*

T	**A**	**B**	**Adder**	**Product register**
0	0101	1011		00000000
1a	0101	1011	0101+0000	01010000
1b shift		0101		00101000
2a	0101	0101	0101+0010	01111000
2b shift		0010		00111100
3a	0101	0010	0000+0011	00111100
3b shift		0001		00011110
4	0101	0001	0101+0001	01101110
		0000		00110111

4.1.1 Booth multiplier

One limitation of the shift-and-add multiplier is that it is not compatible with 2's complement numbers. The Booth multiplier is a slight variation on the shift-and-add multiplier that supports both positive and negative 2's complement numbers. The Booth multiplier performs in the same manner as the shift-and-add multiplier but shifts multiple bits per iteration. For example, there are radix-2 Booth multipliers, which shift two bits during each shift operation. It does this with the following encoding: if multiplier [1:0] = 01 you add the multiplicand, if multiplier [1:0] = 10 you subtract the multiplicand, else you add 0. In order to support this functionality, the Booth multiplier has an additional N-bit register to hold the 2's complement of the multiplicand. Also, the multiplier is initialized into bits [N:1] to start with bit 0 = 0. Consider the multiplier steps for 4 * −11 = −44 shown in Table 4.2.

*Table 4.2 Execution steps of a 4-bit Booth Multiplier for 4 * −11 = −44*

T	A	B	Adder	Product register
0	00100	101010		0000000000
1a	00100	101010	10: subtract	1110000000
1b shift		010101	00000+11100	1111000000
2a	00100	010101	01: add	0001000000
2b shift		001010	11110+00100	0000100000
3a	00100	001010	10: subtract	1110100000
3b shift		000101	00001+11100	1111010000
4	00100	000101	01: add	0001010000
		000010	11110+00100	0000101000
5	00100	000010	10: subtract	1110101000
			00001+11100	1111010100

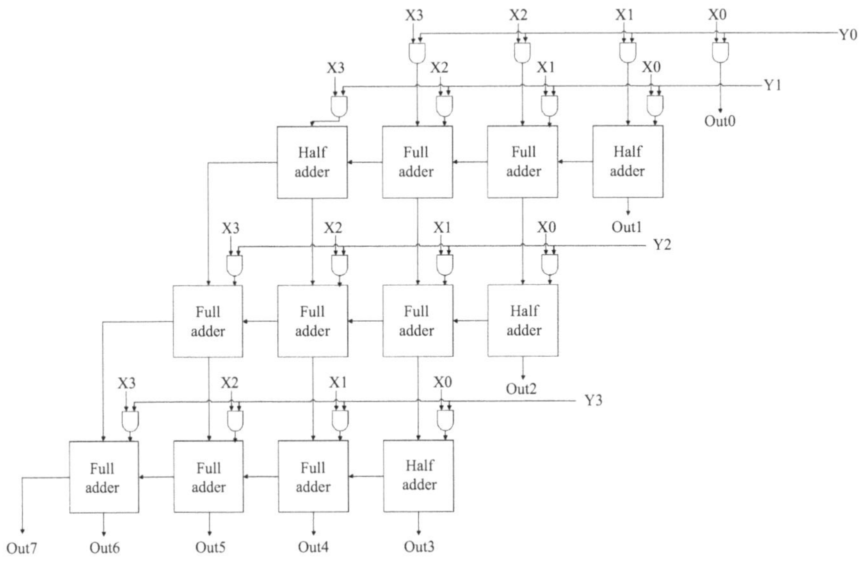

Figure 4.2 4-bit array multiplier

4.2 Array multipliers

Array multipliers use an array of half adders and full adders to accomplish multiplication. Array multipliers extend an $N \times N$ multiplication operation into N additions. This is similar to the functionality of a shift-and-add multiplier except that the additions are pipelined across space to reduce the delay. In order to achieve this pipelining, the adder circuit is replicated N times, creating an array of $(N-1) \times N$ adders. A diagram for a 4×4 array multiplier is given in Figure 4.2.

Execution begins at the top and ripples through to the bottom and to the left in which the final sum and carry-out are produced. Each row is an addition and thus is physically shifted to the left by one cell to account for the weight of each addition. There are $N \times N$ AND gates, which compute the correct input values for that row's addition. Then, the additions begin. The first row adds its two N-bit inputs and propagates its sum and carry-outputs to the next row of addition and so on until the addition completes. The total delay is the longest path through the array from the top right corner to the bottom left corner: an AND gate, one half adder, and $3N - 4$ full adders. The total complexity is N^2 AND gates, $N^2 - 2N - 1$ full adders, and $(N + 1)$ half adders.

4.3 Dadda multipliers

A Dadda multiplier is a popular multiplier for fast computations that uses stages of addition to perform the overall multiplication. Each stage consists of full adders and half adders, each performing independent additions based on inputs from the previous stage. To reduce the number of adders in the design, Dadda multipliers limit the reduction in each stage of the multiplication. Once the reduction has completed and only two rows of operands remain, a standard adder is used to complete the multiplication. This addition can be performed using any adder unit described or others.

An example of a Dadda multiplier reduction for an 8-bit multiplication is given in Figure 4.3.

Each dot in step 0 represents a partial product A_iB_i from the original inputs A and B for a total of N^2 combinations. In the remaining steps, a diagonal line represents a full adder and a diagonal line with a cross-stitch represents a half adder. A standard full adder takes three inputs with the same weight w and produces two outputs, the sum with weight w and the carry-out with weight $w + 1$. In the figure, this can be visualized as each full adder "swallowing" three input dots from one column and producing two output dots. One of the output dots

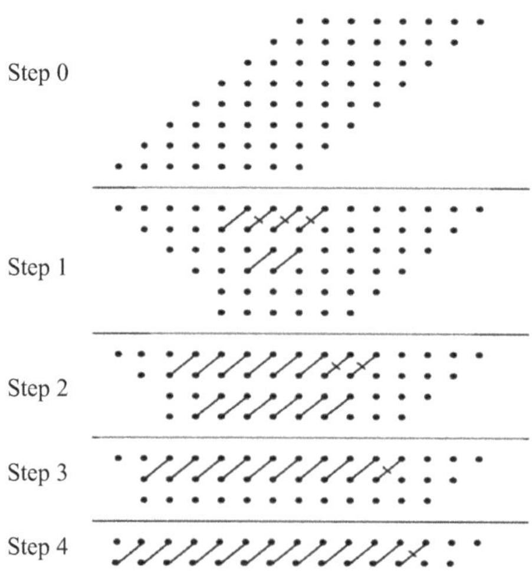

Figure 4.3 Dadda multiplier execution for an 8-bit multiplication

(the sum bit) is in the same column as the inputs and the other (the carry-out bit) is in the next column. A half adder performs the same except it only "swallows" two input dots.

In each stage, the height of the multiplication matrix is reduced. Each stage has a maximum height, represented as

$$D_x = 2, D_{x-1} = \text{floor}(1.5 * D_x)$$

where x is the final addition stage. In Figure 4.3, the final step has a height of two and the earlier stages of heights of three, four, six, and eight, respectively. With this knowledge, the number of half adders and full adders necessary in each step to reduce the height to the maximum height for that stage can be determined. Consider step 1 with a maximum height of six, reduced from eight. Begin from the right-most column in step 0 and progress left through the columns to perform the reduction. For an 8-bit multiplication, this column is C_{15}. Columns C_{15}, C_{14}, C_{13}, C_{12}, C_{11}, and C_{10} all have height six or less. Since this height is equal or less than the desired height for this stage, the columns can be copied directly to this stage. C_9 has a height of seven so a half adder can be used to "swallow" two inputs in the C_9 and produce one output in C_9 and one output in C_8. This reduces the number of dots in C_9 by 1 and increases the number of dots in C_8 by 1 as shown in the figure. C_8 now has nine dots. This can be reduced to six with a full adder and a half adder. In total, these two adders "swallow" five dots and produce two in C_8, which add two dots to C_7. Execution continues through C_1. Each of the subsequent steps is executed in the same manner to achieve the maximum height. After step 4, an adder performs the final 14 bit addition to generate the product.

The total complexity of the Dadda multiplier is dependent on the size of the multiplier because the maximum heights of the stages changes for each size N. For an 8-bit multiplier, there are 64 AND gates for the partial products, 32 full adders, 10 half adders, and a 14-bit adder. If the adder is a standard ripple-carry adder, this is a total complexity of 508 gates. The delay is the delay of the partial products, four steps with a delay of a full adder, and a final 14-bit addition for a total delay of 57 steps.

Chapter 5
Divider arithmetic units

Dividers are commonly found in floating point units. These arithmetic units are not only found in standard computing systems but also in high-performance computing as well. Similar to the presented adders and multipliers in the previous two chapters, design constraints, system characteristics, and expected applications all affect the selection of the optimal divider implementation. Sometimes, a simpler, slower divider is preferred and sometimes a more complex, faster divider is. This chapter presents three dividers that represent varying complexity and speed: binary restoring dividers; Sweeney, Robertson, and Tocher (SRT) dividers; and Goldschmidt dividers. The CMOS design for each of these dividers is explained and analyzed in terms of speed and area.

5.1 Binary restoring dividers

Binary restoring dividers execute serialized subtractions to resolve the quotient and remainder for a given dividend and divisor one bit at a time. This is similar to the way a shift-and-add multiplier executes serialized additions to resolve the product one bit at a time. The divider operates on fixed-point numbers and assumes that the divisor and dividend are normalized.

In the original restoring divider, on each step, the divisor is subtracted from the dividend. If that causes the dividend to be negative then the divisor is added back to "restore" the dividend. This is shown in Figure 5.1.

Here, the nonperforming version of the restoring divider is implemented. It keeps a copy of the dividend, so that if the subtraction gives a negative dividend, no addition is needed since the copy of the original dividend can be used. A flowchart for a nonperforming binary restoring divider with a $2n$-bit dividend and an n-bit divisor is shown in Figure 5.2.

At initialization, the ripple-carry adder performs a 2's complement of the divisor to enable subtraction and stores this into an N-bit register. The dividend P_k is placed in a $2N$-bit shift register and the divisor D is placed in an N-bit register. Then the division iterations begin. First, the dividend shift register is shifted to the left. Then, the divisor is subtracted from the dividend by adding the 2's complement of the divisor to the product.

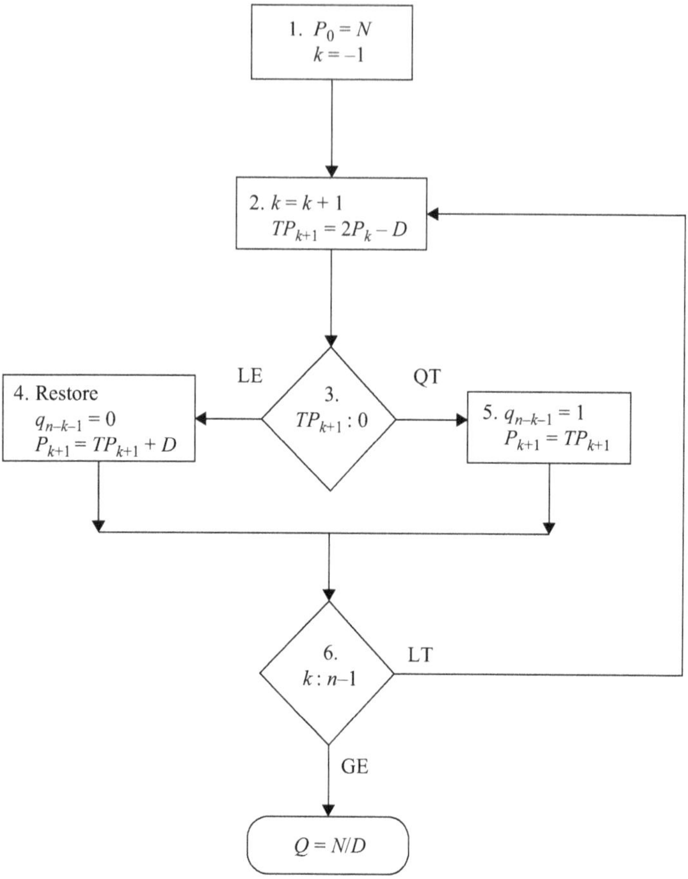

Figure 5.1 Binary restoring divider for an n-bit division and k iterations

If the result of the sum is positive, a "1" is stored as the quotient bit for that iteration. The product register is updated to hold the result of the subtraction. If the result of the sum is negative, "0" is stored as the quotient bit for that iteration. The product register remains untouched, retaining the value from prior to the subtraction, hence the "nonperforming" portion of the divider. To store the quotient bit in each iteration, the quotient bit is shifted left into the least significant bit of the dividend register.

This process repeats for N iterations, at which time the complete quotient is in the lowest bits of the shift register and the remainder is in the higher bits. The total complexity is an N-bit adder, an N-bit multiplexer, a $2N$-bit shift register, and two N-bit registers. The total delay is the initial subtraction, plus N iterations of an addition, a multiplexer, and a shift operation. An example of 5/8 divided by 7/8 is shown in Table 5.1.

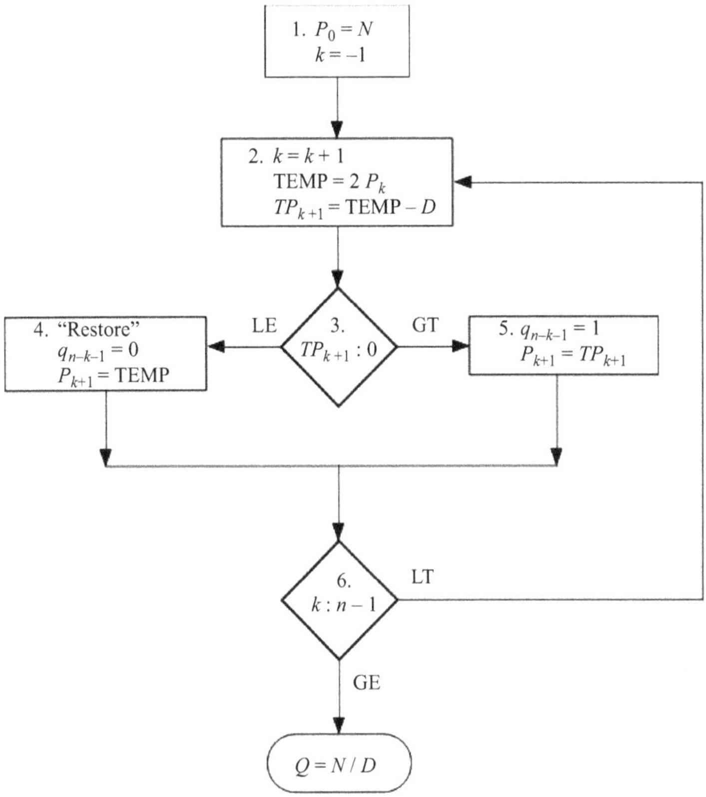

Restoring division
(Nonperforming)

Figure 5.2 Nonperforming binary restoring divider for an n-bit division and k iterations

Table 5.1 Execution steps for binary restoring divider for 5/8 divided by 7/8

T	Execution	Description
0	$P = 0.1010$ $D = 0.1110$ $-D = 1.0010$	Initialization
1	$P_i = 1.0100 + 1.0010$ $P_i = 0.0110$ $Q_i = 1$	Perform $2P - D$ Positive result Set quotient bit Q_i to 1
2	$P_i = 0.1100 + 1.0010$ $P_i = 1.1110$ $Q_i = 0$ $P_i = 0.1100$	Perform $2P - D$ Negative result Quotient bit Q_i to 0 Restore P

(Continues)

Table 5.1 (Continued)

T	Execution	Description
3	$P_i = 1.1000 + 1.0010$	Perform $2P - D$
	$P_i = 0.1010$	Positive result
	$Q_i = 1$	Set quotient bit Q_i to 1
4	$P_i = 1.0100 + 1.0010$	Perform $2P - D$
	$P_i = 0.0110$	Positive result
	$Q_i = 1$	Set quotient bit Q_i to 1
5	$Q = 0.1011$	Final quotient and remainder found
	$P = 0.0000\ 0110$	

5.2 SRT dividers

SRT dividers are similar to restoring dividers but reduce the number of iterations by increasing the radix of operation. Radix-2 SRT dividers have the same number of iterations, radix-4 has half as many, radix-8 has one quarter, and so on. The quotient is represented in binary signed digit form. For this work, radix-2 SRT division is considered.

First, the 2's complement of the divisor is computed. Next, the value $P = 2*$ (dividend) is computed in order to resolve a quotient bit Q_i and the running sum. Q_i is evaluated using the following conditional equations:

$$Q_i = \begin{cases} 1 \text{ if } P > \dfrac{1}{2} \\ -1 \text{ if } P < -\dfrac{1}{2} \\ \text{else } 0 \end{cases}$$

In the first case, when $P > 0.5$, the 2's complement of the divisor D is added to P. In the second case, when $P < -0.5$, the divisor is added to P, else 0 is added. To perform the comparison, a 4-to-1 multiplexer is used to select either the divisor, the 2's complement of the divisor, or 0 to be added to P. The select lines of the multiplexer are the two most significant bits of P. If the select lines are 00 or 11, the value 0 is selected. If the select lines are 01, P is greater than 1/2 and the 2's complement of the divisor is selected. If the select lines are 10, P is less than $-1/2$ and the divisor is selected.

The remaining $N-1$ iterations begin by computing P_{i+1} as twice the current value of P_i. When the division completes, the quotient is converted from binary signed digit to binary. Last, the remainder is checked for sign. If the remainder is negative, the division mistakenly over-subtracted. To correct the result, the divisor is added back to the running sum and 1 is subtracted from the quotient. This undoes the effects of the final subtraction. A flowchart for this is shown in Figure 5.3.

Let N be the dividend, D be the divisor, and Q be the quotient. K is the iteration number. The final step 6 represents the conversion of the result from Binary Signed Digit (BSD) to binary using standard methods. An example of SRT division is shown in Table 5.2.

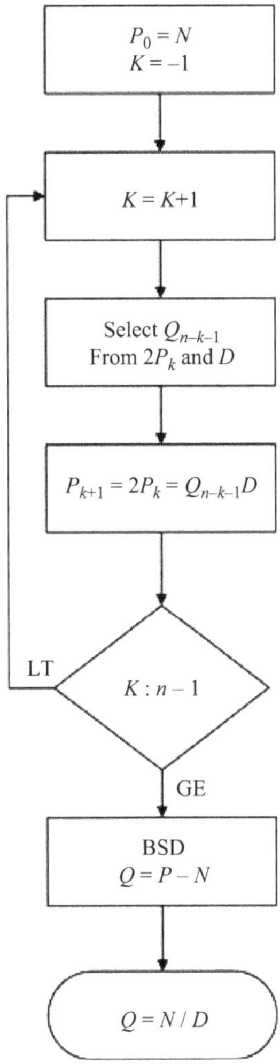

Figure 5.3 Flowchart for a radix-2 SRT divider

Standard binary conversion could be performed after the iterations in an SRT divider as shown in the flow diagram. Alternatively, on-the-fly conversion [52] can be used to compute the binary quotient during the execution of the iterations. This effectively hides the latency of the final quotient conversion. On-the-fly conversion retains information on the results of previous iterations to correctly convert the running binary signed digit quotient as the digits of the quotient are resolved in each iteration. An example of on-the-fly conversion is shown in Table 5.3.

j is the iteration, q_j is the quotient result, $Q[j]$ is an intermediate value, and $QM[j]$ is the final quotient in binary form. In this approach, the final remainder check is still necessary.

Table 5.2 Execution steps for a 4-bit SRT division

Iteration	Execution	Description
0	$P = 0.0110\ 1100$	Initialization
	$D = 0.0101$	
	$-D = 1.1011$	
1	$P_i = 0.1101\ 1000$	Shift P left
	$Q_i = 1$	Set quotient bit
	$P_{i+1} = 0.0001\ 1100$	Subtract $P_i - D = P_{i+1}$
2	$P_i = 0.0011\ 1000$	Shift P left
	$Q_i = 0$	Set quotient bit
3	$P_i = 0.0111\ 0000$	Shift P left
	$Q_i = 0$	Set quotient bit
4	$P_i = 0.1110\ 0000$	Shift P left
	$Q_i = 1$	Set quotient bit
	$P_{i+1} = 0.1001\ 0000$	Subtract $P_i - D = P_{i+1}$
	Q is 1001	
	$P = 0.1001$	

Table 5.3 On-the-fly conversion of BSD to binary for SRT division

j	q_j	$Q[j]$	$QM[j]$
0		0	0
1	1	0.1	0.0
2	1	0.11	0.10
3	0	0.110	0.101
4	1	0.1101	0.1100
5	−1	0.11001	0.11000

The equations that correspond to $Q[j]$ and $QM[j]$ are as follows:

$$Q[j+1] = \begin{cases} (Q[j], q_{j+1}) \text{ if } q_{j+1} \geq 0 \\ (QM[j], (2 - |q_{j+1}|)) \text{ if } q_{j+1} < 0 \end{cases}$$

$$QM[j+1] = \begin{cases} (Q[j], q_{j+1} - 1) \text{ if } q_{j+1} > 0 \\ (QM[j], (1 - |q_{j+1}|)) \text{ if } q_{j+1} \leq 0 \end{cases}$$

where r is the radix of the operation, in this case 2.

The SRT divider requires two N-bit registers, a $2N$-bit shift register, a 4-to-1 mux and N-bit adder and conversion logic. Initialization requires a 2's complement addition, each iteration requires an addition, a mux operation and a shift operation, and the final step requires conversion if standard BSD conversion is used.

5.3 Goldschmidt dividers

Goldschmidt division is a fast division technique that resolves twice as many bits per iteration, significantly reducing execution time for large division widths.

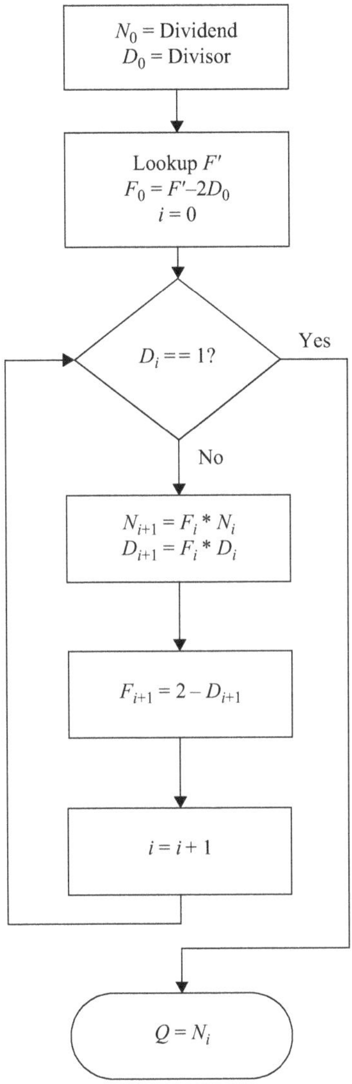

Figure 5.4 Goldschmidt division flowchart

Goldschmidt dividers use repetitive multiplication of the normalized dividend N and divisor D by a factor F until the divisor converges to 1, with the dividend converging to the quotient. At initialization, a factor F' is obtained from a lookup table based on the inputs. Then, F is calculated as $F = F' - 2D_0$. In each iteration, the value F is updated using the equation $F_i = 2 - D_i$. A flowchart can be seen in Figure 5.4.

Table 5.4 Execution steps for an 8-bit Goldschmidt divider for 5/8 divided by 3/4

Step	Execution	Description
0	$P = 5/8$ $D = 3/4$ $F_0 = 2.914 - 2(3/4)$ $F_0 = 1.414$	Initialization
1	$P = 5/8 * 1.414$ $P = 0.88375$ $D = 3/4 * 1.414$ $D = 1.0605$ $F = 0.9395$	Multiply the product by F Multiply the divisor by F Update F for $F = 2 - D$
2	$P = .88375 * 0.9395$ $P = 0.8303$ $D = 1.0605 * 0.9395$ $D = 0.9963$ $F = 1.00366$	Multiply the product by F Multiply the divisor by F Update F for $F = 2 - D$
3	$P = 0.8303*1.00366$ $P = 0.8333$ $D = 0.9963 * 1.00366$ $D = 0.99998$ $F = 1.00001$	Multiply the product by F Multiply the divisor by F Update F for $F = 2 - D$
4	$P = 0.833322 * 1.00001$ $P = 0.83333$ $D = 0.99998*1.00001$ $D = 1.000000$ $F = 1$	Multiply the product by F Multiply the divisor by F Update F for $F = 2 - D$
5	$Q = P = 0.833$	Convergence reached

In total, a division that executes X iterations requires one shift, $2X$ multiplications, and X subtractions. For most Goldschmidt designs, it is assumed that both the dividend and divisor have been normalized a priori. Consider an example of a Goldschmidt divider to perform 5/8 divided by 3/4 with an initial F_0 estimation of 2.914 as shown in Table 5.4.

Note that the precisions of the multiplications are determined by the bit precision of the divider.

Chapter 6
Memristor-based adder designs[1]

This chapter presents various designs of memristor-based adders. IMPLY, hybrid-CMOS, threshold gate, and MAD approaches are employed to implement ripple-carry, carry-select, conditional-sum, and carry-lookahead adders. Each design is described and analyzed in terms of complexity and delay. Completed schematics are also given.

6.1 Ripple Carry Adders[2]

As described in Chapter 3 (Adder arithmetic units), the ripple-carry adder is the simplest of the adders. It has the least logic and fewest components. However, this comes at the cost of higher delay for ripple-carry adders as compared to other popular adders.

Memristor-based ripple-carry adders are presented for the IMPLY, hybrid-CMOS, threshold, and MAD contexts. Each design is explained and optimized for its given context to produce wide variation across the designs.

6.1.1 IMPLY ripple-carry adder

First, a ripple-carry adder is described using the IMPLY approach. This begins with the implementation of an IMPLY full adder and then an explanation of how the full adders can be connected to produce a ripple-carry adder.

Since ripple-carry adders are the simplest adder, many researchers use this unit as a proof of concept for memristor logic designs. In general, memristors are favored for their small form factor. Second, IMPLY designs generally suffer from longer, serialized delays as compared to other design approaches. For these reasons, the majority of the research in designing and optimizing IMPLY-based ripple-carry adders has focused on minimizing memristors rather than delay. This trade-off strives to fully exploit the area reduction benefits of memristor logic.

Most of the research on memristor-based IMPLY ripple-carry adders serializes the IMPLY steps of each bit and performs minimal parallelism. In this way,

[1]This chapter is based on [53,54].
[2]This section is based on [21].

memristors can be maximally reused across steps. After each IMPLY step, the decision is made whether to reset the operand memristors or not. If either of the operand memristors will not be reused, it can be reset and reused. However, this requires a step delay to reset the memristor. Thus, the delay increases, but the complexity remains minimal. By reusing memristors across steps, the number of memristors is minimized. This naturally results in an increased runtime in comparison to a parallelized approach.

The state-of-the-art serial implementation [18] requires only 27 memristors for an 8-bit adder, but it requires 184 IMPLY steps. This delay is significantly higher than the traditional CMOS implementation's delay of 20 steps. However, the complexity is trivial compared to the CMOS implementation. Depending on the application, the memristor-based implementation may be favorable. For example, low-power and microdevices will benefit from the smaller memristor-based design while time-sensitive applications may still prefer the CMOS implementation.

Minimal prior work has focused on parallel implementations of IMPLY ripple-carry adders. Such an implementation was originally shown for an N-bit ripple-carry adder by [16], taking $35N$ memristors and $8N+12$ steps. For an 8-bit adder, this is 280 memristors and a delay of 76 steps. This design was later improved to require $9N$ memristors and a delay of $5N+18$ IMPLY steps [20], or 72 memristors and 58 steps.

The parallel implementation has about three times the number of memristors, but about a third the number of steps as the serial implementation. It is about a one-to-one trade-off. Less reuse is done in favor or performing fewer reset steps and more parallel steps. This is an example of how memristor-based designs are flexible to the application. Depending on the application, a serial or a parallel implementation can be used, each with strengths in either delay or complexity. One consequence of parallel implementations is increased driver complexity in order to maintain independence among the bits' parallel computations.

6.1.1.1 Baseline IMPLY full adder

The IMPLY operation can perform all of the Boolean operations. They can be used to implement the Boolean equations for a full adder:

$$\begin{aligned} \mathrm{Sum} &= \mathrm{A\ XOR\ B\ XOR\ C_{in}} \\ \mathrm{C_{out}} &= \mathrm{(A\ AND\ B)\ OR\ ((A\ OR\ B)\ AND\ C_{in})} \end{aligned}$$

The sum equation consists of two serial XOR operations. A XOR B is implemented with the IMPLY operation as (A IMP B) IMP (($\overline{\mathrm{A}}$ IMP $\overline{\mathrm{B}}$) IMP 0) and requires seven steps and six memristors. The full sum equation becomes

$$\begin{aligned} \mathrm{Sum} = \Big(&\big[\mathrm{(A\ IMP\ B)\ IMP\ \big((\overline{A}\ IMP\ \overline{B})IMP\ 0\big)}\big]\ \mathrm{IMP\ C}\Big)\ \mathrm{IMP} \\ &\Big(\overline{\big[\mathrm{(A\ IMP\ B)\ IMP\ \big((\overline{A}\ IMP\ \overline{B})IMP\ 0\big)}\big]}\ \mathrm{IMP\ \overline{C}}\Big)\ \mathrm{IMP\ 0} \end{aligned}$$

This equation appears much more complicated than the traditional CMOS equation. However, this is partially due to the fact that the CMOS equation is not fully expanded as the IMPLY equation is.

An initial implementation to calculate the sum in a full adder with the IMPLY operation has the execution steps shown in Table 6.1.

The sum operation takes 14 steps and a total of 11 memristors. Note that steps 3 and 10 are used to create copies of the B and C_{in} inputs. This is necessary because they are overwritten in steps 4 and 11. However, if it is known that the B and/or C_{in} inputs are not needed in the future, steps 3 and 10 can be removed. This not only removes two steps from the execution but also removes two memristors. The memristors holding the inputs can be used directly instead of memristors M2 and M6 and these memristors can be removed.

The C_{out} value comprises AND and OR operations. The AND operation is implemented with IMPLY operations as (P IMP (Q IMP 0)) IMP 0, requiring four memristors and three steps. The OR operation is implemented with IMPLY operations as (P IMP 0) IMP Q requiring five memristors and four steps. The full C_{out} equation becomes

$$\begin{aligned} C_{out} = {} & [[(\text{A IMP (B IMP 0))IMP 0] IMP 0]IMP} \\ & [([(\text{A IMP 0)IMP B]IMP } (C_{in} \text{ IMP 0))IMP 0}] \end{aligned}$$

An initial implementation to calculate C_{out} in a full adder with IMPLY operations is shown in Table 6.2.

The C_{out} execution takes 14 steps and 14 memristors. However, when constructing a full adder, the input memristors for A, B, and C_{in} are used in the calculation of both the sum and C_{out}. Thus, these three memristors do not need to be counted twice since they are not overwritten. The total cost for this implementation of a full adder is 28 steps and 25 memristors. This can be compared to the traditional CMOS implementation, which requires six steps and nine gates for a total of 34 MOSFETs. Part of the reason that the CMOS implementation is able to execute in fewer steps is that the gates can perform in parallel. IMPLY operations are

Table 6.1 IMPLY execution steps for a full adder

t	**Operation**	**Description**
1	A IMP 0 → M0	$\overline{\text{A}}$
2	B IMP 0 → M1	$\overline{\text{B}}$
3	M1 IMP 0 → M2	Copy of B
4	A IMP M2 → M2	A IMP B
5	M0 IMP M1 → M1	$\overline{\text{A}}$ IMP $\overline{\text{B}}$ → $\overline{\text{B}}$
6	M1 IMP 0 → M3	($\overline{\text{A}}$ IMP $\overline{\text{B}}$) IMP 0
7	M2 IMP M3	A XOR B
8	M3 IMP 0 → M4	$\overline{\text{(A XOR B)}}$
9	C_{in} IMP 0 → M5	$\overline{C_{in}}$
10	M5 IMP 0 → M6	Copy of C_{in}
11	M3 IMP M6 → M6	(A XOR B) IMP C_{in}
12	M4 IMP M5 → M5	$\overline{\text{(A XOR B)}}$ IMP $\overline{C_{in}}$ → $\overline{C_{in}}$
13	M5 IMP 0 → M7	[$\overline{\text{(A XOR B)}}$ IMP $\overline{C_{in}}$] IMP 0
14	M6 IMP M7	[(A XOR B) IMP C_{in}] IMP ([$\overline{\text{(A XOR B)}}$ IMP $\overline{C_{in}}$] IMP 0) = A XOR B XOR C

Table 6.2 Execution steps to calculate C_{out}

t	Operation	Description
1	B IMP 0 → M0	$\overline{B}$
2	M0 IMP 0 → M1	Copy of B
3	A IMP 0 → M2	$\overline{A}$
4	M2 → M1	A OR B
5	B IMP 0 → M3	$\overline{B}$
6	A IMP M3 → M4	A IMP $\overline{B}$
7	M4 IMP 0 → M5	A AND B
8	C_{in} IMP 0 → M6	$\overline{C_{in}}$
9	M1 IMP M6 → M6	A OR B IMP $\overline{C_{in}}$
10	M6 IMP 0 → M7	(A OR B) AND C_{in}
11	M7 IMP 0 → M8	$\overline{((A\ OR\ B)\ AND\ C_{in})}$
12	M8 IMP 0 → M9	Copy of ((A OR B) AND C_{in})
13	M5 IMP 0 → M10	$\overline{(A\ AND\ B)}$
14	M10 → M9	(A AND B) OR ((A OR B) AND C_{in})

performed in series. However, this serialization limitation can be overcome in many situations.

6.1.1.2 Optimizing the critical path

Now consider a few optimizations to this implementation that reduce the number of execution steps and reduce the number of memristors. First, any redundancy in steps can be eliminated. For example, steps 1 and 5 in the resolution of C_{out} are identical. Step 5 and memristor M3 can be eliminated. Instead of M3, the value in M0 is reused for step 6.

Second, copy steps for intermediate results are not necessary. For example, in a stand-alone OR operation, the second operand must be copied to allow its use in the future. This requires two steps and two memristors. For example, the copy operations of operand B in steps 2 and 3 of the sum execution are needed because B is used later for the resolution of the C_{out} signal. However, this is not always necessary. Consider the expression (A OR B) AND C_{in} used in (A AND B) OR ((A OR B) AND C_{in}). This expression is used only once. It is not the same as operand B which is needed later. Steps 11 and 12 and memristors M8 and M9 used to copy this expression are unnecessary. They can be eliminated, and memristor M7 can be directly used in step 14.

Third, A XOR B is computed in the first steps of the sum resolution. The equation for C_{out} can be rewritten as

$$C_{out} = (A\ AND\ B)\ OR\ ((A\ XOR\ B)\ AND\ C_{in})$$

Now the design can reuse the result from A XOR B that has already been computed rather than computing the OR operation. This removes the need for steps 2–4 and memristors M1 and M2 for the C_{out} resolution. Now the result of step 7 in the sum resolution can be used directly in place of M1 in step 9.

Table 6.3 Optimized execution steps to calculate C_{out}

t	**Operation**	**Description**
1	B IMP 0 → M0	$\overline{\text{B}}$
2	A IMP M0 → M4	A IMP $\overline{\text{B}}$
3	M4 IMP 0 → M5	A AND B
4	C_{in} IMP 0 → M6	$\overline{C_{in}}$
5	(M3 from sum steps) IMP M6 → M6	A XOR B IMP $\overline{C_{in}}$
6	M6 IMP 0 → M7	(A XOR B) AND C_{in}
7	M5 IMP 0 → M10	$\overline{\text{(A AND B)}}$
8	M10 → M7	(A AND B) OR ((A XOR B) AND C_{in})

All these optimizations together reduce the number of steps for C_{out} from 14 to 8 and the number of memristors from 11 to 6, almost 2× improvements in both metrics. These steps are shown in Table 6.3.

In the previous section, a series of incremental optimizations were performed to reduce the number of steps in an IMPLY full adder. The most crucial optimization is the removal of redundant operations, unnecessary copy operations, and similar extraneous operations. However, it is often the case that Boolean logic needs an entire ground-up transformation from the CMOS domain to the memristor domain. This is because most, if not all, traditional Boolean expressions have been simplified and optimized with the assumption that the AND, OR, and NOT gates are the smallest unit of logic and have a delay of one. This assumption no longer holds in the memristor domain. In the memristor domain, the IMPLY and NOT operations are the smallest units of logic and have a delay of one. Thus, Boolean expressions need to be rewritten, simplified, and optimized in terms of IMPLY and NOT operations.

Now consider the use of a full adder in the context of its main application: an *N*-bit ripple-carry adder. Given this knowledge, directed optimizations can be performed on the IMPLY operations. In general, the C_{in}-to-C_{out} delay lies on the critical path and the C_{in}-to-sum delay does not. Thus, the C_{in}-to-C_{out} path should be performed first. Also, optimizations should favor the C_{out} path over the sum path.

In a ripple-carry adder, all of the inputs arrive to the full adders at the same time *t* for a given computation. Only the C_{in} bits will be staggered from stage to stage as they are computed and propagated. Thus, any steps in the full adder, which do not require C_{in}, should be executed as early as possible. This reduces the number of idle steps before C_{in} arrives and the number of active steps after it does.

Given this information, the goal is to minimize the C_{in}-to-C_{out} delay through each adder and try to minimize its steps. The equation for C_{out} is

$$C_{out} = (\text{A AND B}) \text{ OR } ((\text{A XOR B}) \text{ AND } C_{in})$$

It is possible that (A AND B) and (A XOR B) could be computed in the full adder prior to C_{in} arriving. This was shown in the previous section. The final AND and OR operation must wait until C_{in} arrives. In traditional CMOS, this is equivalent to two gate delays. However, in IMPLY operations with memristors, both the AND and OR operation take multiple steps. It takes five steps (steps 4–8) to compute these two operations and resolve C_{out}.

This delay implies that for each full adder, it will take five IMPLY steps to propagate the carry to the next full adder. Thus, if there are N full adders in a chain forming an N-bit ripple-carry adder, it will take at least $5N$ steps to complete the addition. Can this delay be reduced?

The values of (A AND B) and (A XOR B) can be produced before the carry-in signal arrives. Thus, consider the delay of the final AND operation and the final OR operation. Note that the result of the AND operation is an input to the final OR operation. The operations can be rewritten as

$$C_{out} = (C_{in} \text{ AND G}) \text{ OR F}$$

where F = A AND B and G = A XOR B. In terms of IMPLY operations, this becomes

$$C_{out} = (((C_{in} \text{ IMP (G IMP 0))IMP 0)IMP 0)IMP F}$$

This requires five steps and three additional memristors. However, notice that two "IMP 0" operations occur consecutively. This occurs because the AND operation completes with a final "IMP 0" operation. Then, the result of the AND operation is used as the first operand in the OR operation, which requires another "IMP 0" operation. If the AND and OR operation are performed in separate contexts, both require the "IMP 0" step. However, when done consecutively, this is equivalent to a copy operation. Since neither of the intermediate results is used elsewhere in execution, the copy operation is entirely unnecessary. Thus, an AND followed by an OR operation can be simplified as

$$C_{out} = (C_{in} \text{ IMP (G IMP 0))IMP F}$$

This requires only three steps and one additional memristor. Now, the operands for the C_{out} equation can be considered:

$$C_{out} = (C_{in} \text{ IMP ((A XOR B) IMP 0)) IMP (A AND B)}$$

This breaks down to the steps in Table 6.4.

Notice that steps 1–3 do not require the C_{in} input. Thus, all these three steps can be performed before the carry-in signal arrives. It is also an option to remove

Table 6.4 Simplified IMPLY steps for C_{out} execution in a full adder

Step	Execution	Description
1	Q IMP 0 → M0	(A XOR B) IMP 0 == A XNOR B
2	P IMP 0 → M1	(A AND B) IMP 0 == A NAND B
3	M1 IMP 0 → M2	Copy of (A AND B)
4	C_{in} IMP M0 → M0	(A XNOR B) NAND C
5	M0 IMP M2 → M2	(A AND B) OR (C_{in} AND (A XOR B)) = C_{out}

steps 2 and 3 entirely if it is found that (A AND B) is never needed again during the execution of the full adder. This reduces the total cost of the C_{out} logic to just three steps and two memristors. Also, the carry-propagation delay from C_{in} to C_{out} is only two steps, steps 4 and 5.

This optimization is an example of the sensitivity of IMPLY operations to the operand selection. By strategically selecting the equation for C_{out} as

$$C_{out} = (C_{in} \text{ AND } G) \text{ OR } F$$

This produces a design with minimal delay and complexity requirements. If C_{out} had been written as

$$C_{out} = F \text{ OR } (C_{in} \text{ AND } G)$$

or

$$C_{out} = (G \text{ AND } C_{in}) \text{ OR } F$$

The solution would not have been as optimal. It is left as an exercise to the reader to calculate the steps and memristors required for the variant expressions of C_{out}.

6.1.1.3 Optimized IMPLY ripple-carry adder

The optimizations for C_{out} can be combined with the previously discussed steps for the sum signal. The execution steps for a full adder implementation optimized for both delay and component counts are shown in Table 6.5. It requires eight memristors and has a delay of 21.

Steps 1–6 are identical to steps 1–6 in the initial scheme for the sum resolution. Steps 7, 8, 9, and 11 are used to compute A AND B, A XOR B, and A XNOR B, which are required for C_{in} resolution. Step 10 is included as an optimization to reduce the memristor count. Step 10 essentially resets memristors M0, M1, and M2 to 0 since their current values are no longer needed. This allows them to be reused in later steps in the full adder. Alternatively, Step 10 could be removed and an additional three memristors could be used in the full adder. Steps 12 and 13 are identical to the C_{out} steps that were derived above. Steps 14 and 17 are only required if the full adder is part of a larger *N*-bit adder and the C_{out} bit is required by a subsequent adder. If this is the case, it is vital for the next full adder to use the C_{out} immediately to reduce the critical path. Otherwise, these steps can be eliminated. The reason that C_{out} is sensed in steps 14 and 17 rather than 14 and 15 is because each full adder uses two steps after initially reading the value of C_{out} to produce its own C_{out}. Then, it reads the value again to assist in resolving its sum output, which is off the critical path. The remaining steps are to resolve the final XOR operation with C_{out} to obtain the sum bit.

The implementation was designed to ensure that the C_{in} memristor was always used as the first operand in each IMPLY operation in which it is directly used (steps 14 and 17). This removes the need to copy the value of C_{in} before its use, removing two steps from the carry-in to carry-out critical path. Given this implementation, the delay for an *N*-bit adder is 11 steps $+ 2{*}N + 8 = 2N + 19$ steps. This is close to

Table 6.5 Steps for an optimized IMPLY parallel add operation

Step	Functionality	Goal
1	A IMP 0	A→M2
2	B IMP 0	B→M3
3	B	M3→M1
4	(A IMP 0) IMP (B IMP 0)	M2→M3
5	A IMP B	A→M1
6	((A IMP 0) IMP (B IMP 0)) IMP 0	M3→M0
7	B IMP (A IMP 0)	B→M2
8	B IMP (A IMP 0) IMP 0	M2→M4
9	A XOR B	M1→M0
10		FALSE(M1,M2,M3)
11	(A XOR B) IMP 0	M0→M2
12	C_{in} IMP ((A XOR B) IMP 0)	C_{in}→M2
13	C_{in} AND (A XOR B) + AB→ C_{out}	M2→M4
14	*Next bit is reading C_{in}*	
15	C_{in} IMP 0	C_{in} → M1
16	C_{in} OR (A XOR B)	M1→M0
17	*Next bit is reading C_{in}*	
18	(C_{in} OR (A XOR B)) IMP 0	M0→M3
19	(C_{in} NAND (AXORB)) IMP ((C_{in} OR (A XOR B)) IMP 0)	M2→M3
20		False M1
21	(C_{in} NAND (AXORB)) IMP ((C_{in} + (A XOR B)) IMP 0) IMP 0 = SUM	M3→M1

the delay of a traditional CMOS ripple-carry adder, which has a delay of $2N+4$ gate delays. In fact, the critical carry-in to carry-out delay is two in both designs, indicating that they converge as N increases. However, note that a CMOS gate delay and a memristor IMPLY step may not be equivalent units of time.

The implementation of a full adder requires eight memristors: inputs A, B, and C_{in} and five working memristors M0 to M4. However, when this full adder is used in the context of a larger arithmetic unit, the C_{in} bits are shared across chained full adders. In other words, memristor M4 in full adder FA_i is the same as memristor C_{in} in full adder FA_{i+1}. For the IMPLY steps that use C_{in}, an IMPLY operation will logically occur across the full adders, driving V_{cond} on M4 in FA_i and V_{set} on M2 or M1, respectively, in FA_{i+1}. For all FA_i such that $i>0$, the full adder requires seven memristors for a total of $7N+1$ memristors for an N-bit ripple-carry adder.

To complete the circuitry for the full adder, the drivers for the IMPLY signals must be added to each memristor to apply V_{cond} and V_{set}. The full circuitry required for a 1-bit IMPLY full adder is shown in Figure 6.1.

The pull-down resister R_g is selected as 2 kΩ in this example and low resistance = 1 kΩ and high resistance = 100 kΩ.

For an N-bit adder, N of the full adders are placed in a chain. In order to allow individual bits to operate in parallel, there are voltage-controlled switches on the

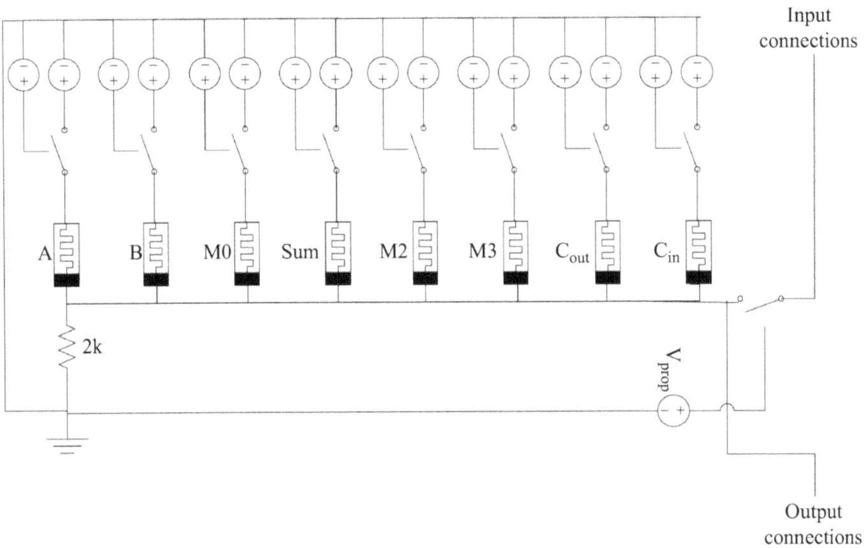

Figure 6.1 1-bit IMPLY adder schematic

lines between the bits. The full circuitry required for an 8-bit IMPLY adder is given in Figure 6.2.

Each column is a full adder and execution begins from the right and traverses to the left. The first column, corresponding to bit 0 has eight memristors, whereas the rest have seven. This is because the C_{in} for bits 1–7 lies in the C_{out} memristor of the previous adder as described previously. Notice each full adder is connected via a switch to the next full adder. This allows each full adder to perform IMPLY operations in parallel without interference. When the C_{out} memristor is resolved in bit i, the switch between full adder i and full adder $i + 1$ is activated and the IMPLY operation can be performed across the adders as described. Then the switch is turned off and execution proceeds in both adders.

6.1.1.4 Pipelining

An additional benefit of the IMPLY approach is that operations can be pipelined through the circuitry. In traditional CMOS, the gates in an adder are combinational and the data are flowing through the circuitry as voltages. Pipelining is only possible in this context if latches and additional logic are introduced into the adder structure to stage values. In the memristor domain, the values are inherently "stored" in the memristors after each operation, similar to a latch. This unique property allows each IMPLY stage to be pipelined so memristors can operate in parallel on disjoint tasks.

Consider the N-bit adder that has been discussed. In the current implementation, as soon as the sum bit i has been resolved, the circuitry for bit i sits idle until the next addition. Each bit i is busy during steps 1 to 11 and steps $12 + 2i$ to $21+2i$

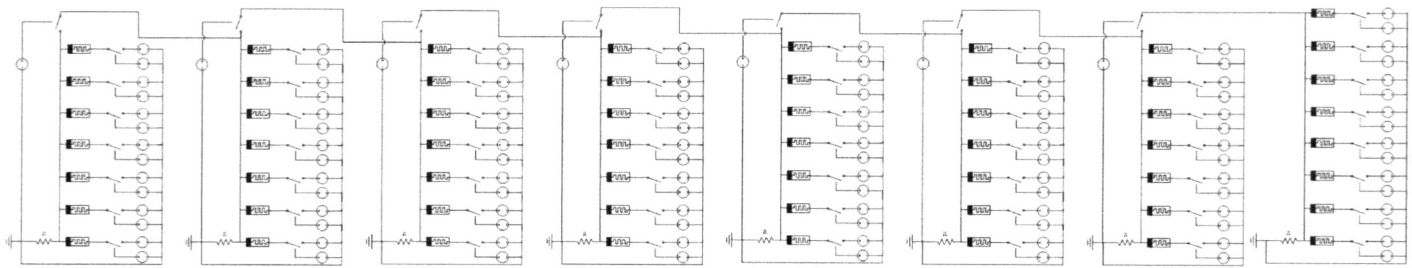

Figure 6.2 8-bit IMPLY ripple-carry adder

for a total of 21 steps. The adder executes for a total of $21 + 2N$ steps. Thus, each full adder is idle for $2N$ steps. Now remove these idle steps from the design.

Let the ripple-carry adder begin a new addition as soon as the first full adder, FA0, has completed. Full adder FA0 completes in step 21 and can begin operating on the next inputs in the subsequent step rather than waiting $2N$ idle steps. The circuitry already supports the full adders executing IMPLY operations in parallel. All full adders execute steps 1–11 in parallel. There is no limitation that requires these parallel operations to be identical operations. The circuitry does not even require that these IMPLY operations be part of the same N-bit addition.

The overhead to perform pipelining is two steps in each full adder: one for resetting all of the memristors in the full adder and the other for loading the next inputs A_i and B_i into the input memristors for bit i. The only other modification to the execution steps in Table 6.5 is that the result of the sum should be stored in a separate memristor, outside of the adder so that this memristor can be reset and reused. The memristor could exist in the subsequent unit in which the sum is expected to be used as an operand or in a new result memristor in the system. The final sum resolution step can now occur in step 20 since a result memristor does not need to be reset. The revised steps for a pipelined full adder are in Table 6.6.

A new addition can begin every 22 steps in this IMPLY approach. This improves the throughput for the IMPLY ripple-carry adder over alternatives and

Table 6.6 Steps for a pipelined IMPLY full adder

Step	Functionality	Goal
1	A IMP 0	A→M2
2	B IMP 0	B→M3
3	B	M3→M1
4	(A IMP 0) IMP (B IMP 0)	M2→M3
5	A IMP B	A→M1
6	((A IMP 0) IMP (B IMP 0)) IMP 0	M3→M0
7	B IMP (A IMP 0)	B→M2
8	B IMP (A IMP 0) IMP 0	M2→M4
9	A XOR B	M1→M0
10		FALSE(M1,M2,M3)
11	(A XOR B) IMP 0	M0→M2
12	C_{in} IMP ((A XOR B) IMP 0)	C_{in}→M2
13	C_{in} AND (A XOR B) + AB→ C_{out}	M2→M4
14	*Next bit is reading C_{in}*	
15	C_{in} IMP 0	C_{in} → M1
16	C_{in} OR (A XOR B)	M1→M0
17	*Next bit is reading C_{in}*	
18	(C_{in} OR (A XOR B)) IMP 0	M0→M3
19	(C_{in} NAND (AXORB)) IMP ((C_{in} OR (A XOR B)) IMP 0)	M2→M3
20	(C_{in} NAND (AXORB)) IMP ((C_{in} + (A XOR B)) IMP 0) IMP 0 = SUM	M3→SUM (outside of adder)
21		FALSE(ALL)
22		Load A, B inputs

CMOS because CMOS must wait for an entire addition to complete before beginning a new one, waiting the full $2N+4$ cycles. The IMPLY approach must wait 22 cycles, regardless of N. The benefits of the pipelining scheme grow as the utilization of the adder increases.

For visual assistance, a chart showing the resolution of the C_{out} and the sum in each bit of an 8-bit ripple-carry adder for three pipelined additions is shown in Table 6.7.

Every 22 steps, a full adder resolves the carry-out signal for an addition. Seven steps later, the sum is resolved. Two steps are required to reset the adder and then the next addition begins. In this way, a sum is fully resolved in step 7 every 22 steps.

Table 6.7 Resolution of the C_{out} and the sum signals for a pipelined 8-bit IMPLY ripple-carry adder

Step	**Bit 0**	**Bit 1**	**Bit 2**	**Bit 3**	**Bit 4**	**Bit 5**	**Bit 6**	**Bit 7**
1	Start 0	Start 0	Start 0	Start 0	Start 0	Start 0	Start 0	Start 0
2								
3								
4								
5								
6								
7								
8								
9								
10								
11								
12								
13	C_{out0}							
14								
15		C_{out0}						
16								
17			C_{out0}					
18								
19				C_{out0}				
20	Sum0							
21					C_{out0}			
22		Sum0						
23	Start 1					C_{out0}		
24			Sum0					
25		Start 1					C_{out0}	
26				Sum0				
27			Start 1					C_{out0}
28					Sum0			
29				Start 1				
30						Sum0		
31					Start 1			

Table 6.7 (Continued)

Step	Bit 0	Bit 1	Bit 2	Bit 3	Bit 4	Bit 5	Bit 6	Bit 7
32							Sum0	
33						Start 1		
34								Sum0
35	C_{out1}						Start 1	
36								
37		C_{out1}						Start 1
38								
39			C_{out1}					
40								
41				C_{out1}				
42	Sum1							
43					C_{out1}			
44		Sum1						
45	Start 2					C_{out1}		
46			Sum1					
47		Start 2					C_{out1}	
48				Sum1				
49			Start 2					C_{out1}
50					Sum1			
51				Start 2				
52						Sum1		
53					Start 2			
54							Sum1	
55						Start 2		
56								Sum1
57	C_{out2}						Start 2	
58								
59		C_{out2}						Start 2
60								
61			C_{out2}					
62								
63				C_{out2}				
64	Sum2							
65					C_{out2}			
66		Sum2						
67	Start 3					C_{out2}		
68			Sum2					
69		Start 3					C_{out2}	
70				Sum2				
71			Start 3					C_{out2}
72					Sum2			
73				Start 3				
74						Sum2		
75					Start 3			
76							Sum2	
77						Start 3		
78								Sum2

6.1.2 Hybrid-CMOS ripple-carry adder

Minimal prior work exists in the design of hybrid-CMOS ripple-carry adders due to the signal degradation issues inherent in hybrid-CMOS gates. Signal restoration buffers are a prominent design concern and current simulation and modeling tools do not have clean support for the detection and installation of buffers for hybrid-CMOS design. This poses challenges for researchers interested in developing large-scale memristor-based designs with hybrid-CMOS gates.

The two works that have explored hybrid-CMOS designs require $16N$ memristors, $18N$ MOSFETs, and $10N$ vias for an N-bit adder with the same gate delay as the traditional CMOS implementation [23]. For an 8-bit adder, this is 128 memristors, 144 MOSFETs, and 80 vias. The vias are used to transfer from the CMOS to nanoscale memristor domain and vice versa.

This section presents the design of a ripple-carry adder constructed from hybrid-CMOS gates. It improves upon previous works and offers insight into the necessity and placement of strategic signal restoration buffers. Traditional CMOS gates are used for the NOT gates while memristor-based AND and OR gates are used. When necessary, CMOS signal restoration buffers are included to maintain signal integrity.

6.1.2.1 Optimized hybrid-CMOS ripple-carry adder

Recall the XOR gate introduced in Chapter 2 and shown again in Figure 6.3.

The XOR gate is crucial to addition, occurring two times in each full adder in the ripple-carry adder design. The XOR gates are used to resolve the sum output bits. It is possible to improve the implementation of the hybrid-CMOS XOR circuit introduced earlier to reduce the number of CMOS MOSFETs needed. A revised hybrid-CMOS XOR gate is shown in Figure 6.4.

This proposed circuit requires one fewer CMOS NOT gate. Since CMOS gates are expensive in hybrid-CMOS designs, it is preferential to remove them when possible.

When used in a full adder, the revised XOR design has even greater benefits. Recall that the full adder uses the value of A AND B in order to compute C_{out}. The new XOR design produces the A AND B result as a by-product of the existing circuitry, whereas the original XOR design did not. The C_{out} logic can use this value directly and can save the additional logic that would be needed otherwise.

Similarly, the C_{out} expression uses (A XOR B AND C_{in}) to resolve the final OR operation. Since the second XOR in the sum output uses (A XOR B) and C_{in} as its two inputs, the expression (A XOR B AND C_{in}) is produced as a by-product of the second XOR gate. Thus, both expressions required for the final OR in the resolution of C_{out} are produced as by-products of the XOR gates for the sum output.

Now, the only logic in the full adder besides the two XOR gates for the sum output is a single OR gate for the final resolution of $C_{out.}$ This reuse can be seen in the schematic of a hybrid-CMOS full adder in Figure 6.5.

Two hybrid-CMOS XOR gates are concatenated to produce the sum output. The outputs of the first AND gate in each XOR gate are used in the final OR gate for the C_{out} signal. The reuse of these signals removes the need for two hybrid-CMOS AND gates from the design. This is a total savings of four memristors per

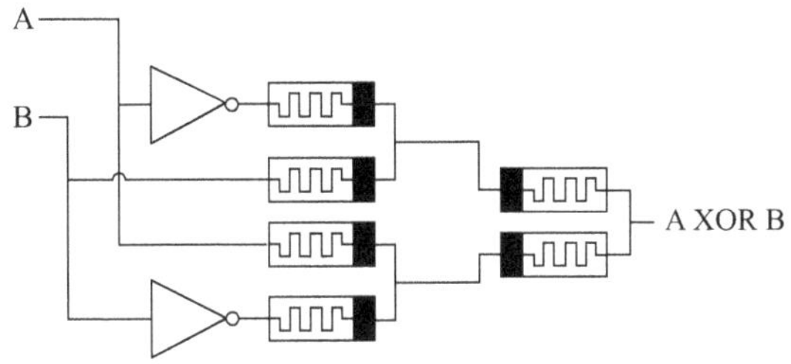

Figure 6.3 Hybrid-CMOS implementation of an XOR gate

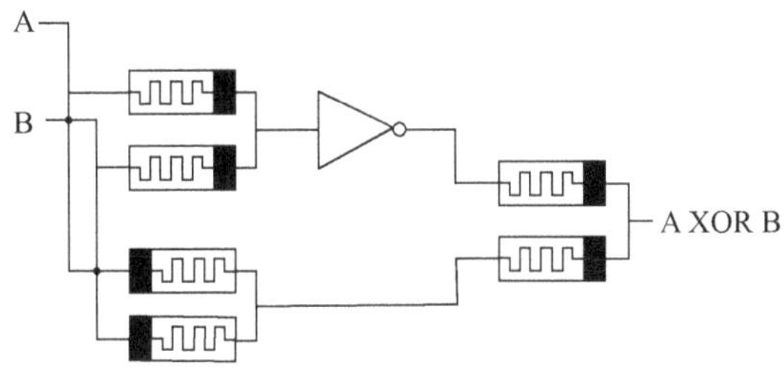

Figure 6.4 Optimized Hybrid-CMOS XOR gate schematic

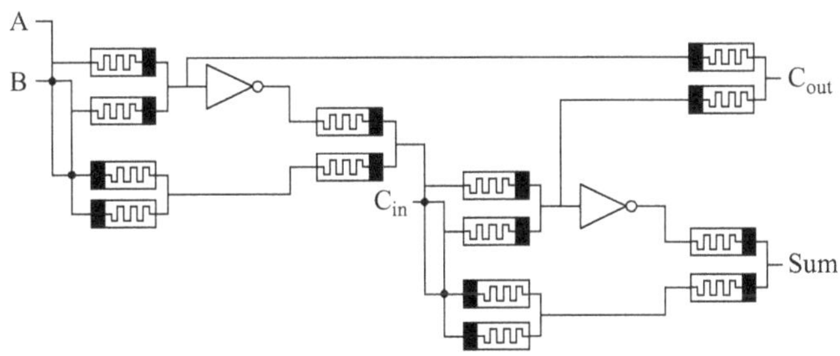

Figure 6.5 Hybrid-CMOS full adder based on the optimized XOR

full adder when the optimized XOR is used. As the complexity of the ripple-carry adder grows, these benefits become significant. A comparison of the componential requirements for the hybrid-CMOS full adder in comparison to a CMOS full adder is shown in Table 6.8.

At first glance, the delay of an N-bit hybrid-CMOS ripple-carry adder should be identical to CMOS, $(2N+4)$, since they have a 1-to-1 mapping of gates. The propagation of the carry signal takes two steps through each full adder, through an AND gate and an OR gate. There is also a constant delay for the first carry-out and final sum.

However, one consequence of the hybrid-CMOS approach is that the signals suffer from signal degradation as they propagate through the adder. Often times, it

Table 6.8 Complexity comparison for CMOS and hybrid-CMOS full adders

	Full adder		***N*-bit adder**	
	Memristors	MOSFETs	Memristors	MOSFETs
Hybrid-CMOS	14	4	14*N*	4*N**
CMOS		34		34*N*

*Does not include signal restoration buffers.

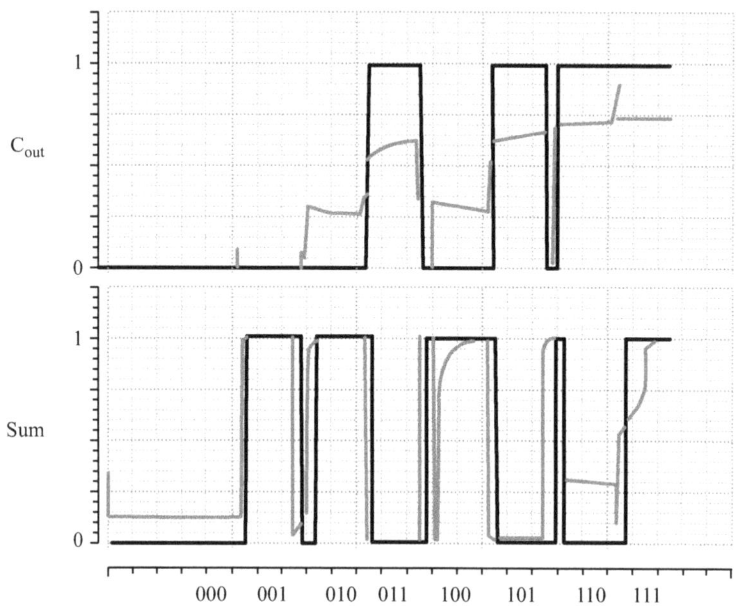

Figure 6.6 Hybrid-CMOS full adder execution with and without buffering

is necessary to add signal restoration buffers to maintain the signal's integrity. The strength of the output signals before and after signal restoration buffers are added is shown in Figure 6.6 for each input combination for A, B, and C_{in}, starting with 000 and finishing with 111.

The gray line is the strength of the output signal before signal restoration. The black line is the strength of the output signal after signal restoration. Initially, one would expect that the C_{out} and Sum signals would suffer about equally from signal degradation. However, it is clear from Figure 6.6 that the Sum signal exhibits nearly no degradation while the C_{out} signal has significant degradation.

CMOS buffers are added to the design to restore the signals with long propagation paths. Consider the carry-in to carry-out path. Within a single full adder, this path is only two gates and can be buffered at the output as it feeds into the next logic block. There is no need for internal signal restoration within the full adder. In an *N*-bit adder, signal restoration buffers are needed at each carry-propagation stage

Table 6.9 Complexity comparison for CMOS and hybrid-CMOS full adders

	Full adder		***N*-bit adder**	
	Memristors	MOSFETs	Memristors	MOSFETs
Hybrid-CMOS	14	4	$14N$	$12N$*
CMOS		34		$34N$

*Includes signal restoration buffers.

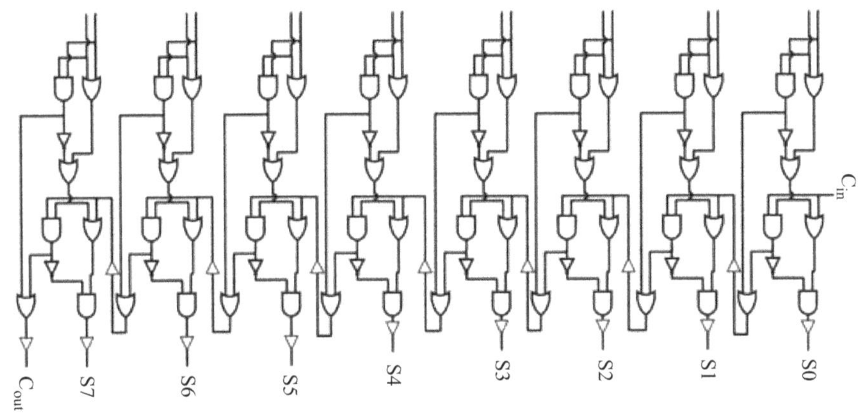

Figure 6.7 8-bit hybrid-CMOS ripple-carry adder

to preserve the carry signals. This adds 4 MOSFETs per bit for a total of $4N$ MOSFETs.

Refer back to the schematic for the hybrid-CMOS full adder. Notice that one of the two inputs to the final gate on the sum output path has a NOT gate inverting its value. This inverter essentially acts as a signal restorative inverter, as it outputs either a full "0" or a full "1". Thus, the Sum output can only be negatively impacted by one of its inputs whereas the C_{out} path does not contain any inverters. This is a conclusion with far-reaching effects. Many times in hybrid-CMOS designs, signal restoration buffers can be removed when they are on a path with a NOT gate or other CMOS buffer. It is crucial to use this knowledge when optimizing hybrid-CMOS designs, especially as the complexity of the design grows. The total complexity of a hybrid-CMOS full adder including signal restoration buffers is shown in Table 6.9.

Each buffer also adds one step to the delay of each full adder. Thus, the total delay and complexity of an N-bit becomes $3N + 4$ rather than the traditional CMOS delay of $2N + 4$ steps. The total complexity is $14N$ memristors + $12N$ MOSFETs as shown in Table 6.9. This is still less than half the number of MOSFETs in the CMOS design. The final gate-level schematic is shown in Figure 6.7.

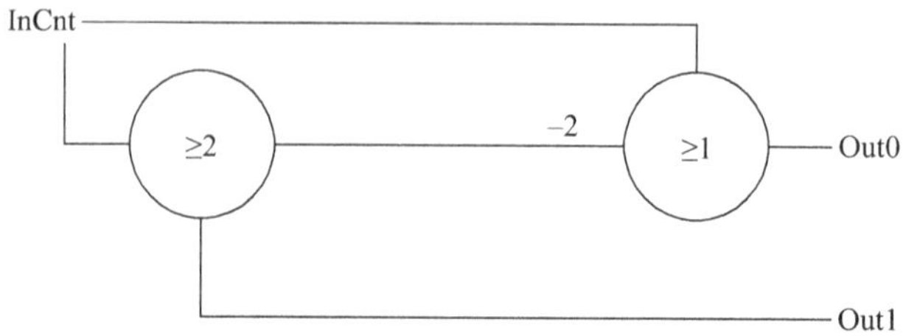

Figure 6.8 Minnick and Kautz methods for a 3-to-2 counter with threshold logic

At a gate level, the design is identical to the CMOS design with the addition of signal buffers on each output.

6.1.3 Threshold gate ripple-carry adder

A third implementation of an N-bit ripple-carry adder uses threshold gates as the fundamental block. These threshold gates can be constructed from either GOTO pairs or CSTG gates. There is no known prior work on the development of memristor-based ripple-carry adder units using threshold gates. However, there has been research on the general use of threshold gates in the development of arithmetic units.

In the context of threshold gates, a full adder takes a new form. The full adder can be thought of as a 3-to-2 counter that takes three inputs and counts how many inputs are a "1." This count is stored as a 2-bit output, the sum is the least significant bit, and C_{out} is the most significant bit. The Kautz/Minnick methods describe how to construct 3-to-2 counters from threshold gates [55–57]. A visual representation for a 3-to-2 counter with threshold gates using these methods is shown in Figure 6.8.

A 3-to-2 counter is identical in both Kautz and Minnick methods. Out_0 and Out_1 are the output bits, with Out_0 as the least significant bit and Out_1 as the most significant bit. Thus, Out_0 maps to the sum bit, and Out_1 maps to the carry-out bit in a full adder. Input InCnt is a 3-bit input, represented as a single value. InCnt represents how many inputs are 1.

If InCnt is greater than or equal to 2, the output Out_1, or the carry-out, is a 1. This is equivalent to the traditional C_{out} equation, which uses Boolean operations to represent the requirement that at least two inputs must be a "1."

If [InCnt − 2*Out_1] is greater than or equal to 1, the output Out_0, or the sum, is a 1. The counter has the following truth table in Table 6.10, which is identical to a 3-to-2 counter.

Each threshold detector node in Figure 6.8 corresponds to a threshold gate. Output Out_1 only requires one threshold gate and is computed one cycle after InCnt arrives. Output Out_0 requires two threshold gates and is computed two cycles after InCnt arrives. A full adder requires only two threshold gates. Now that the basic form of a threshold gate ripple-carry adder has been discussed, consider its implementation using both GOTO pairs and CSTG gates.

Table 6.10 Truth table for a Kautz/Minnick 3-to-2 counter

InCnt	**[InCnt – 2*Out$_1$]**	**Out$_1$**	**Out$_0$**
0	0	0	0
1	1	0	1
2	0	1	0
3	1	1	1

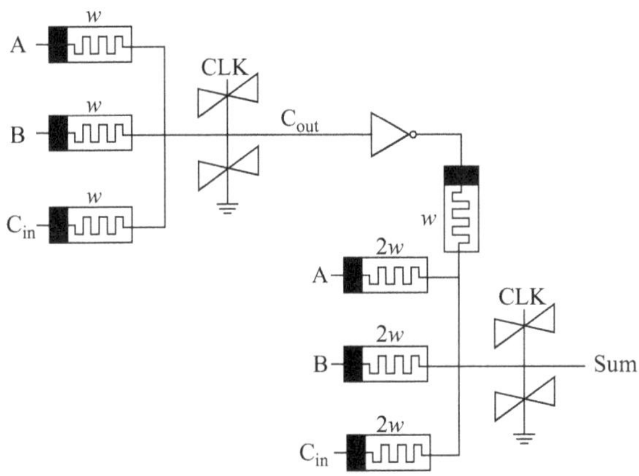

Figure 6.9 Threshold gate full adder using GOTO pairs

6.1.3.1 Ripple-carry adder with GOTO pairs

An implementation of a full adder constructed from threshold gates using GOTO pairs is shown in Figure 6.9.

The first GOTO threshold gate takes the three inputs A, B, and C_{in} with equal weights w for inputs to implement a true majority gate. The second GOTO threshold gate takes the three inputs A, B, and C_{in} with weight w, and the output of the first threshold gate negated with a weight 2. Recall that there is no direct representation for negative weights with GOTO pair threshold gates. Thus, in order to represent the negative weight from Figure 6.8, a CMOS NOT gate is used at the output of the first threshold gate. This effectively implements the functionality in Figure 6.8.

The GOTO full adder can be replicated and concatenated to form a ripple-carry adder. The delay for C_{in} to C_{out} is one gate delay. With the NOT gate introduced, the delay from C_{in} to the sum output in this design is three gate delays. The total delay for an N-bit ripple-carry adder constructed from this full adder is $N+2$ threshold gate delays. The implementation has a complexity of 7N memristors, 2N MOSFETs, and 2N GOTO pairs.

6.1.3.2 Ripple-carry adder with CSTG gates

Alternatively, CSTG gates can be used instead of GOTO pairs to construct a full adder. There are still two threshold gates and the inputs and outputs are identical. However, the CSTG implementation of a full adder does not require an inverter as the GOTO implementation did. Recall from the initial schematic of a CSTG that both the threshold gate output and its inverse are produced during normal operation. Thus, only the two threshold gates are required per full adder in the CSTG design.

The first threshold gate has three inputs with equal weight w. The inputs each require one memristor, one resistor, and two MOSFETs. The output requires four MOSFETs. In total, the first threshold gate requires three resistors, three memristors, and ten MOSFETs. The second threshold gate has four inputs. In total, it requires 4 resistors, 4 memristors, and 12 MOSFETs.

The full adder can be instantiated multiple times to create a ripple-carry adder. No additional hardware is needed besides the full adders themselves. The total complexity is $7N$ memristors, $7N$ resistors, and $22N$ MOSFETs. The delay from C_{in} to C_{out} remains one gate delay. Since there is no inverter in the CSTG implementation, the delay from C_{in} to the sum output is only two gate delays. The total delay is $N+1$ threshold gate delays.

6.1.3.3 Ripple-carry adder schematic

Figure 6.10 shows an 8-bit adder constructed from CSTG threshold gate full adders.

A and B are two 8-bit inputs, and C_{in} is the initial carry-in signal. As described in the introduction of threshold gates, the output of a threshold gate TG is written as

$$\text{Output} = \left(\sum_{i=1}^{n} w_i * X_i \right) \geq T$$

where T is the threshold of the gate as shown in the name of the gate, w_i is the weight of input i, and X_i is the value of input i. In the ripple-carry adder schematic, the first gate in each full adder has a threshold of two and the second gate has a threshold of one.

The schematic is nearly identical to the schematic constructed from GOTO threshold gate full adders. The only difference is that if the schematic uses GOTO gates, inverters would be required on each carry-out output.

6.1.4 MAD gate ripple-carry adder

Similar to IMPLY gates, MAD gates use memristors for both logical operation and memory. The result of a MAD operation is stored into a result memristor in a similar manner to the IMPLY operation. This is in contrast to hybrid-CMOS and threshold gates, which use memristors for logic but not storage.

6.1.4.1 Baseline MAD full adder

The introduction to MAD gates showed that the gates are capable of implementing all Boolean operations. These principles are extended to a stand-alone full adder as shown in Figure 6.11.

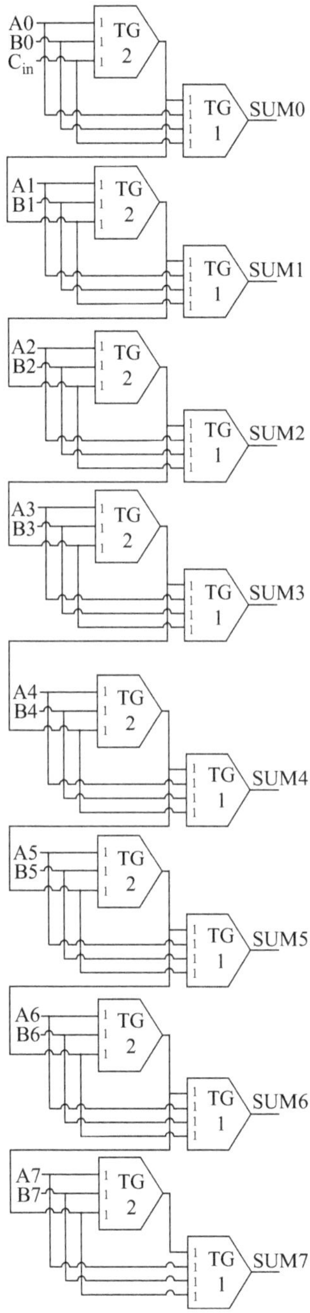

Figure 6.10 Optimized threshold gate 8-bit adder

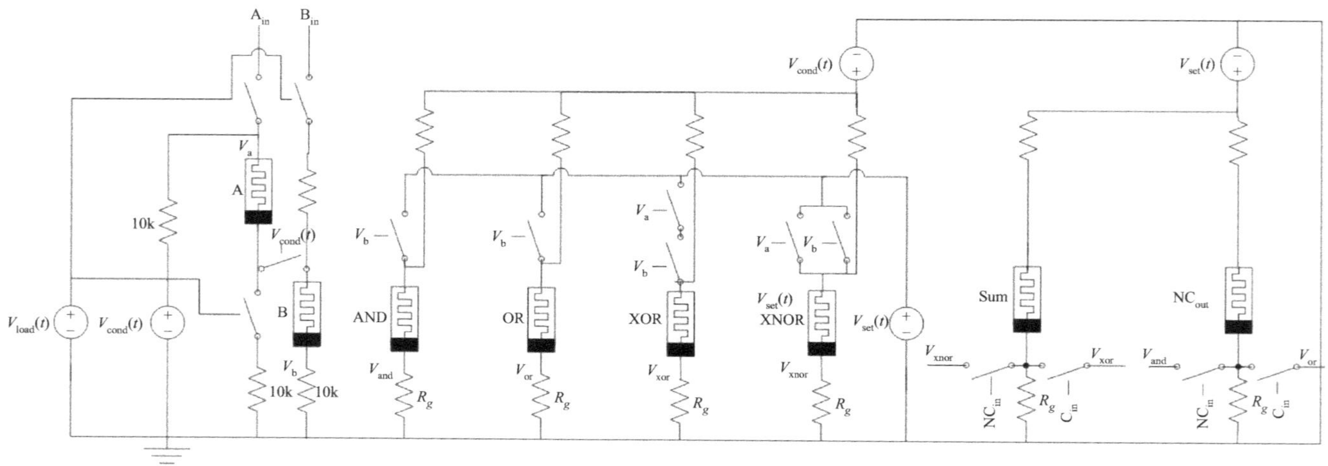

Figure 6.11 MAD full adder

The full adder requires 5 drivers, 8 memristors, and 14 switches. Two of the memristors A and B are the input memristors, and two of the memristors Sum and NC_{out} are the output memristors. The carry-out signal NC_{out} is named as such to indicate that it is the inverse carry-out signal. The remaining four memristors are named according to the Boolean operation they store.

Each driver is labeled with the voltage it drives. One of the drivers drives the V_{load} voltage. This driver is connected to the two switches that gate the input values A_{in} and B_{in} to their corresponding input memristors A and B. Let the threshold voltage on the input switches be V_{th}. At initialization, V_{load} is driven such that $V_{load} > V_{th}$, and the input switches close. This allows the inputs to be stored into the input memristors. For the remainder of the execution, the V_{load} driver remains inactive. Two drivers drive the V_{cond} voltage. These drivers are used to read memristor values, the first for the input memristors and the second for the intermediate memristors. The final two drivers drive the V_{set} voltage. These drivers are used to set the intermediate memristors and the final output memristors.

As described, at $t = 0$, inputs of A and B are initialized into the operand memristors through previous operations or direct write operations with the V_{load} driver set to high. This operation gates the inputs A_{in} and B_{in} into the n-terminals of the input memristors.

Next, the four memristors for the A OR B, A AND B, A XOR B, and A XNOR B operations are resolved in parallel, a key benefit of MAD methodologies. Each of these operations is resolved into the corresponding intermediate memristor, labeled accordingly. V_{cond} is driven high which closes the switch between the input memristors and places the memristors in series. The V_{cond} voltage is also driven to the input memristors themselves so their values can be read. The input circuitry now takes the same form as a standard MAD gate cell. Each operation is resolved by reading the corresponding voltage(s) at V_a and V_b. V_{set} is driven to all of the intermediate Boolean output memristors, and voltages at V_a and V_b are used to drive the gates on each memristor. For each operation, the V_{set} signal will only set the output memristor to "1" if the inputs A and B are the appropriate values as described in the introduction to MAD Boolean operations.

Last, the sum and inverse carry-out results are resolved. In this step, V_{cond} is applied to the intermediate results so their values can be read. V_{set} is applied to the sum and inverse carry-out memristors so these memristors can be updated. The carry-in signal and its inverse, C_{in} and NC_{in}, are also used to resolve the output memristors. The carry-in and its inverse drive switches, which gate the intermediate values to the sum and carry-out memristors. Consider the logic on the C_{out} memristor and recall the equation for a carry-out bit:

$$C_{out} = (\text{A AND B}) \text{ OR } ((\text{A OR B}) \text{ AND } C_{in})$$

This can be rewritten as

$$C_{out} = (\text{NOT}(C_{in}) \text{ AND } (\text{A AND B})) \text{ or } ((\text{A OR B}) \text{ AND } C_{in})$$

This expression is logically equivalent to the original expression. Although this adds an unnecessary expression, it is simpler for understanding the MAD circuitry in the full adder. Consider the different scenarios.

Note that exactly one of C_{in} and NC_{in} will be a "1" and the other a "0" since they are inverses. Thus, only one of the switches will close at the p-terminal of the NC_{out} memristor. If the inverse carry-in bit is a 1, the switch will gate the voltage at V_{and} to the p-terminal of the NC_{out} memristor. If the carry-in bit is a "1," the switch will gate the voltage at V_{or} to the p-terminal.

When the carry-in bit is a "0" and the inverse carry-in bit is a "1," the switch to gate the value V_{or} to the p-terminal will remain open and will have no effect on the output memristor. The switch to gate the value V_{and} to the p-terminal will close. Consider the voltage division circuitry with the AND memristor when V_{cond} is applied to the intermediate memristors. If the value in the AND memristor is a "0," or high resistance state, then the voltage at node V_{and} will be near 0 V when V_{cond} is applied. This means that the voltage at the p-terminal of the NC_{out} memristor will be near 0 V. The voltage-drop across the NC_{out} output memristor will have a magnitude near V_{set}, setting it to "1." If the value in the AND memristor is a "1," or low resistance state, then the voltage at V_{and} will be near V_{cond}. The NC_{in} switch will drive this voltage at the p-terminal of the NC_{out} memristor creating a voltage-drop of magnitude $\sim V_{set} - V_{cond}$ across the NC_{out} memristor. This magnitude is too small to set the memristor to "1," and it will remain a "0."

For the case that the carry-in bit is a "1" and the inverse carry-in bit is a "0," the switch to gate the value V_{and} to the p-terminal will remain open and will have no effect on the output memristor. The switch to gate the value V_{or} to the p-terminal will close. Consider the voltage division circuitry with the OR memristor when V_{cond} is applied to the intermediate memristors. It follows the same logic as discussed for the AND memristor above. If the value in the OR memristor is a "0," then the voltage-drop across the NC_{out} memristor will have a magnitude near V_{set}, setting it to "1." If the value in the OR memristor is a "1," then the voltage-drop across the NC_{out} memristor will have a magnitude $\sim V_{set} - V_{cond}$, which is too small to set the memristor to "1," and it will remain a "0."

Table 6.11 represents the full set of scenarios.

Table 6.11 Truth table for the NC_{out} output memristor in a MAD full adder

ABC_{in}	A OR B	A AND B	V_{and}	V_{or}	V_{drop}	NC_{out}
000	0	0	0	N/A	V_{set}	1
001	0	0	N/A	0	V_{set}	1
010	1	0	0	N/A	V_{set}	1
011	1	0	N/A	$\sim V_{cond}$	$\sim V_{set} - V_{cond}$	0
100	1	0	0	N/A	V_{set}	1
101	1	0	N/A	$\sim V_{cond}$	$\sim V_{set} - V_{cond}$	0
110	1	1	$\sim V_{cond}$	N/A	$\sim V_{set} - V_{cond}$	0
111	1	1	N/A	$\sim V_{cond}$	$\sim V_{set} - V_{cond}$	0

This is identical to the truth table for C_{out} as shown later including the value of the inverse NC_{out} for completeness shown in Table 6.12.

The NC_{out} memristor contains the inverse of the C_{out} signal after this step.

The sum output memristor works in a similar manner except it does not hold an inverse output. Exactly one of the switches on the p-terminal of the sum memristor will be closed when the value of the carry-in and its inverse are read. If the inverse carry-in bit is a 1, the switch will gate the voltage at V_{xnor} to the p-terminal of the sum memristor. If the carry-in bit is a "1," the switch will gate the voltage at V_{xor} to the p-terminal.

Now consider the case that the carry-in bit is a "0." The switch that gates V_{xor} will remain open and will not affect the output of the sum memristor. The switch that gates V_{xnor} will close, and the voltage at node V_{xnor} will be applied to the p-terminal of the sum memristor. If the value in the XNOR memristor is a "0," then the voltage at node V_{xnor} will be near 0 V when V_{cond} is applied. The NC_{in} switch will drive this voltage at the p-terminal of the sum memristor creating a voltage-drop of magnitude near V_{set} across the sum memristor, setting it to "1." If the value in the XNOR memristor is a "1," then the voltage at node V_{xnor} will be near V_{cond}. The NC_{in} switch will drive this voltage at the p-terminal of the sum memristor creating a voltage-drop of magnitude $\sim V_{set} - V_{cond}$ across the sum memristor, which is too small to set the memristor to "1." In other words, the sum output will be set to "1" if $C_{in} = 0$ and A XNOR B = 0. Equivalently, the sum output will be set to "1" if $C_{in} = 0$ and A XOR B = 1. This matches with the expression for the sum output of a full adder,

$$\text{Sum} = \text{A XOR B XOR C}$$

Now consider the case that the carry-in bit is a "1." The switch that gates V_{xnor} will remain open and will not affect the output of the sum memristor. The switch that gates V_{xor} will close, and the voltage at node V_{xor} will be applied to the p-terminal of the sum memristor. If the value in the XOR memristor is a "0," then the voltage at V_{xor} will be near 0 V when V_{cond} is applied. The C_{in} switch will drive this voltage at the p-terminal of the sum memristor creating a voltage-drop of magnitude near V_{set} across the sum memristor, setting it to "1." If the value in the XOR memristor

Table 6.12 Truth table for C_{out} in a MAD full adder

$\mathbf{ABC_{in}}$	$\overline{\mathbf{C_{out}}}$	$\mathbf{C_{out}}$
000	1	0
001	1	0
010	1	0
011	0	1
100	1	0
101	0	1
110	0	1
111	0	1

is a "1," then the voltage at V_{xor} will be near V_{cond}. The C_{in} switch will drive this voltage at the p-terminal of the sum memristor creating a voltage-drop of magnitude $\sim V_{set} - V_{cond}$ across the sum memristor, which is too small to set the memristor to "1." In other words, the sum output will be set to "1" if $C_{in} = 1$ and A XOR B = 0. This too matches with the expression for the sum output. Together, these scenarios complete the input combinations for a sum output of "1." A complete table showing the voltages at the various nodes, and the result of the sum memristor is shown in Table 6.13.

By inspection, this truth table matches with the expected output for the sum of a full adder. Thus, the full adder introduced in this section correctly resolves both the inverse carry-out and sum memristors in step 2 of the adder.

In all, this implementation has a delay of two and requires eight memristors. If the drivers are included, the adder requires 8 memristors, 5 drivers, and 14 switches. This delay is significantly lower than alternative implementations, especially the IMPLY operation, at the cost of more complexity.

At first glance, it may appear that an extra step is required to apply a NOT operation to the NC_{out} memristor to obtain the value of C_{out}. However, this is not necessary. The value of the inverse of the carry-out is as informative as the carry-out. Since MAD designs are based on reading voltages nodes in the circuit, the value of a memristor and its inverse can be obtained from the same single memristor. Depending on which terminal of the memristor is sensed, the result will correspond to the value in the memristor or the inverse. This was demonstrated with the NOT and COPY operations in the introduction to MAD operations. To obtain the carry-out signal, a read voltage is applied to the NC_{out} memristor and the n-terminal should be sensed. If the inverse of the carry-out is desired, the p-terminal of the memristor will be sensed. This will be seen in later sections. In general, to sense the value of a memristor, the p-terminal should be sensed, and to sense the inverse of the value of a memristor, the n-terminal should be sensed. Thus, this adder not only resolves the sum and inverse carry-out signal, but also the inverse sum and carry-out signals.

Table 6.13 Truth table for the sum output memristor in a MAD full adder

ABC$_{in}$	**A XNOR B**	**A XOR B**	V_{xnor}	V_{xor}	V_{drop}	**Sum**
000	1	0	$\sim V_{cond}$	N/A	$\sim V_{set} - V_{cond}$	0
001	1	0	N/A	0	V_{set}	1
010	0	1	0	N/A	V_{set}	1
011	0	1	N/A	$\sim V_{cond}$	$\sim V_{set} - V_{cond}$	0
100	0	1	0	N/A	V_{set}	1
101	0	1	N/A	$\sim V_{cond}$	$\sim V_{set} - V_{cond}$	0
110	1	0	$\sim V_{cond}$	N/A	$\sim V_{set} - V_{cond}$	0
111	1	0	N/A	0	V_{set}	1

6.1.4.2 Optimized MAD full adder

It is possible to implement a more optimized MAD full adder. Although this implementation of a MAD full adder in the previous section achieved low delay and component count, there are more optimizations that can be made. The MAD full adder can be further optimized to perform in a single step rather than two. The complexity is also reduced to 4 memristors, 13 switches, and 4 drivers. The circuit diagram for the revised MAD full adder is shown in Figure 6.12.

Names and parameters were selected to match those in Figure 6.11 to make the adaptations more clear. In the original design, intermediate results for A AND B, A OR B, A XOR B, and A XNOR B were computed in the first step and the carry-out and sum values were resolved in the second. This design removes the intermediate memristors in exchange for switches, producing the carry-out and sum signals in a single step. This modification accounts for the changes in the number of switches and memristors in the design. The voltage division circuitry for the input memristors remains unchanged.

Similar to the original design, a read signal, V_{cond}, is applied to the input memristors in series, and the voltages at nodes V_a and V_b are sensed to drive the switches on the neighboring memristors. However, these nodes are no longer used to set intermediate memristors as they were in the original design. Instead, they are used to set switches in a more complex circuitry on the output memristors. The switches that use nodes V_a and V_b are labeled accordingly. The remaining switches are driven by either the value of the carry-in signal C_{in} or its inverse NC_{in}.

Consider the logic on the C_{out} memristor. Specifically, the circuitry for the carry-out memristor is implementing

$$C_{out} = (C_{in} \text{ AND } (\text{A OR B})) \text{ OR } (\text{A AND B})$$

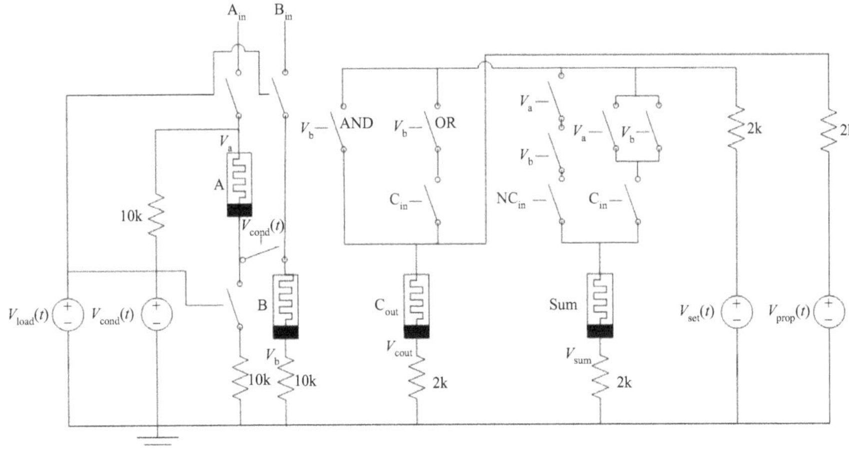

Figure 6.12 Optimized MAD-based full adder

This is seen in the circuitry in which the two paths from V_{set} to the carry-out memristor are. The left path corresponds to the expression (A AND B) and the right path corresponds to the expression C_{in} AND (A OR B). Since these paths lie in parallel, either one can create a path between the V_{set} voltage and the C_{out} memristor. In other words, if either the C_{in} AND (A OR B) path OR the (A AND B) path has its switches closed, the output will be set to a "1." This implements the final OR operation in the C_{out} expression.

Consider each path. The left-most path will only gate V_{set} through to the memristor if the voltage at V_b is above the V_{th} for an AND operation (as labeled). Essentially, the left path is a standard MAD AND gate operation. This aligns with the fact that its intention is to implement (A AND B).

The right-most path has two switches between the V_{set} voltage and the output memristor. Thus, two conditions must be satisfied for this path to set the output memristor to a "1." The right path will only gate V_{set} to the C_{out} memristor if the voltage at V_b is above the V_{th} for an OR operation (as labeled) and the carry-in signal is above the V_{th} of its switch. Essentially, the right path will gate V_{set} to the C_{out} memristor if C_{in} AND (A OR B). Now, if either path's expression is true, V_{set} will have a closed circuit to the carry-out memristor and the preceding Boolean equation is accomplished.

A similar methodology is used for the sum memristor, implementing

$$\text{Sum} = (\text{NC}_{\text{in}}\ \text{AND}\ (\text{A XOR B}))\ \text{or}\ (\text{C}_{\text{in}}\ \text{AND}\ (\text{A XNOR B}))$$

Although this equation takes a different form, it produces the traditional sum equation as shown in Table 6.14.

This equation is implemented via the left and right paths, respectively. The left path implements the expression (NC_{in} AND (A XOR B)) and the right path implements the expression $-$(C_{in} AND (A XNOR B)).

For the left path, the switches driven by V_a and V_b are identical to the switches in the standard XOR MAD operation. Their thresholds are selected in the same manner as described for a standard XOR Boolean operation. The third switch is driven by the inverse of the carry-out signal. In other words, the left path will

Table 6.14 Truth table for the alternative sum equation in a full adder

ABC$_{in}$	**Sum**
000	0
001	1
010	1
011	0
100	1
101	0
110	0
111	1

only create a closed path from V_{set} to the sum memristor if the XOR operation is satisfied and the carry-out is a "0." This effectively implements (NC_{in} AND (A XOR B)).

The right path also has two switches driven by V_a and V_b. However, on this path, the switches are designed to be identical to the switches in the standard XNOR MAD operation. Rather than being placed in series, they are placed in parallel. Their thresholds are selected to match the thresholds of the switches driven by V_a and V_b in a standard MAD XNOR operation. In series with these switches is a third switch driven by the carry-in signal. In other words, the right path will only create a closed path from V_{set} to the sum memristor if the XNOR operation is satisfied and the carry-out is a "1." This effectively implements (C_{in} AND (A XNOR B)). Since these two paths lie in parallel, the complete circuit implements the desired sum equation above.

All of this logic can and must be performed in the same step. Thus, everything computes in a single step. The voltage division circuitry is driven with a read voltage V_{cond} at the same time the C_{out} and sum memristors are driven with a set voltage V_{set}. The voltages at nodes V_a and V_b are coupled with the carry-in signal and its inverse to drive the switches that set the output memristors to "1" or retain their reset value of "0."

This approach to a MAD full adder reduces the need for the intermediate Boolean results by incorporating them into the circuitry for the output memristors themselves. This removes a step from the addition process and removes one driver, four memristors, and one switch from the design. It also improves the conceptualization and readability of the circuit, reformatting the switch logic to a 1-to-1 mapping with the logic equations for the outputs. Note, this design still produces the inverses of the sum and carry-out since this is purely a matter of node selection at the time the outputs are read.

6.1.4.3 Baseline MAD ripple-carry adder

The original MAD full adder design can be extended with minimal circuitry to implement an *N*-bit ripple-carry adder as shown in Figure 6.13.

Although a 3-bit adder is shown, any adder width *N* can be constructed by concatenating *N* full adders in the same manner. The full adder is simply replicated *N* times. There is no change to the full adder circuitry. The only change is the connections between the full adders. Now, the carry signals and their inverses are driven by the previous full adder in the chain.

At $t = 0$, the V_{load} driver for the input memristors is applied to each bit as described for the full adder, loading the inputs A_i and B_i into their corresponding full adders in parallel. Then, in step 1, all of the intermediate Boolean operations are resolved by applying V_{cond} and V_{set}, respectively. This is done in the same manner as the individual full adder. Now it is simply done in parallel across all of the full adders. Last, the resolution of the output memristors can begin.

One full adder at a time performs the resolution step, beginning with the first full adder, FA0. The full adder resolves its sum and inverse carry-out as described for the stand-alone full adder. In the next step, the next full adder in the ripple-carry

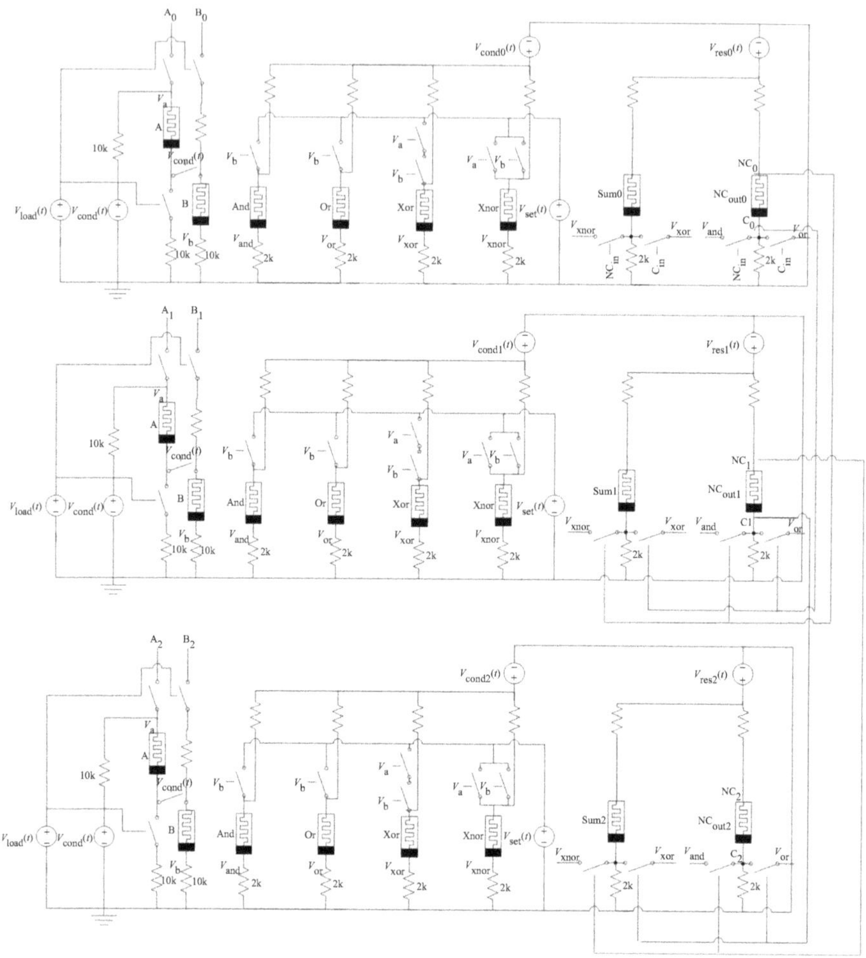

Figure 6.13 3-bit MAD ripple-carry adder

adder chain, FA1, resolves its sum and carry-out. The inverse carry-out memristor in FA0 is applied to a V_{cond} signal so that the voltage at the n-terminal represents the value of the carry-out, signal and the voltage at the p-terminal represents the inverse. The voltages at these two nodes drive the gates on the corresponding switches in FA1 in the same way the carry-in bit did for FA0. At the same time, a read voltage V_{cond} is driven on the intermediate memristors of FA1 to sense the values at their corresponding nodes. Together, this resolves the output memristors. The process then repeats for the subsequent bits until the final sum and carry-out are resolved. Each of the bits resolves their sum and carry signals one step at a time so an N-bit addition requires $N + 1$ steps. A table showing the resolution of each bit in an 8-bit ripple-carry adder is shown in Table 6.15.

Table 6.15 Resolution of the C_{out} and the sum signals for an 8-bit MAD ripple-carry adder

Step	Bit 0	Bit 1	Bit 2	Bit 3	Bit 4	Bit 5	Bit 6	Bit 7
0	Load	Load	Load	Load	Load	Load	Load	Load
1								
2	Resolve							
3		Resolve						
4			Resolve					
5				Resolve				
6					Resolve			
7						Resolve		
8							Resolve	
9								Resolve

The "Load" stage is when the inputs are loaded in the input memristors, and the "Resolve" stage is when the output memristors are resolved.

The total complexity is $8N$ memristors, $14N$ switches, and $2N+3$ drivers. Notice that five drivers are not needed for each full adder. The three drivers, which set the initial input memristors, read the input memristors and set the intermediate memristors can be shared across all of the bits since these operations happen in the same step in every full adder. Only the two drivers for the outputs are unique across each bit since they happen in different cycles.

6.1.4.4 Optimized MAD ripple-carry adder

The optimized MAD full adder can be used to implement an N-bit ripple-carry adder as shown in Figure 6.14.

Again, the full adder is simply replicated N times. There is no change to the full adder circuitry. The only change is the connections between the full adders. Now, the carry signals and their inverses are driven by the C_{out} memristor in the previous full adder in the chain.

The inputs A_i and B_i are initialized into every full adder's input memristors at $t=0$ using the V_{load} driver. Now the resolution phase can begin. In the same manner as the baseline ripple-carry adder, this phase propagates through the full adder chain, resolving one full adder at a time.

In the first step, the first full adder, FA0, resolves its sum and C_{out} memristors. V_{cond} is applied to the input memristors A_0 and B_0, V_{set} is applied to the sum and output memristors in FA0, and V_a, V_b, and the carry-in signal are driven to their corresponding switches. The output memristors in FA0 are now resolved. In the next step, the next full adder, FA1, can resolve its output memristors. V_{prop} applies a read signal, V_{cond}, to the C_{out} memristor in FA0. As done in the original design, the n-terminal is sensed to represent the inverse of the carry-out, and the voltage at the p-terminal is sensed to represent the carry-out signal. These nodes are denoted as VN_{cout} and VC_{out}, respectively. At the same time, a read voltage V_{cond} is driven on the input memristors of FA1 to sense the values at their corresponding nodes in

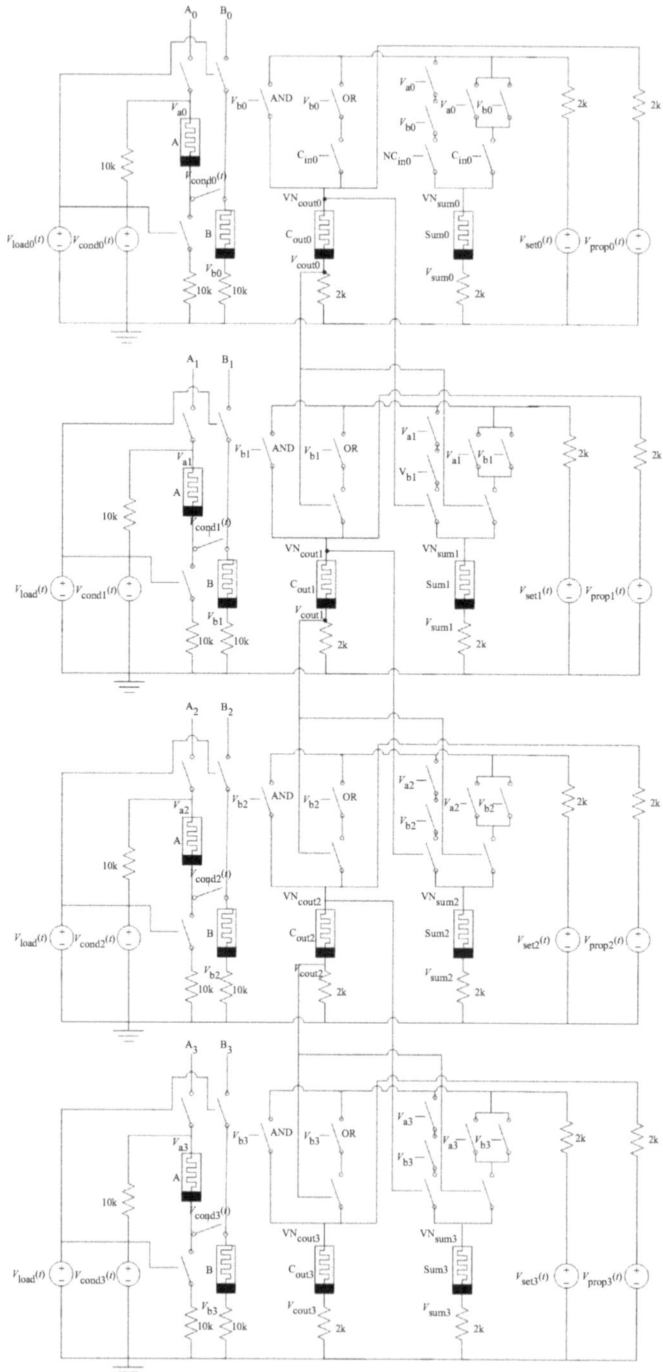

Figure 6.14 Optimized MAD design for a 4-bit ripple-carry adder

the voltage division circuit. Together, all these values are used in the same manner as FA0 to drive the respective switches in FA1. The process repeats through each bit of the adder, resolving the final carry-out and sum in the Nth step. This is one less step than the original MAD design. A table showing the resolution of each bit in an 8-bit ripple-carry adder is shown in Table 6.16.

Again, the "Load" stage is when the inputs are loaded in the input memristors and the "Resolve" stage is when the output memristors are resolved.

The total complexity for the N-bit adder is $4N$ memristors, $13N$ switches, and $3N+1$ drivers. Again, the number of drivers is fewer than the number of memristors because the initialization driver has identical behavior for all bits and can be reused across them. The component and delay comparisons for the original MAD adder and revised MAD adder are summarized in Table 6.17.

The revised MAD adder requires half the number of memristors that the original MAD full adder requires. Even with the additional switch circuitry on the revised MAD full adder, there are still fewer switches and drivers due to the removal of the intermediate operations.

6.1.4.5 Pipelining

The MAD design can further improve delay by pipelining independent additions, similar to the IMPLY approach. MAD operations perform through the use of

Table 6.16 Resolution of the C_{out} and the sum signals for an optimized 8-bit MAD ripple-carry adder

Step	Bit 0	Bit 1	Bit 2	Bit 3	Bit 4	Bit 5	Bit 6	Bit 7
0	Load	Load	Load	Load	Load	Load	Load	Load
1	Resolve							
2		Resolve						
3			Resolve					
4				Resolve				
5					Resolve			
6						Resolve		
7							Resolve	
8								Resolve

Table 6.17 Delay and complexity improvements for the MAD adders

	Original MAD full adder	Optimized MAD full adder	Original MAD N-bit adder	Optimized MAD N-bit adder
Complexity	8 memristors	4 memristors	$8N$ memristors	$4N$ memristors
	14 switches	13 switches	$14N$ switches	$13N$ switches
	5 drivers	4 drivers	$2N+3$ drivers	$3N+1$ drivers
Delay	2	1	$N+1$	N

drivers, and results are stored in memristors at the end of each step. This is similar to a "latch" mechanism, storing the value of each operation. This can be used to pipeline the design.

In the context of a ripple-carry adder, it is desirable to pipeline across the full adders such that two full adders can operate on two different additions in parallel. This removes the idle steps in the design. Consider the nonpipelined optimized ripple-carry adder design. The design has a delay of $N+1$ steps but the first full adder is complete after two steps. The full adder remains idle for the remaining $N-1$ steps of the addition. Similarly, the last full adder is only active during step 1 and step $N+1$. This adder is also idle for $N-1$ steps of the addition. Pipelining removes these idle steps.

Pipelining allows a full adder to begin a new addition as soon as it completes its addition and propagates its outputs to the next adder. The overhead is two steps: one for the subsequent adder to sense the result and one for resetting the memristors. Consider FA0 for bit 0. At $t=1$, the sum and carry-out are resolved. At $t=2$, the carry-out value is read by bit 1 with V_{prop} and the outputs are no longer needed. At $t=3$, all of the memristors are reset to logical "0." At $t=4$, the next addition's A_0 and B_0 operands are loaded into the adder. The remaining full adders can continue normal execution during these steps, logically separated from the operations in FA0. Thus, every four cycles a new addition can begin.

For visual assistance, a chart showing the pipelining of an 8-bit MAD ripple-carry adder for three pipelined additions is shown in Table 6.18.

The "Load" stage loads the inputs into the input memristors, the "Resolve" stage sets the output memristors, the "Prop" stage reads the output memristor voltages for use in the next adder, and the "Reset" stage sets the memristors back to "0." Now, X additions can complete in $8+4(X-1)$ steps.

In traditional CMOS and the other approaches, additions cannot overlap and must wait for the previous addition to complete. Even in the IMPLY approach, which supported pipelining additions, a new addition could start only after at least 22 cycles. The MAD approach has a much higher throughput. The only consequence of this optimization for the MAD designs is that the full adders do not all load their inputs at the same step anymore. Thus, the load driver cannot be shared in parallel across the full adders. Each full adder needs a unique V_{load} driver, increasing the number of drivers from $3N+1$ to $4N$.

6.1.4.6 Translation into a crossbar

It is possible to implement the presented MAD full adder in a memristor crossbar. This is useful for logic-in-memory applications and systems. However, the design must be transformed similarly to the way that the basic MAD Boolean gates were transformed for the crossbar context. The structure of the circuitry must change to adapt to the logic-in-memory framework, but the overall functionality remains the same.

The input circuitry of a MAD gate can be mapped into a crossbar to perform a Boolean operation. The full adder is a basic extension on these principles. The input circuitry is identical to the Boolean operations so the transformations necessary on

Table 6.18 Resolution of the C_{out} and the sum signals for a pipelined 8-bit ripple-carry adder

Step	Bit 0	Bit 1	Bit 2	Bit 3	Bit 4	Bit 5	Bit 6	Bit 7
0	Load	Load	Load	Load	Load	Load	Load	Load
1	Resolve							
2	Prop	Resolve						
3	Reset	Prop	Resolve					
4	Load	Reset	Prop	Resolve				
5	Resolve	Load	Reset	Prop	Resolve			
6	Prop	Resolve	Load	Reset	Prop	Resolve		
7	Reset	Prop	Resolve	Load	Reset	Prop	Resolve	
8	Load	Reset	Prop	Resolve	Load	Reset	Prop	Resolve
9	Resolve	Load	Reset	Prop	Resolve	Load	Reset	Prop
10	Prop	Resolve	Load	Reset	Prop	Resolve	Load	Reset
11	Reset	Prop	Resolve	Load	Reset	Prop	Resolve	Load
12		Reset	Prop	Resolve	Load	Reset	Prop	Resolve
13			Reset	Prop	Resolve	Load	Reset	Prop
14				Reset	Prop	Resolve	Load	Reset
15					Reset	Prop	Resolve	Load
16						Reset	Prop	Resolve
17							Reset	Prop
18								Reset

this part of the circuitry are identical too. The remaining changes are on the output circuitry.

Instead of performing a single Boolean operation, the full adder performs two complex operations. One operation computes C_{out} and the other computes the sum. The same nodes V_a and V_b are needed on the input circuitry, but the driver logic on the outputs is now more complicated. Luckily, this does not drastically impact the crossbar structure itself. Recall the logic presented for the MAD crossbar to select which operation was being performed in the memory each step. The driver selection logic is shown in Figure 6.15.

The switch logic is entirely removed from the crossbar itself. Thus, no changes to the crossbar are needed. Instead the single driver logic that selects the operation must be updated. In order to enable the C_{out} and sum memristor logic, optional paths need to be added in this multiplexer. Now, these paths can be chosen to gate the voltage V_{set} to the selected output memristor in the crossbar. A revised operation selection for the drivers is shown in Figure 6.16.

Now, the operation select lines will choose either a traditional MAD Boolean operation, or the more complex set operations for the C_{out} memristor and sum memristor.

The timing for the adder must change. Only a single operation can perform on the crossbar at a time. Thus, the read steps for the input memristors and the carry-in signal must be serialized and the set steps for each of the output memristors must be serialized. Now, in step $t = 1$, the node voltages at V_a and V_b are read as usual for a

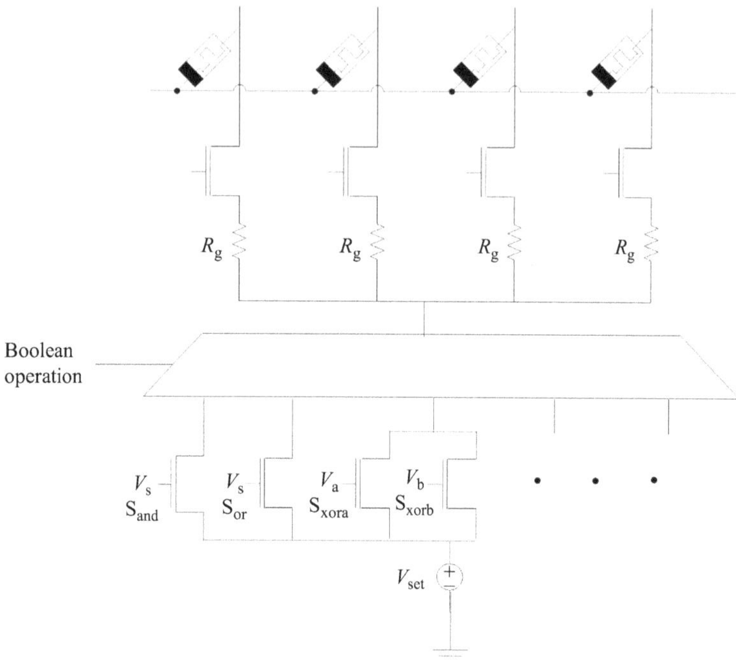

Figure 6.15 Driver logic for selecting MAD crossbar Boolean operations

crossbar operation. At $t = 2$, the C_{in} value is read from its memristor in the crossbar. At $t = 3$, the C_{out} memristor is set, and at $t = 4$, the sum memristor is set.

The full adder delay increases from two steps to four due to the serial behavior in the crossbar context. Operations can no longer be performed in parallel. The delay for an N-bit adder is now $4N$ steps. This delay of $4N$ steps is a significant increase compared to the delay of N steps for a MAD full adder outside of the crossbar. However, it is still significantly less than the alternative logic-in-memory approach: IMPLY.

6.1.5 Ripple-carry adder analysis and comparison

A summary of the complexity and delay of the presented memristor-based ripple-carry adders in comparison to a traditional CMOS ripple-carry adder is given in Table 6.19.

In terms of delay, all proposed implementations have comparable or better results relative to the CMOS approach. The IMPLY methodology maintains a delay of two for carry propagation (the same as CMOS) and GOTO pairs and MAD gates improve this to one. When buffering is considered for the hybrid CMOS design, the delay is increased to three steps per carry propagation for a total delay of $3N + 4$.

It is important to note that the units of delay are not equivalent across approaches. For the IMPLY and MAD designs, a unit delay is a step in which voltages are applied to the circuit. In the hybrid-CMOS and threshold gate designs,

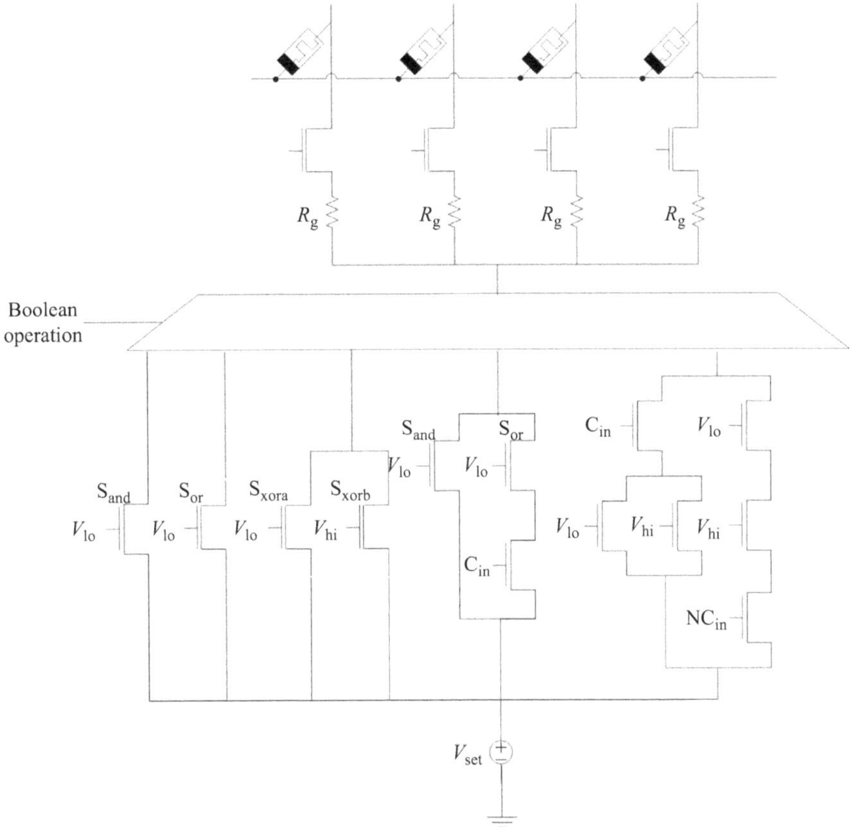

Figure 6.16 Revised driver logic for selecting MAD crossbar operations

it is the gate delay through the respective memristor gate. For the CMOS design, a unit delay is the delay through a traditional CMOS gate. Direct apples-to-apples comparisons cannot be made between these delays, but general comparisons can be.

In terms of area, all discussed implementations improve upon the CMOS design. The hybrid-CMOS design has about a 38 percent reduction in the number of devices as compared to previous work. The IMPLY implementation decreases the component count to $7N + 1$ memristors and their drivers, significantly decreasing the number from $34N$ MOSFETs in CMOS. The GOTO-based designs see one of the greatest improvements in area, decreasing the requirements from $34N$ MOSFETs to $7N$ memristors and $2N$ GOTO pairs. The MAD design requires only $4N$ memristors, $3N + 1$ drivers, and $13N$ switches as compared to $34N$ MOSFETs for a traditional CMOS design.

Table 6.19 Complexity and delay comparisons for the ripple-carry adders

	CMOS	IMPLY	Hybrid-CMOS	Threshold-GOTO	MAD
Delay	$2N+4$	$2N+19$	$3N+4$	$N+2$	N
Complexity	$34N$ MOSFETs	$7N+1$ memristors	$14N$ memristors	$7N$ memristors	$4N$ memristors
		$7N$ drivers	$12N$ MOSFETs	$2N$ MOSFETs	$3N+1$ drivers
		$8N-1$ switches		$2N$ GOTO pairs	$13N$ switches

Table 6.20 Throughput comparisons for X additions in an N-bit ripple-carry adder

	CMOS	IMPLY	Hybrid-CMOS	Threshold	MAD
Delay	$2NX+4X$	$2N+22X-3$	$3NX+4X$	$NX+2X$	$N+4X-4$

Recall that MAD and IMPLY implementations have the additional benefit of overlapped operations, or pipelining. Since their functionality is controlled by drivers, they are able to implement various stages of numerous operations in parallel, pipelining execution and improving throughput as shown in Table 6.20.

In the MAD design, a new addition can begin every four steps. In IMPLY, a new addition can begin every 22 steps. In traditional CMOS and the remaining approaches, a new addition cannot begin until the full execution has completed on the previous addition. Thus, the throughput of the IMPLY and MAD adder design is significantly higher than alternative designs, performing X additions in $X+O(N)$ rather than $O(NX)$. As the number of additions increases, the MAD design establishes a significant advantage over the IMPLY approach. As X increases, the MAD design has a total latency of $4X$, less than 1/5 of the IMPLY latency, $22X$. One consequence of this optimization for the MAD designs is that the drivers cannot be shared across bits, increasing the number of drivers from $3N+1$ to $4N$.

6.2 Carry Lookahead adders[3]

Memristor-based carry-lookahead adders are presented for the IMPLY, hybrid-CMOS, threshold, and MAD contexts. Carry-lookahead adders are the most complex of the adders discussed in this textbook. They also generally have the fastest delay.

This chapter introduces an IMPLY, hybrid-CMOS, threshold gate, and MAD implementation of the carry-lookahead adder. The focus of these designs is on

[3]This section is based on [58].

16-bit carry-lookahead adders using 4-bit lookahead logic blocks, but the design principles can be extended to any configuration. Each design is implemented, optimized, and analyzed in terms of performance and delay. A full comparison of the designs in comparison to CMOS is given at the end of the chapter.

6.2.1 IMPLY carry-lookahead adder

Prior work implemented an IMPLY-based carry-lookahead adder that reduced the delay for a 4-bit adder to 25 steps for the carry-out and 32 steps for the sum [16]. Area and complexity numbers were not given for this implementation.

The IMPLY implementation can simplify the logic for the final carry-out bit and sum bit significantly by recognizing Boolean-based simplifications in the IMPLY domain. In carry-lookahead adders, the AND operation is often followed by the OR operation or a second AND operation. As for the ripple-carry adder, an AND and OR operation can together be simplified to

(P AND Q) OR R = (P IMP (Q IMP 0)) IMP R

This requires only three steps and four memristors. Similarly, two consecutive AND operations can be simplified. A naïve implementation of two AND operations would serialize the individual operations as follows:

1. B IMP 0
2. A IMP (B IMP 0)
3. (A IMP (B IMP 0)) IMP 0 = A AND B
4. C IMP 0
5. AB IMP (C IMP 0)
6. (AB IMP (C IMP 0)) IMP 0 = A AND B AND C

This implementation requires six steps and seven memristors. By treating these operations as single operational units, the IMPLY operations can be simplified as

1. B IMP 0
2. A IMP (B IMP 0)
3. C IMP (A IMP (B IMP 0))
4. (C IMP (A IMP (B IMP 0))) IMP 0 = A AND B AND C

This implementation only requires four steps and five memristors. It also does not overwrite any of the inputs. It relies on the assumption that the intermediate result of (A AND B) is not needed, which is correct in the context of a carry-lookahead adder.

Both of these Boolean expressions are used extensively in carry-lookahead adders. The calculations for the group propagate (GP) and generate signals and the carry-out and sum values use chained instances of them. This leads to a multiplicative improvement in the overall delay and area of the design.

Numerous Boolean expressions are calculated inside of a carry-lookahead adder block. Each bit produces its individual generate and propagate bits and the 4-bit block produces the group generate (GG) and GP bits. Then, the block produces each bit's sum and carry-out bits. Recall that for a 4-bit block, these equations are

$$
\begin{aligned}
P_i &= A_i \text{ XOR } B_i \\
G_i &= A_i \text{ AND } B_i \\
GP &= P_0 * P_1 * P_2 * P_3 \\
GG &= G_3 + G_2 * P_3 + G_1 * P_2 * P_3 + G_0 * P_1 * P_2 * P_3 \\
\text{Cout0} &= G_0 + P_0C_{in} \\
\text{Cout1} &= G_1 + P_1G_0 + P_1P_0C_{in} \\
\text{Cout2} &= G_2 + P_2G_1 + P_2P_1G_0 + P_2P_1P_0C_{in} \\
\text{Cout3} &= G_3 + P_3G_2 + P_3P_2G_1 + P_3P_2P_1G_0 + P_3P_2P_1P_0C_{in} \\
&= GG_0 + GP_0C_{in} \\
\text{Sum}_0 &= A_0 \text{ XOR } B_0 \text{ XOR } C_{in} \\
\text{Sum}_1 &= A_1 \text{ XOR } B_1 \text{ XOR } [G_0 + P_0C_{in}] \\
\text{Sum}_2 &= A_2 \text{ XOR } B_2 \text{ XOR } [G_1 + P_1G_0 + P_1P_0C_{in}] \\
\text{Sum}_3 &= A_3 \text{ XOR } B_3 \text{ XOR } [G_2 + P_2G_1 + P_2P_1G_0 + P_2P_1P_0C_{in}]
\end{aligned}
$$

The GG and GP bits are on the critical path of the design. The GG and GP bits are used to resolve the carry-out signal to the next block to allow forward progress. To obtain GG and GP, the adders must first resolve their individual G and P bits. Once GG and GP are available, the carry-out for the signal is resolved. Once the carry-out signal is propagated, the remaining carry-out and sum signals can be resolved since they are not on the critical path. The execution steps for a 4-bit block in a carry-lookahead adder are given in Table 6.21. The sum resolution is identical to a ripple-carry adder once the carry signals are available.

Steps 1 through 10 produce the propagate P and generate G bits as well as their inverses. These steps are performed in parallel across each bit in the block. The steps that are performed in parallel are denoted with a "*." All four of these outputs are used later in the block's execution. The rest of the execution determines GG, GP, and then the carry-out and sum signals.

6.2.1.1 Multiset IMPLY operation

Steps 9 and 10 produce the G and inverse of P results by inverting the values of P and NOT G. Step 9 is a multiset operation to produce multiple copies of the NOT P value. Step 9 expands to each of the steps shown in Table 6.22 for the various blocks.

Each block performs step 9 at the same time. In Block0, there are two destination memristors, M4 and M6. M5 is not used because it is the destination memristor in step 10. To perform a multiset operation, V_{set} is applied to both output memristors. The rest of the IMPLY operation is identical. So, V_{cond} is applied to M3 and V_{set} is applied to both M4 and M6. The result of the IMPLY operation ends up in both M4 and M6. The same operation occurs in the other three blocks with three destinations in Block1 and four destinations in Block2 and Block3.

The multiset operation enables more parallelism in the design. Without the multiset operation, each (P IMP 0) operation would be serialized. This would take two steps in Block0, three steps in Block1, and four steps in Block2. The

Table 6.21 Steps for an optimized 4-bit block in an IMPLY carry-lookahead adder

Step	Functionality	Goal
1*	A IMP 0	A → M0
2*	B IMP 0	B → M1
3*	B	M1 → M2
4*	(A IMP 0) IMP (B IMP 0)	M0 → M1
5*	A IMP B	A → M2
6*	((A IMP 0) IMP (B IMP 0)) IMP 0	M1 → M3
7*	A NAND B = NOT G	B → M0
8*	A XOR B = P	M2 → M3
9*	NOT P (2 P0, 3 P1, 4 P2, 4 P3)	M3 → M4
10*	G	M0 → M5
11	P2 IMP (P3 IMP 0)	P2 NAND P3
12	P2 IMP (P3 IMP 0)	P2 NAND P3 (second copy)
13	P0 IMP (P1 IMP 0)	P0 NAND P1
14	(P0 NAND P1) IMP 0	P0 AND P1
15	G0 IMP (P1 IMP 0)	G0 NAND P1
16	(G0 NAND P1) IMP 0	G0 AND P1
17	(G0 AND P1) IMP (P2 NAND P3)	G0P1 NAND P2P3
18	G1 IMP (P2 NAND P3)	G1 NAND P2P3
19	G2 IMP (P3 IMP 0)	G2 NAND P3
20	(G2 IMP (P3 IMP 0)) IMP G3	P3G2 + G3
21	G1P2P3 IMP (G3 + P3G2)	G1P2P3 + G2P3 + G3
22	((G0 AND P1) IMP (P2 NAND P3)) IMP (G1P2P3 + (G3 + P3G2)) = GG	G0P1P2P3 + G1P2P3 + G2P3 + G3
23	GG sensed by the 16-bit block	
24	(P0 AND P1) IMP (P2 NAND P3)	P0P1 NAND P2P3
25	((P0 AND P1) IMP (P2 NAND P3)) IMP 0 = GP	P0P1 AND P2P3
26	G1 IMP (P2 IMP 0)	G1 NAND P2
27	P1 IMP (P2 IMP 0)	P1 NAND P2
28	P1 IMP (P2 IMP 0)	P1 NAND P2 (second copy)
29	(G0 IMP (P1 IMP 0)) IMP G1	P1G0 + G1
30	(G1 IMP (P2 IMP 0)) IMP G2	P2G1 + G2
31	G0 IMP (P1 IMP (P2 IMP 0))	G0 NAND P1P2
32	G0P1P2 IMP (G2 + P2G1)	G0P1P2 + (G2 + P2G1)
33	C0 IMP (P0 IMP 0)	C0 NAND P0
34	(C0 IMP (P0 IMP 0)) IMP G0 = C1	P0C0 + G0
35	C0 IMP (P0 IMP (P1 IMP 0))	C0 NAND P0P1
36	(C0 NAND P0) IMP 0	C0 AND P0
37	(C0 AND P0) IMP (P1 NAND P2)	NAND C0P0*P1P2
38	C0P1P0 IMP (G1 + P1G0) = C2	C0P1P0 + P1G0 + G1
39	((C0 AND P0) IMP (P1 NAND P2)) IMP (G0P1P2 + (G2 + P2G1)) = C3	C0P0P1P2 + G0P1P2 + (G2 + P2G1)
40*	C IMP 0 → M0	NOT C
41*	(NOT P) IMP (C IMP 0)	
42*	((NOT P) IMP (C IMP 0)) IMP 0 → M1	
43*	P IMP C	
44*	(P IMP C) IMP ((NOT P) IMP (C IMP 0)) IMP 0 = Sum	(A XOR B) XOR C

Table 6.22 Expansion of step 9 in carry-lookahead adder

STEP	Op	Block0	Block1	Block2	Block3
9*	(P IMP 0)	M3 → M4, M6	M3 → M4,M6, M7	M3 → M4,M6,M7, M8	M3 → M4,M6,M7, M8
Result	NOT P	$\overline{P}$ in M4, M6	$\overline{P}$ in M4,M6,M7	$\overline{P}$ in M4,M6,M7, M8	$\overline{P}$ in M4,M6,M7, M8

Table 6.23 Example of steps benefiting from multiset IMPLY operation

Step	Functionality	Goal
11	P2 IMP (P3 IMP 0)	P2 NAND P3
12	P2 IMP (P3 IMP 0)	P2 NAND P3 (second copy)
17	(G0 AND P1) IMP (P2 NAND P3)	G0P1 NAND P2P3
18	G1 IMP (P2 NAND P3)	G1 NAND P2P3

consequence is that the multiset operation increases the number of memristors in the design. Block0 requires nine memristors, Block1 requires 10 memristors, Block2 requires 11 memristors, and Block3 requires 11 memristors. In total, the individual adders require 41 memristors.

Multiset operations also save numerous steps later in execution. For example, consider steps 11 and 12 results of which are used in steps 17 and 18. These four steps are condensed in Table 6.23.

The value of (P3 IMP 0) sits in memristors M4, M6, M7, and M8 in Block3 as shown in the expansion of step 9. Select M4 for use in step 11. Step 11 performs P2 IMP (P3 IMP 0), overwriting M4 with the result. The original value of M4, P3 IMP 0, is no longer available. However, it is necessary in step 12. Since the multiset operation was performed, this is not an issue. If it had not been, it would have to be reproduced.

Although it is relatively inexpensive to recreate the value of (P3 IMP 0), which is not always the case. For more intricate execution, the value being overwritten in an IMPLY step may require multiple steps to recreate. Thus, it is valuable to perform multiset operations on memristors, which are used multiple times.

6.2.1.2 Group propagate and group generate production

Steps 11 to 25 produce the GG and GP bits by using the outputs from the four individual bits. Since the necessary multiset operations have been performed in step 9, many of the memristors for these steps are already included in the design. Only three additional memristors are needed.

The GG bit is read in step 23 after it is resolved and the GP bit is set in step 25. The GG bit is read as soon as possible so that the 16-bit logic in the carry-lookahead adder can begin immediately. The GP bit is directly stored into the 16-bit logic in step 25 by applying V_{set} to a memristor outside of the 4-bit block. This is not done for

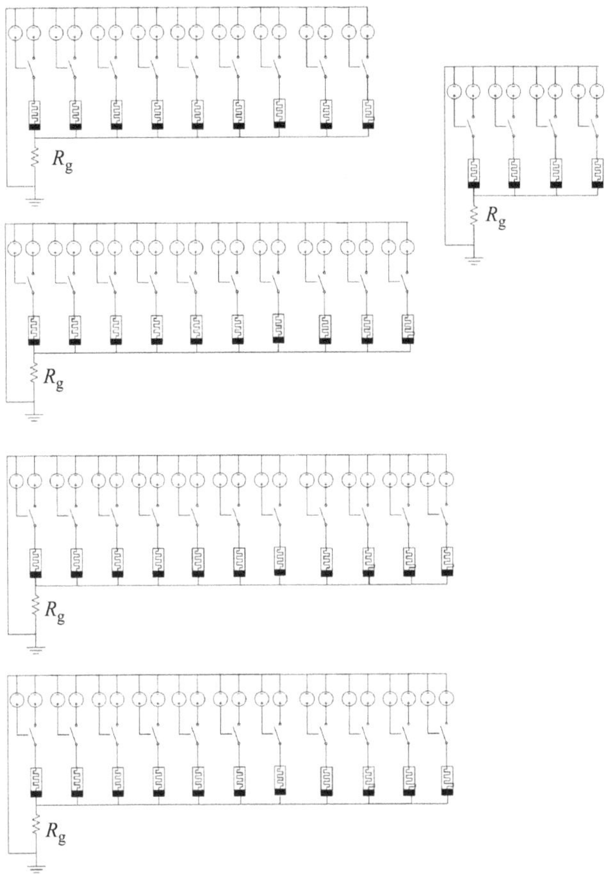

Figure 6.17 4-bit IMPLY block for a carry-lookahead adder

the GG bit because its final step is not a simple IMPLY NOT operation. The details for the 16-bit logic will be discussed in further detail later in the section.

6.2.1.3 Sum and carry-out production

The remaining steps are dedicated to resolving the internal carry-out and sum signals, which are not on the critical path. The carry-out signals are computed by the block itself using the G and P bits from each of the bits. In all, the operations to compute the carry signals in steps 26–39 only introduce one additional memristor. This again shows the benefits of the multiset operation. It is also partially due to the strategic order of use for the memristors in the IMPLY steps.

Once the carry signals are resolved, their values are propagated back to the individual bits. Finally, each bit resolves its sum in parallel as shown in steps 40 to 44. These steps require three memristors in each bit. However, recall that the first ten steps were performed in the individual bits. The memristors used in these steps, namely, M0, M1, and M2, can be reused. During the steps when the block is executing steps 11–39 and the bits are idle, the bits can reset these memristors. Then, they can be reused in steps 40–44. Thus, the sum computations do not require any additional memristors. The total number of memristors in the entire 4-bit block is 45 memristors: 41 for the individual bits, 3 for the GG and GP bits, and 1 for the sum and carry-out signals. The 4-bit block is shown in Figure 6.17.

The top four columns are the four bits in the block. Notice that all of them do not have the same number of memristors, corresponding to the different number of P memristors in the multiset operation. The four additional memristors are for the GG, GP, and carry logic in the 4-bit block.

6.2.1.4 16-bit logic

In a 16-bit carry-lookahead adder, all four of the 4-bit blocks perform the steps in Table 6.23 in parallel. All of the blocks resolve their GG and GP bits in the first 25 steps. Then, the blocks can propagate these values to the 16-bit logic in the adder. The blocks can then continue in parallel, completing steps 26 to 32. The blocks cannot complete steps 33 to 44 until they receive their carry-in signal. For the first block, this signal is available immediately. For the remaining blocks, they receive their carry-in signal from the 16-bit logic.

The 16-bit logic is responsible for taking the GG and GP bits from each of the four 4-bit blocks and producing the carry-in signals to each of the blocks (besides the first). These carry signals use the same equations as the carry-in signals in the individual blocks but use the GG and propagate signals rather than the individual generate and propagate signals:

$$\begin{aligned} C4_{in} &= GG_0 + GP_0C_{in} \\ C8_{in} &= GG_1 + GP_1GG_0 + GP_1GP_0C_{in} \\ C12_{in} &= GG_2 + GP_2GG_1 + GP_2GP_1GG_0 + GP_2GP_1GP_0C_{in} \end{aligned}$$

$C4_{in}$ is the carry-in signal to Block1, $C8_{in}$ is the carry-in signal to Block2, and $C12_{in}$ is the carry-in signal to Block3. These operations can begin executing as soon as the group signals are available in step 23. The full list of execution steps for the 16-bit logic in a 16-bit carry-lookahead adder is shown in Table 6.24.

In step 22, the carry-in signal to the adder is inverted. This can occur in any step before the GG and GP signals arrive. All of the GG and GP bits are read and copied from the blocks in parallel in steps 23–25. Recall that the execution steps for the individual bits showed that GG was read in step 23, and GP was stored directly into the 16-bit logic in step 25. This is corroborated by steps 23 and 25 in Table 6.24.

The carry-in bits each take three steps to resolve. $C4_{in}$ and $C8_{in}$ resolve in steps 27 and 30. Recall that the IMPLY operations were ordered in such a way that the 4-bit blocks do not use the carry-in bit until steps 33 and 35. Therefore, the latency to compute $C4_{in}$ is hidden. In steps 33 and 35, Block1 reads its carry-in. Interlaced with these steps is the resolution of $C12_{in}$. Finally, Block2 and Block3 alternate reading their carry-in signals in steps 36 to 49.

Based on the timing of the carry-in signals, Block1 completes its sum resolution in 44 steps, Block2 in 47 steps, and Block3 in 48 steps for a total delay of 48 steps for the 16-bit carry-lookahead adder. The 16-bit logic introduces three memristors for

Table 6.24 Execution steps for 16-bit logic in a carry-lookahead adder

22	C_{in} IMP 0	NOT C_{in}
23	GG IMP 0	NOT GG
24	(GG IMP 0) IMP 0	Copy of GG
25		GP bits are set from 4-bit blocks
26	GP1 IMP (C_{in} IMP 0)	GP1 NAND C_{in}
27	(GP1 NAND C_{in}) IMP GG1 = $C4_{in}$	GG1 + (GP1 AND C_{in})
28	$C4_{in}$ IMP 0	NOT $C4_{in}$
29	GP2 IMP ($C4_{in}$ IMP 0)	GP2 NAND $C4_{in}$
30	(GP2 NAND $C4_{in}$) IMP GG2 = $C8_{in}$	GG2 + (GP2 AND $C4_{in}$)
31	$C8_{in}$ IMP 0	NOT $C8_{in}$
32	GP3 IMP ($C8_{in}$ IMP 0)	GP3 NAND $C8_{in}$
33		Block1 reads $C4_{in}$
34	(GP3 NAND $C8_{in}$) IMP GG3 = $C12_{in}$	GG3 + (GP3 AND $C8_{in}$)
35		Block1 reads $C4_{in}$
36		Block2 reads $C12_{in}$
37		Block3 reads $C12_{in}$
38		Block2 reads $C12_{in}$
39		Block3 reads $C12_{in}$

each of Blocks 1, 2, and 3 for the carry logic: two for the copy operation of GG and one for holding GP. There are three additional memristors used to hold the inverses of C_{in}, $C4_{in}$, and $C8_{in}$. The total complexity is 192 memristors.

A full schematic of a 16-bit IMPLY based carry-lookahead adder is shown in Figure 6.18.

The four 4-bit blocks are shown in the symmetric four blocks across the top of the schematic. The remaining memristors are the 16-bit logic, which requires four memristors per carry-in signal, $C4_{in}$, $C8_{in}$, and $C12_{in}$.

6.2.1.5 Pipelining and ordering optimizations

Unlike other IMPLY designs, the IMPLY carry-lookahead adder cannot be easily pipelined. This is because most of the units in the adder are busy for the majority of the execution. For example, the individual bits are active in steps 1–10 and from steps 40 to the end of execution. The 4-bit logic is active from steps 11 to 39 and the 16-bit logic is active from steps 23 to 40. The blocks cannot accept a new input until they are done processing their first. Thus, the carry-lookahead adder cannot be directly pipelined.

In the traditional Boolean equations for a carry-lookahead adder, nearly all of the outputs use the same intermediate expressions but in different steps. This implies that each step must be performed in a serial manner or the values must be copied first to ensure values are not overwritten. However, both of these options result in a significant increase in delay and/or complexity.

The proposed implementation circumvents both of these options by translating the traditional Boolean expressions. The updated expressions are rewritten and ordered such that expressions minimize overwriting values and maximize the reuse of intermediate values. For example, once an intermediate result has only one use

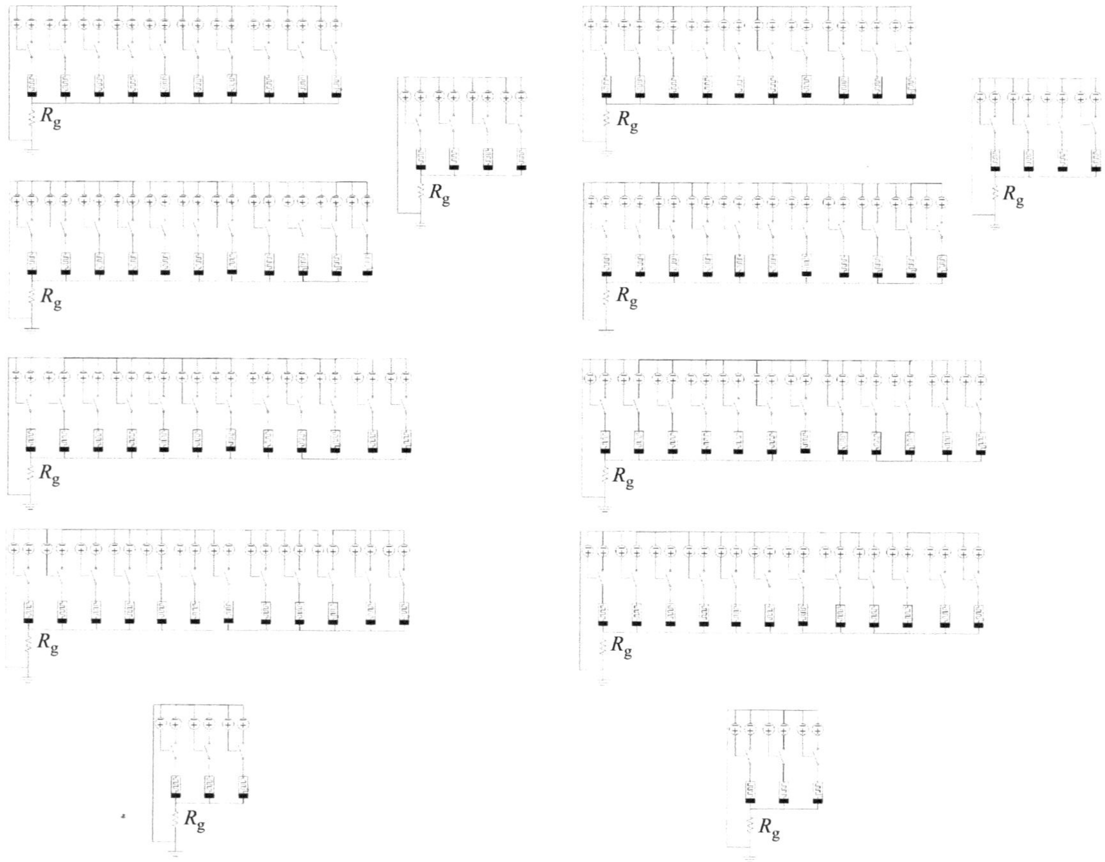

Figure 6.18 16-bit IMPLY carry-lookahead adder schematic

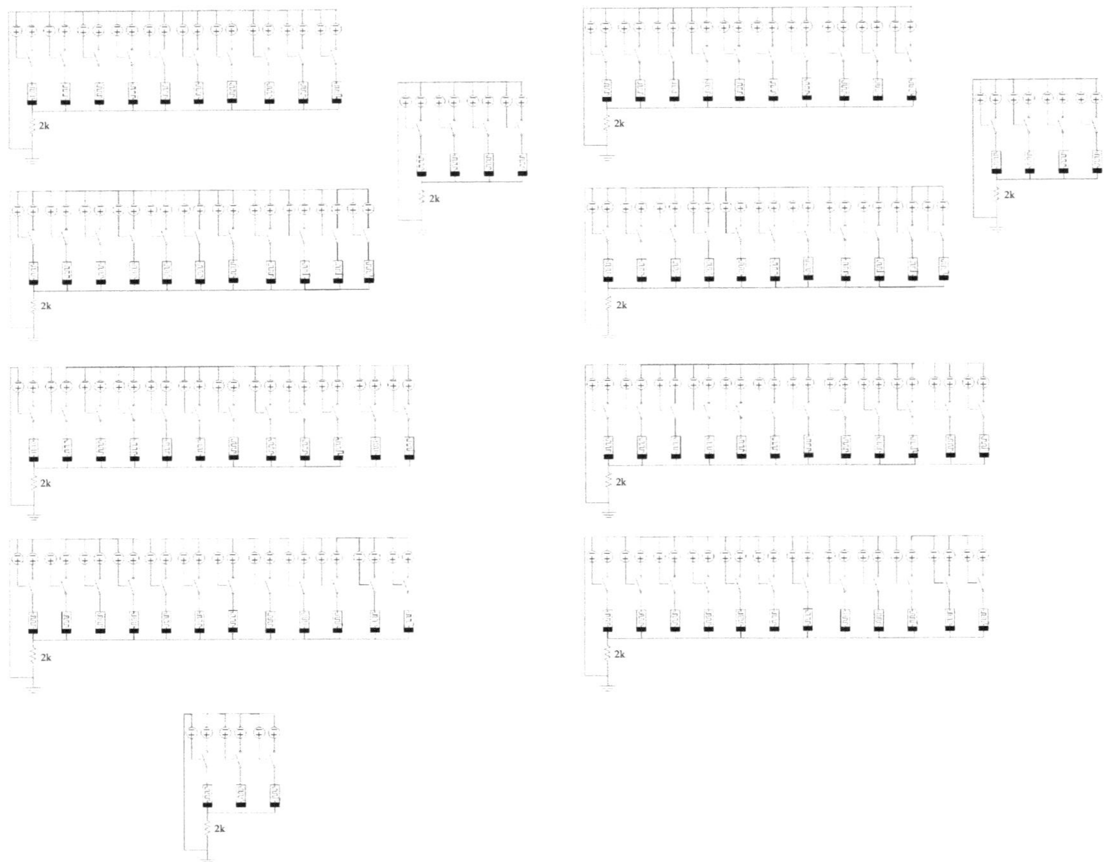

Figure 6.18 (Continued)

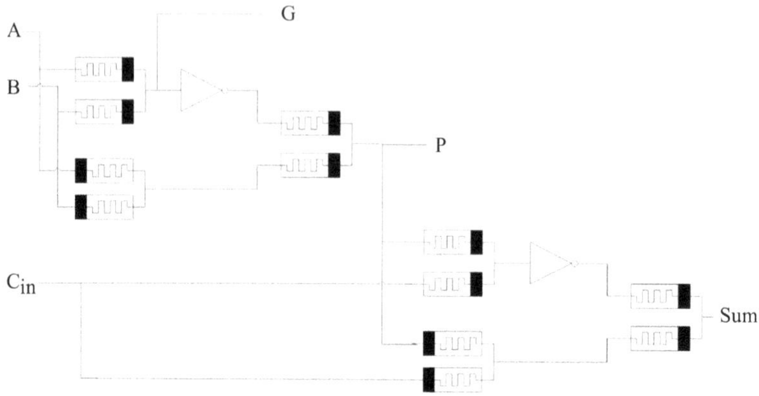

Figure 6.19 Hybrid-CMOS adder for the carry-lookahead adder

remaining, it can be overwritten. Thus, this operand will be used as the first operand in the IMPLY operations until its last operation. Then it will be used as the second operand.

Similarly, opportunities for multiset operations are leveraged. Overall, this has the benefit of reducing the total number of steps and memristors in the design.

6.2.2 Hybrid-CMOS carry-lookahead adder

This section presents an optimized implementation of a 16-bit hybrid-CMOS carry-lookahead adder.

One consequence of hybrid-CMOS designs is their reliance on CMOS NOT gates. However, carry-lookahead adders use very few instances of the NOT gate, since all of the generate and propagate logic require only AND and OR operations. The majority of the logic can be implemented using only memristors. This observation makes carry-lookahead adders a prime candidate for implementation in the hybrid-CMOS domain.

Start with the construction of the 4-bit blocks. The 4-bit block holds four adders and the logic for GG, GP, and the internal carry signals. The adders in a carry-lookahead adder are slightly modified versions of a traditional full adder. The schematic is shown in Figure 6.19.

The adder no longer requires the two gates for computing the carry-out signal. It only needs the two XOR gates for computing the sum. There are also two output probes for the G and P signal. However, no additional memristors are needed since these are naturally produced in a standard hybrid-CMOS full adder. Each adder requires 12 memristors and 4 MOSFETs to compute the generate, propagate, and sum signals. Buffers are not shown and will be discussed in the remainder of the section.

Four of these adders are included in the 4-bit block in a carry-lookahead adder. A gate-level schematic of this block is shown in Figure 6.20.

First, the four full adders in the block produce the G and P bit. Then, the GP, GG, and carry-out bits are resolved. The carry-out signals feed back into the

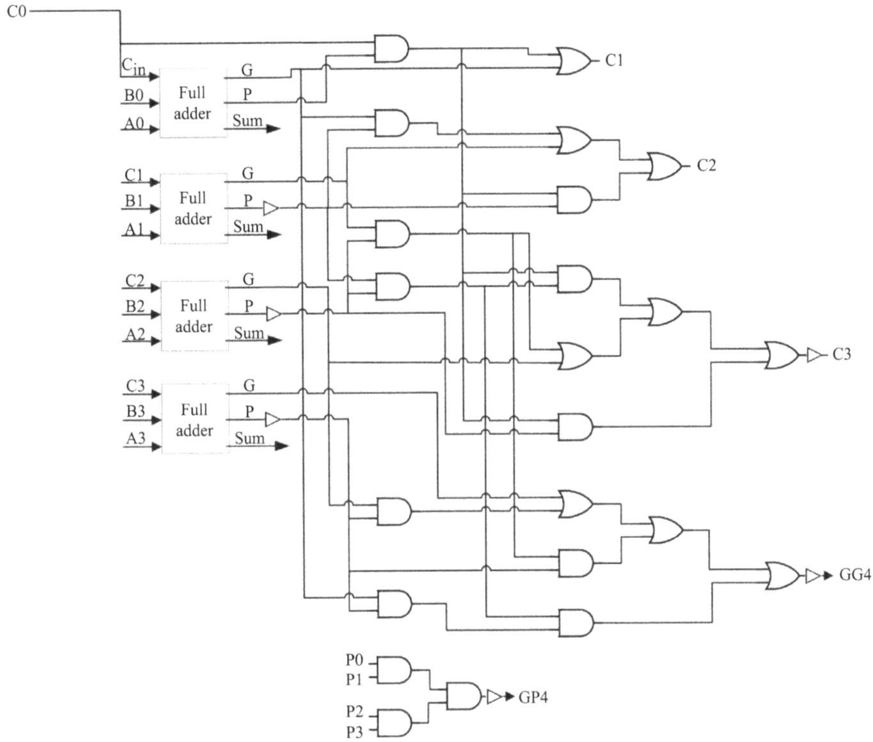

Figure 6.20 4-bit Hybrid-CMOS carry-lookahead adder schematic

corresponding full adders to complete the sums. Although the wires are not shown in the schematic for this, it is indicated by the labels.

In this case, the propagate signals from bits 1, 2, and 3 and C3 must be buffered. These signals need buffering due to their locations on long gate chains. Additionally, the GP and GG signals are buffered in preparation for their use in the 16-bit logic for the adder.

Once the buffers are added, the GP signal is ready after five steps and the GG is reave after seven. The last internal carry-out signal in the block, C_3, is ready after seven steps and is fed back into full adder 3. Full adder 3 needs three cycles to compute the XOR operation for the sum and is ready at $t = 10$.

The 4-bit block comprises 4 adders, 23 gates for the carry propagation, GG and GP logic, and 6 buffers. The 6 buffers contribute 24 MOSFETs to the design. Thus, the total complexity of the 4-bit block is 94 memristors and 40 MOSFETs.

When the 4-bit blocks are extended to a 16-bit adder, additional signal restoration buffers must be added because both the fan-out and the concatenation of gates increase for the outputs. The 16-bit carry-lookahead adder schematic is shown in Figure 6.21.

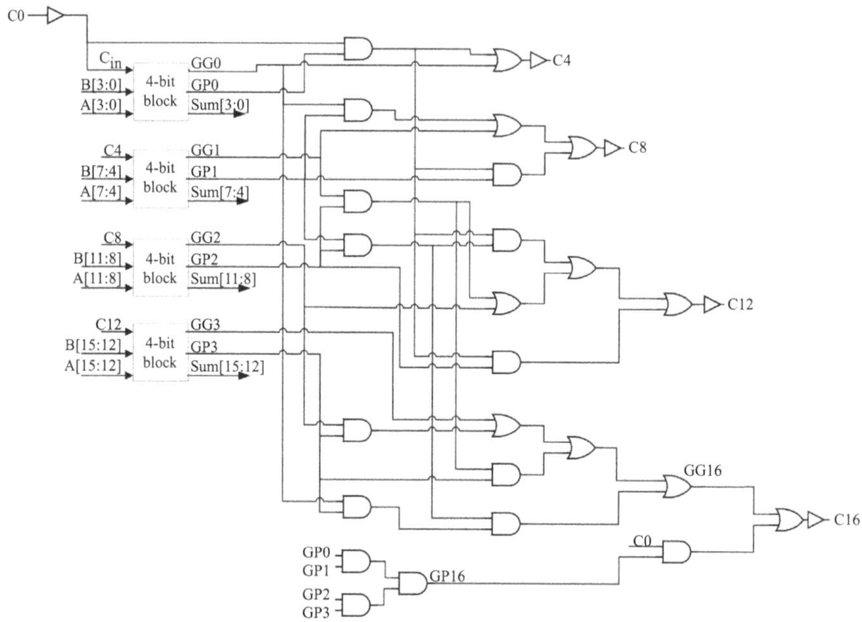

Figure 6.21 16-bit Hybrid-CMOS carry-lookahead adder schematic

Notice that the schematic looks identical to the schematic for the 4-bit block except the 1-bit adders have been replaced with 4-bit blocks. This is indicative of the hierarchical nature of the carry-lookahead adder. As the adder grows logarithmically, the logic takes the same form at each level. The last step in the 16-bit logic must complete the final carry-out, C16, for the entire addition.

The 16-bit logic adds 25 gates and 4 buffers for a total of 50 memristors and 8 MOSFETs. The total complexity is 426 memristors and 180 MOSFETs. Note that buffers are not required on the outputs of the 4-bit blocks since these were included in the blocks themselves. The only necessary buffers are on the carry signals.

The delay for the carry bit and sum bit are 14 and 23 steps. The GG signals from the 4-bit blocks are not resolved until step 7, and there are additional six-gate delays to resolve C_{12}. As a result, C_{in} to the final block does not arrive until step 13. As shown for the 4-bit block, the delay for the sum is 10 steps so the final sum is resolved at step 23.

6.2.3 Threshold gate carry-lookahead adder

This section implements a 16-bit carry-lookahead adder constructed from threshold gates. A carry-lookahead adder constructed from threshold gates achieves lower depth and lower gate count than alternative implementations. Threshold gates allow the traditional equations for propagate and generate signals to be rewritten to produce a smaller delay [59].

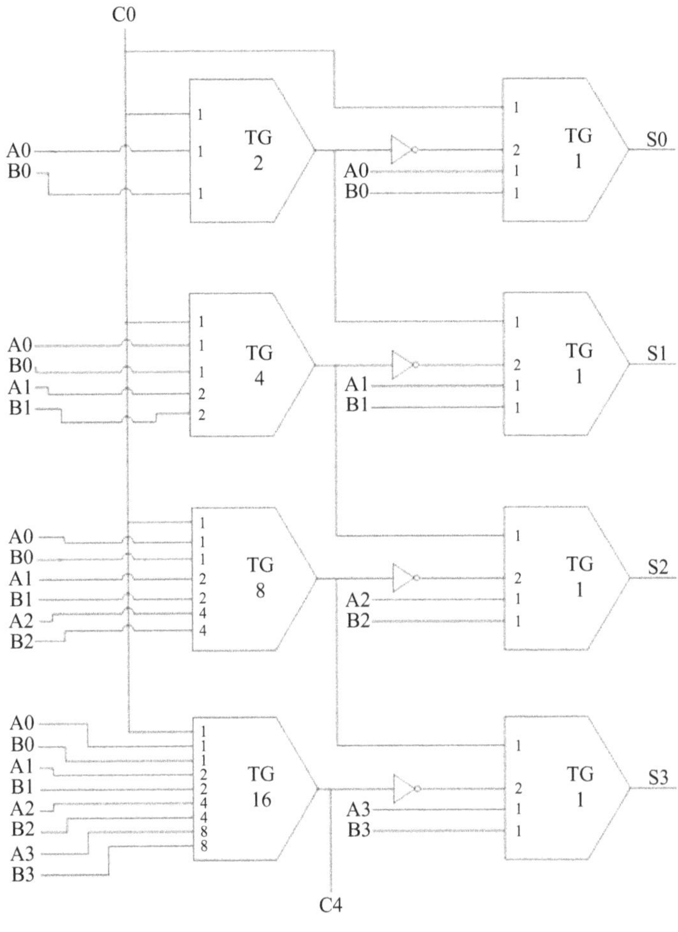

Figure 6.22 4-bit threshold gate carry-lookahead adder

A 4-bit carry-lookahead adder block can be constructed from eight threshold gates and four NOT gates as shown in Figure 6.22.

It is not necessary to generate the generate or propagate signals for individual bits directly since they are only used as intermediate results to produce GG, GP, and the carry signals. Similarly, it is not necessary to calculate the internal carry signals. These signals are not used outside of the 4-bit block. Only the GG and GP signals are needed by the 16-bit logic, and only the sums are used for the final result. Thus, the design does not have to produce the individual G and P bits or the internal carry signals if they are not needed.

In the threshold gate design, it is optimal to forego the resolution of these internal signals. Instead, increased-input counts and varying input weights on the threshold gates achieve the same result without the intermediate results. The sum and block carry-out are produced directly with the threshold gates.

6.2.3.1 Gate expressions per block

The circuit works by assigning a pair of threshold gates to each bit in the adder. Each row in the schematic has one pair of gates. For a given bit b, the first threshold gate essentially resolves the carry-out bit by internally processing the bit's P and G bits with the block's carry-in signal. The second threshold gate uses this carry-out bit in combination with the bit's inputs to resolve the final sum.

Bit 0 is identical to a threshold gate full adder, using two threshold gates in series to produce the sum. Its equivalent threshold equation is

$$\text{Sum}_0 = \text{A}_0 + \text{B}_0 + \text{C}_{\text{in}} - 2\left(\overline{\text{C0}_{\text{out}}}\right) \geq 1$$

Consider bit 1. The first threshold gate for bit 1 performs the Boolean equation for its carry-out:

$$\text{C1}_{\text{out}} = \text{G}_1 \text{ OR } (\text{P}_1 \text{ AND } \text{C0}_{\text{out}})$$

The threshold gate has the following equation:

$$\text{Out} = 1 \text{ if } 2\text{A}_1 + 2\text{B}_1 + \text{A}_0 + \text{B}_0 + \text{C}_{\text{in}} \geq 4$$

$$\text{else } 0$$

As indicated by the weights and labels on the gate inputs. To confirm the correctness of this gate equation, consider all the cases that will make the output "1" as shown in the truth table in Table 6.25.

From the truth table, the equation for the gate's output can be written as

$$\begin{aligned}\text{Out} = {} & (\text{A}_1 \text{ AND } \text{B}_1) \text{ OR } (\text{A}_1 \text{ AND } \text{MAJ}(\text{A}_0, \text{B}_0, \text{C}_{\text{in}})) \text{ OR} \\ & (\text{B}_1 \text{ AND } \text{MAJ}(\text{A}_0, \text{B}_0, \text{C}_{\text{in}}))\end{aligned}$$

The "Maj" notation means "majority" so the second term in the expression can be read as "A_1 and at least two of (A_0, B_0, and C_{in})." Note that $\text{MAJ}(\text{A}_0, \text{B}_0, \text{C}_{\text{in}})$ is equivalent to C0_{out}, so the equation can be simplified to

$$\text{Out} = (\text{A}_1 \text{ AND } \text{B}_1) \text{ OR } ((\text{A}_1 \text{ OR } \text{B}_1)\text{AND } \text{C0}_{\text{out}})$$

The first expression is equivalent to G_1 and that the second expression is equivalent to P_1. Thus, the first threshold gate for Bit 1 produces the carry-out signal for C1_{out} with the Boolean equation

$$\text{C1}_{\text{out}} = \text{G}_1 \text{ OR } (\text{P}_1 \text{ AND } \text{C0}_{\text{out}})$$

which produces the carry-out signal for C1_{out}. The second threshold gate for bit 1 calculates the sum for bit 1 with the following Boolean equation

$$\text{Sum}_1 = \text{A}_1 + \text{B}_1 + \text{C0}_{\text{out}} - 2\overline{\text{C1}_{\text{out}}} \geq 1$$

This is the same as the threshold gate for Sum_0 for the updated signal names. This equation calculates (A_1 XOR B_1 XOR C0_{out}). In fact, all of the threshold gates, which compute the sum outputs, use the same Boolean equation, using the corresponding A and B inputs, the incoming carry-in signal, and the computed carry-out

Table 6.25 Truth table for bit 1 threshold gate 1 in carry-lookahead adder

A_1	B_1	A_0	B_0	C_{in}	Threshold	Output
0	0	0	0	0	0	0
0	0	0	0	1	1	0
0	0	0	1	0	1	0
0	0	0	1	1	2	0
0	0	1	0	0	1	0
0	0	1	0	1	2	0
0	0	1	1	0	2	0
0	0	1	1	1	3	0
0	1	0	0	0	2	0
0	1	0	0	1	3	0
0	1	0	1	0	3	0
0	1	0	1	1	4	1
0	1	1	0	0	3	0
0	1	1	0	1	4	1
0	1	1	1	0	4	1
0	1	1	1	1	5	1
1	0	0	0	0	2	0
1	0	0	0	1	3	0
1	0	0	1	0	3	0
1	0	0	1	1	4	1
1	0	1	0	0	3	0
1	0	1	0	1	4	1
1	0	1	1	0	4	1
1	0	1	1	1	5	1
1	1	0	0	0	4	1
1	1	0	0	1	5	1
1	1	0	1	0	5	1
1	1	0	1	1	6	1
1	1	1	0	0	5	1
1	1	1	0	1	6	1
1	1	1	1	0	6	1
1	1	1	1	1	7	1

signal from the first threshold gate. All of the gates in the second column take the identical form.

The carry-out signals for bits 2 and 3 are computed in a similar fashion to bit 1, with the intention of encapsulating the bit's G OR (P AND C) equations. For bit 2, the equation is

$$\text{Out} = 1 \text{ if } 4A_2 + 4B_2 + 2A_1 + 2B_1 + A_0 + B_0 + C_{in} \geq 8$$
$$\text{else } 0$$

This equation is similar to the equation for bit 1 except the terms for A_2 and B_2 have been added with weights of four and the threshold has changed from four to eight. In doing so, the expression $(2A_1 + 2B_1 + A_0 + B_0 + C_{in})$ alone is incapable of meeting the threshold requirement for the gate. At least A_2 or B_2 must be a "1."

The equation can be thought of as

$$\text{Out} = (A_2 \text{ AND } B_2)\text{OR } ((A_2 \text{ OR } B_2) \text{ AND } (\text{bit 1's } C_{out} \text{ threshold gate} == 1))$$

Or

$$\text{Out} = G_2 \text{ OR } (P_2 \text{ AND } C1_{out})$$

Similarly, for bit 3 the first threshold gate's equation is

$$\text{Out} = 1 \text{ if } 8A_3 + 8B_3 + 4A_2 + 4B_2 + 2A_1 + 2B_1 + A_0 + B_0 + C_{in} \geq 16$$
$$\text{else } 0$$

This equation is again similar to the equation for bit 2 except the terms for A_3 and B_3 have been added and the threshold has increased to 16. Now, A_3 and B_3 together satisfy the threshold or one of them coupled with the remainder of the expression. In other words,

$$\text{Out} = (A_3 \text{ AND } B_3)\text{OR } ((A_3 \text{ OR } B_3) \text{ AND } (\text{bit 2's } C_{out} \text{ threshold gate} == 1))$$

Or

$$\text{Out} = G_3 \text{ OR } (P_3 \text{ AND } C2_{out})$$

This design reduces the carry-out delay to one and the sum delay to three due to the NOT gate. Note that it is possible to feed the output carry-out signal from the first threshold gate of each bit b as an input to the first threshold gate of bit b+1 to reduce the number of inputs. However, this would create a dependency chain and each carry-out bit would resolve one cycle after the previous. This would increase the final carry-out delay to four and the final sum delay to six.

6.2.3.2 16-bit logic block

This 4-bit block design can be extended to a 16-bit carry-lookahead adder in a hierarchical manner. Figure 6.23 shows a schematic for a 16-bit design.

The top four rows are the first 4-bit block as described. The remaining rows represent the other three 4-bit blocks in the design. Their equations can be obtained with a similar methodology as described for the 4-bit block, using their respective carry-in signal and input signals.

In the first cycle, the first block executes its first column of threshold gates, resolving all of its carry-out signals, $C0_{out}$, $C1_{out}$, $C2_{out}$, and $C3_{out}$. $C3_{out}$ is passed as the carry-in signal to the second block. In step 2, Block2 can begin. It executes its first row of threshold gates using its inputs and its carry-in signal from Block1. Now the carry-out signals $C4_{out}$, $C5_{out}$, $C6_{out}$, and $C7_{out}$ are resolved. In step 3, Block3 executes its first column of gates, and in step four, the fourth block does. In each block, it executes the second column of threshold gates in the two steps after the first column executes. Thus, Block1 resolves its sums in step 3, Block2 in step 4, Block3 in step 5, and Block4 in step 6.

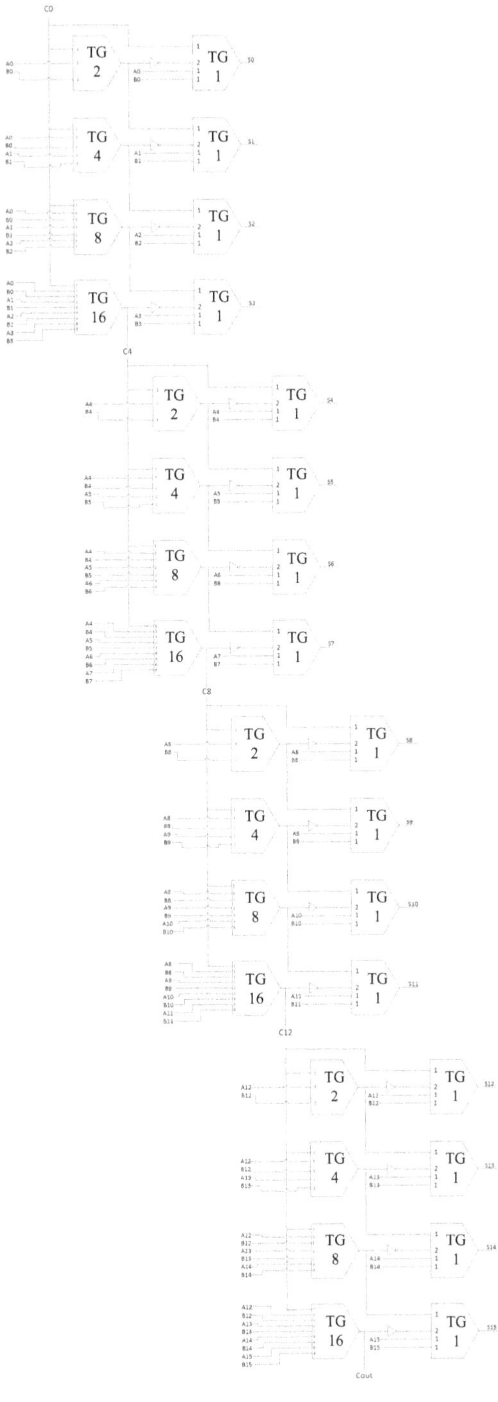

Figure 6.23 16-bit threshold gate carry-lookahead adder

This design reduces delay by leveraging increased input counts on the threshold gates. The entire design only requires 32 threshold gates and 32 MOSFETs. The total component requirements can be computed as follows. For a GOTO implementation, each gate requires one memristor per input and one GOTO pair per output. In each 4-bit block, there are 40 inputs and 8 outputs. Thus, the total complexity for a GOTO implementation of a 16-bit carry-lookahead adder is 160 memristors, 32 GOTO pairs, and 32 MOSFETs. For a CSTG implementation, each gate requires one memristor, one resistor, and two MOSFETs per input and four MOSFETs per output. Thus, the total complexity for a CSTG implementation of a 16-bit carry-lookahead adder is 160 memristors, 160 resistors, and 480 MOSFETs.

6.2.4 MAD gate carry-lookahead adder

This section presents a 16-bit carry-lookahead adder constructed from MAD gates. First, the construction of 4-bit lookahead logic blocks is shown, and then the logic necessary to connect them into a 16-bit complete adder design is shown.

The 4-bit blocks have four adders, each responsible for resolving its P and G bits. The propagate bit is equivalent to A OR B or A XOR B, and the generate bit is equivalent to A AND B. For the MAD design, the propagate bit is A OR B. The OR operation is performed with MAD gates.

To obtain these Boolean values, a V_{cond} signal is applied to the inputs A and B in series within each full adder, while a V_{set} signal is applied to the G and P memristors. The voltage at the p-terminal of the B memristor would be sensed to determine whether V_{set} was gated to the G or P memristor. The P memristor would be gated by a switch with a threshold voltage corresponding to an OR operations and the G memristor would be gated by a switch with threshold voltage corresponding to an AND operation. The circuitry can be seen in the schematic in Figure 6.24.

The P and G logic requires six switches, four memristors, and four drivers per bit. The drivers can be shared across all the bits if they execute in lockstep. The top row shows the inputs memristors in series for the four bits on the left and the G and P bits for the four bits on the right. The switches on the output memristors are labeled according to the operation they perform, and the memristors are labeled according to the bit they correspond to.

Once the G and P values are in their respective memristors, the GP and GG group signals can be resolved for the block. This circuitry is shown in the bottom row of Figure 6.24. Recall that the equation for the GP signal is

$$GP = P_0 \text{ AND } P_1 \text{ AND } P_2 \text{ AND } P_3$$

This is essentially a 4-input AND. To resolve the GP memristor, a V_{cond} signal would be applied to each of the P memristors at once and the voltages at the p-terminals would be sensed (labeled VP0, VP1, VP2, and VP3). The GP memristor would be applied a V_{set} signal that is gated by a series of four switches, each driven by the voltage at one of the P memristors' p-terminal node. In this way, the GP

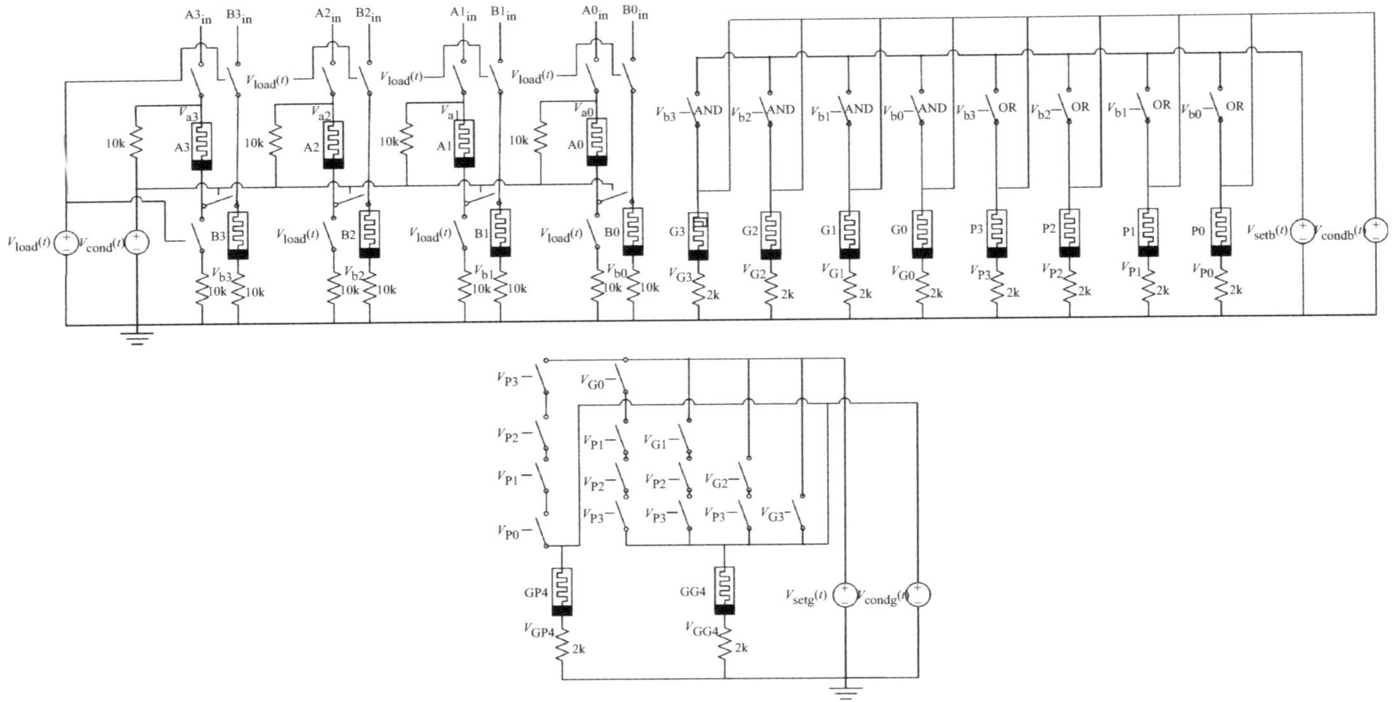

Figure 6.24 Baseline implementation of MAD GG and GP logic for a 4-bit block in a carry-lookahead adder

memristor will only be set to "1" if all of the P memristors satisfy the necessary threshold voltage.

The necessary threshold voltage is selected such that a value of P = "1" will meet the threshold and a value of P = "0" will not. This threshold voltage can be computed. In Figure 6.24, each P memristor is part of a voltage division circuit with its pull-down resistor. The resistor has a value of 2 kΩ. The resistance in the P memristor will either be "0" (100 kΩ) or "1" (1 kΩ). Thus, the voltage at node V_P when V_{cond} is applied is $(V_{cond}/51)$ V when P is a "0" and $(2V_{cond}/3)$ V when P is a "1." Thus, the threshold voltage V_{th} on each of the switches gated to the GP memristor is selected to satisfy

$$\frac{V_{cond}}{51}\ \mathrm{V} < V_{th} < \frac{2V_{cond}}{3}\ \mathrm{V}$$

then the gate will only close when P is a "1."

The GG signal has the equation

$$GG = G_3 \text{ OR } G_2P_3 \text{ OR } G_1P_2P_3 \text{ OR } G_0P_1P_2P_3$$

This expression can be implemented in the MAD context with parallel paths of switches. The equation for GG has four different expressions that can set it to "1." Each of these expressions is represented by a path in the circuitry for GG. Each expression consists of a single term or multiple terms in an AND operation. Each of these expressions is represented by a series of switches, each driven by a term in the AND operation. This logic can be seen in Figure 6.24 for the GG memristor.

The rightmost path from V_{set} to the GG memristor has a single switch on it driven by G_3. This represents the first term in the equation for GG. The value of G_3 is sensed at its p-terminal, VG3. If G_3 = "1," V_{set} will be gated to the output memristor, regardless of the other paths. The next path has two switches, one driven by G_2 and the other by P_3. The values of G_2 and P_3 are sensed at their p-terminals, labeled as VG2 and VP3. The next path has three switches, driven by G_1, P_2, and P_3, and the leftmost path has four switches driven by G_o, P_1, P_2, and P_3.

A path will only gate V_{set} to the output memristor if all of the switches are closed, meaning their switch drivers are "1." If any of the paths are fully closed, the GG signal memristor will be set to 1, achieving the OR operation among the terms. In all, the presented GP and GG logic adds 14 switches, 2 memristors, and 2 drivers to the circuitry for a total of 38 switches, 18 memristors, and 6 drivers.

6.2.4.1 Group signal optimizations

Instead, consider the design with the P and G signals removed. This is similar to the motivation behind the optimizations in the threshold gate implementation. The P and G signals are only used inside of the 4-bit block. If the design can compute the outputs for the adder without these signals, there is no need to compute them.

In the original design, the input memristors are read to resolve P and G, and then P and G are read to resolve the GG and GP bits. Instead, the input memristors can be read to directly set the GG and GP signals and carry-out signals. This bypasses the need for the individual P and G memristors while maintaining identical functionality. This also removes a step from the overall delay because it is not necessary to set these intermediate results. The optimized circuitry for a 4-bit block is shown in Figure 6.25.

The input circuitry remains unchanged. For each bit, a V_{cond} signal is applied to the memristors A and B in series and the voltage at the p-terminals of each B memristor are read. The architecture of the switches on the GP and GG memristors does not change but the drivers do.

Now, the voltages at the p-terminals of the B memristors are used as the drivers for the gates on the GP and GG memristors. Each switch that was driven by the voltage at VP_i or VG_i in the original design is now driven by the voltage at V_a or V_b in the manner that its previous P_i or G_i memristor had been driven by. This is a 1-to-1 mapping. The labels on the switches for GP and GG are labeled accordingly. Table 6.26 shows how each of the switches on the GP and GG memristors have mapped their drivers.

Before	Equation	After
VP0	A_0 OR B_0	$VB0_{or}$
VP1	A_1 OR B_1	$VB1_{or}$
VP2	A_2 OR B_2	$VB2_{or}$
VP3	A_3 OR B_3	$VB3_{or}$

Consider the left most path of GG for example. This path has four switches in series in both the original design and the optimized design. In the original design, the path sends the values at nodes VG0, VP1, VP2, and VP3 to drive the four switches. These values had been set in the previous step by their corresponding equations, A_0 AND B_0, A_1 OR B_1, A_2 OR B_2, and A_3 OR B_3.

In the optimized designs, the values of these four equations are used directly by sending the value of node V_b. If the original equation was an AND operation, the threshold on that gate is changed to be the threshold voltage for a MAD AND operation. If the original equation was an OR operation, the threshold on the gate is changed to be the threshold voltage for a MAD OR operation. This process is performed for each switch on the GP and GG logic.

These optimizations reduce the total complexity of the design by removing the G and P memristors and their switches and drivers. This removes eight memristors, eight switches, and two drivers. Now, the 4-bit block only requires 30 switches, 10 memristors, and 4 drivers. The optimization also removes the intermediate step to compute them. Now the block has a delay of just one step.

6.2.4.2 Carry-signal optimizations

The internal carry bits inside the 4-bit block can be resolved in parallel to the GG and GP signals by probing the same terminal on the input B memristors. These can

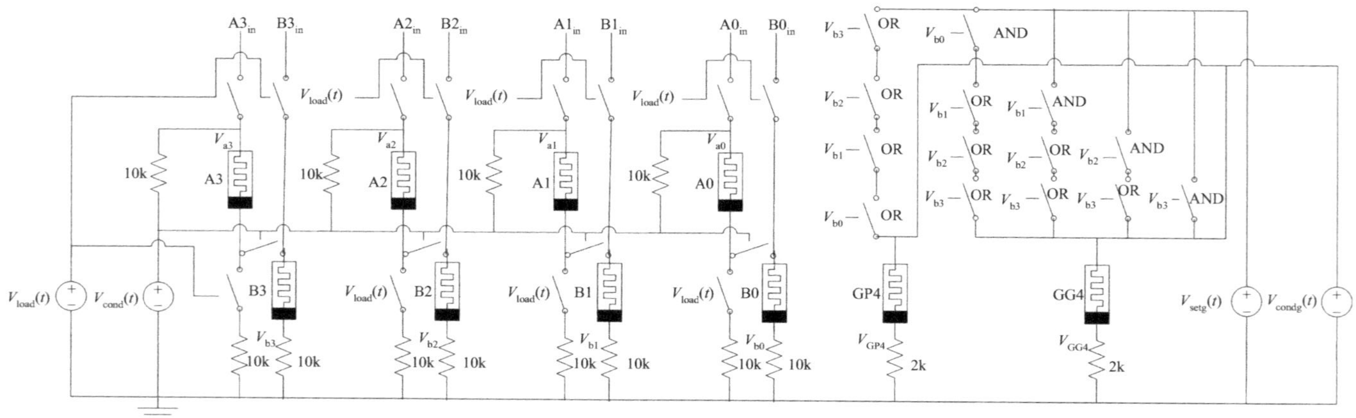

Figure 6.25 Optimized implementation of MAD GG and GP logic for a 4-bit block in a carry-lookahead adder

Table 6.26 Switch transformations on optimized GG and GP logic block for the (a) GG and (b) GP signal

Left path			**Path 2**			**Path3**			**Path4**		
Before	Equation	After	Before	Equation	After	Before	Equation	After	Before	Equation	After
VG0	A_0 AND B_0	$VB0_{and}$	VG1	A_1 AND B_1	$VB1_{and}$	VG2	A_2 AND B_2	$VB2_{and}$	VG3	A_3 AND B_3	$VB3_{and}$
VP1	A_1 OR B_1	$VB1_{or}$	VP2	A_2 OR B_2	$VB2_{or}$	VP3	A_3 OR B_3	$VB3_{or}$			
VP2	A_2 OR B_2	$VB2_{or}$	VP3	A_3 OR B_3	$VB3_{or}$						
VP3	A_3 OR B_3	$VB3_{or}$									

be constructed with the MAD methodology in the same manner as described for the GG and GP signals. The internal G and P bits are no longer necessary.

Consider the internal carry-out equations for the 4-bit block below:

$$
\begin{aligned}
C_{0out} &= G_0 + P_0C_{in} \\
C_{1out} &= G_1 + P_1G_0 + P_1P_0C_{in} \\
C_{2out} &= G_2 + P_2G_1 + P_2P_1G_0 + P_2P_1P_0C_{in}
\end{aligned}
$$

Instead, replace the G_x and P_x signals with the MAD logic that set their internal output memristors in the original design. Recall that the G and P memristors were set by sensing the voltage at the V_b nodes for the four bits. Generate bits required the voltage to be above the threshold for a MAD AND operation, and the propagate bits required the voltage to be above the threshold for a MAD OR operation. The carry-out equations can be rewritten as

$$
\begin{aligned}
C_{0out} &= (V_{b0} == \text{AND}) + (V_{b0} == \text{OR})C_{in} \\
C_{1out} &= (V_{b1} == \text{AND}) + (V_{b1} == \text{OR})\ (V_{b0} = \text{AND}) \\
&\quad + (V_{b1} == \text{OR})(V_{b0} == \text{OR})C_{in} \\
C_{2out} &= (V_{b2} == \text{AND}) + (V_{b2} == \text{OR}\ )(V_{b1} == \text{AND}) + (V_{b2} == \text{OR})(V_{b1} == \text{OR}) \\
&\quad (V_{b0} == \text{AND}) + (V_{b2} == \text{OR})(V_{b1} == \text{OR})(V_{b0} == \text{OR})C_{in}
\end{aligned}
$$

where (V_x == OP) means that the voltage at V_x satisfies the threshold voltage of a switch for the operation OP. These equations are nearly identical to the paths for the GG logic. Figure 6.26 shows the resulting circuitry for these carry signals in a 4-bit block.

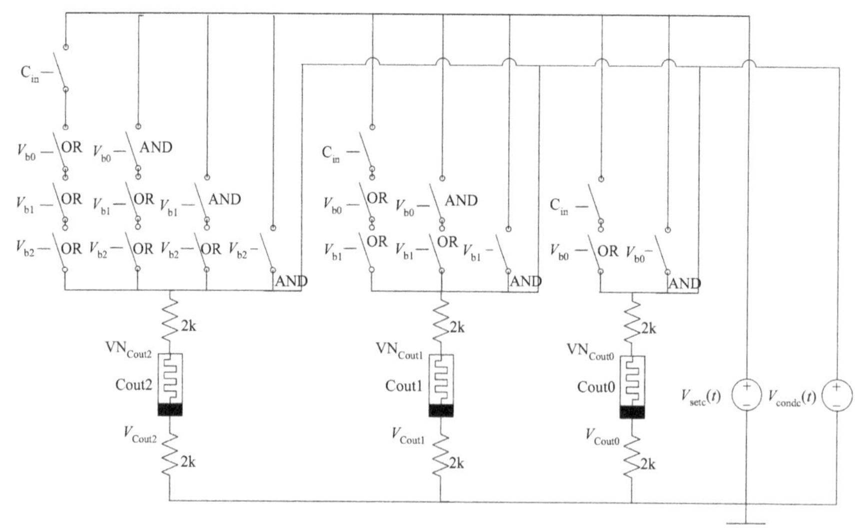

Figure 6.26 Optimized implementation of MAD carry logic for a 4-bit block in a carry-lookahead adder

This logic is constructed using the same reasoning as described for the GG and GP bits. Last, the sum logic for the 4-bit block is implemented. The sum logic presented for the MAD full adder is reused in this design and replicated for each bit in the block. The schematic, including the carry logic, can be seen in Figure 6.27.

Each of the sum bits senses the voltages for their corresponding input and carry signals, as indicated by the labels on the switches.

Note that the optimization performed to remove the need for G and P memristors cannot be done to remove the carry signals. Even though these carry signals are not outputs to the remainder of the 16-bit adder, the carry signals need to be stored in an intermediate memristor. This is because the sum resolution for each bit requires both the carry signal and its inverse.

In order to obtain a memristor's value and its inverse at the same time, the p- and n-terminal of the memristor are probed when V_{cond} is applied. Otherwise, a NOT operation is required. Either way, the value needs to be stored into a memristor. Thus, the carry signals need to be stored into memristors. For the GG and GP signals, the inverses of the G and P signals were not required so their memristors could be removed.

The carry-out bits resolve once the carry-in signal to the block is available and the sums are resolved in the next step. The carry-out and sum logic adds 43 switches, 7 memristors, and 4 drivers. The total complexity of the 4-bit block is 73 switches, 17 memristors, and 8 drivers.

6.2.4.3 16-bit logic block

The principles for the 4-bit block are extended to the 16-bit block to complete the adder design. As described, in the first block, the carry signals and the GG and GP signals are resolved in parallel in the first step of execution. These signals can resolve in the first step because they only need the values of the input memristors and the carry-in signal.

Similarly, the GG_{16} and GP_{16} signals and the carry signals in the 16-bit block ($C4_{in}$, $C8_{in}$, and $C12_{in}$) can all resolve in parallel. They all only require the GG and GP signals from the 4-bit blocks and the input carry-in signal to the adder. Thus, they can resolve all these values in the second step of execution.

The carry bits and GG_{16} and GP_{16} bits are resolved using the same circuitry as shown for the 4-bit block but with the switch drivers modified to represent the respective inputs to the 16-bit logic block. In the third step, $C4_{in}$, $C8_{in}$, and $C12_{in}$ are passed into their respective 4-bit blocks and the carry signals internal to these blocks are resolved. In the fourth and final step, all remaining sum signals for the 16-bit addition are resolved. In general, for an N-bit addition with block size $= \log(N)$ blocks, the MAD implementation takes $\log(N)$ steps. This is much faster than the other approaches.

The complete schematic for the 16-bit adder is shown in Figure 6.28.

The total delay for a 16-bit carry-lookahead addition is four steps. The total complexity is of the four blocks, the logic for GG_{16} and GP_{16} and the 16-bit logic for the $C4_{in}$, $C8_{in}$, and $C12_{in}$ signals. In all, there are 325 switches, 73 memristors, and 24 drivers. All of the 4-bit blocks share four drivers for the G and P bits, but

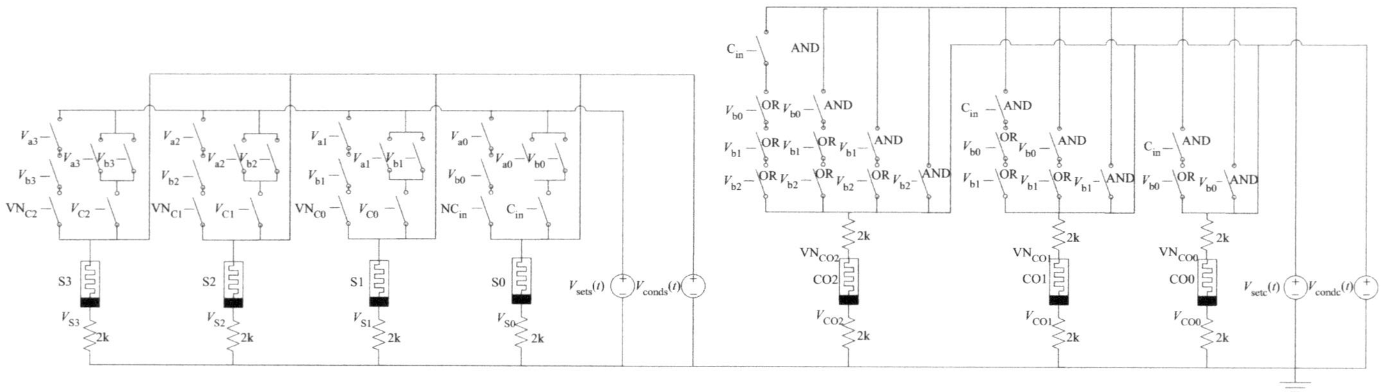

Figure 6.27 Optimized implementation of MAD sum logic for a 4-bit block in a carry-lookahead adder

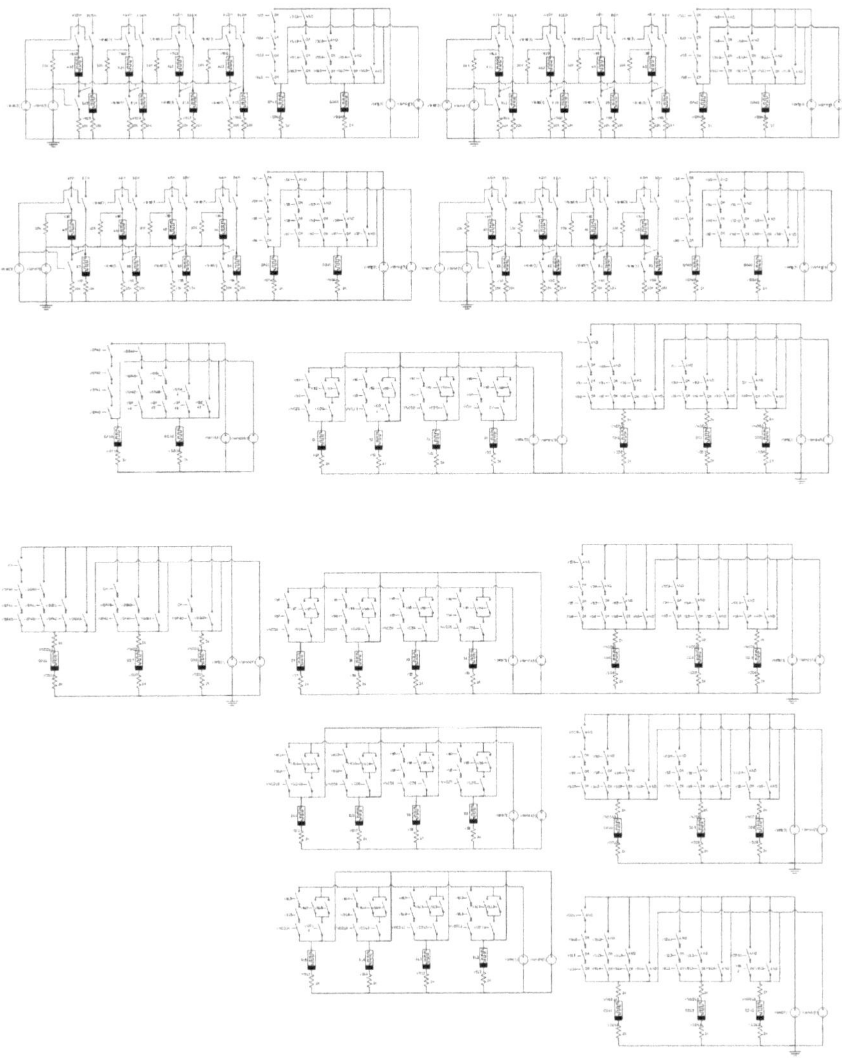

Figure 6.28 Optimized 16-bit MAD carry-lookahead adder

each block has their own four drivers for their carry and sum signals. The 16-bit logic requires four drivers, two for the group signals and two for the carry signals.

6.2.5 Carry-lookahead adder analysis and comparison

This section discusses the complexity and delay comparisons for the various carry-lookahead implementations in comparison to each other and the traditional CMOS design. A comparison of the carry-out and sum delays and the complexities for both the 4- and 16-bit blocks are given in Table 6.27.

Table 6.27 4-bit and 16-bit delay and complexity comparisons for carry-lookahead adders

4-bit block	**Delay-C_{out}**	**Delay-sum**	**Complexity**
Conventional CMOS	6	7	300 MOSFETs
IMPLY	27	44	45 memristors 45 drivers
Hybrid-CMOS	7	10	94 memristors 40 MOSFETs
Threshold gate—GOTO	1	3	40 memristors 8 GOTO pairs 8 MOSFETs
Threshold gate—CSTG	1	3	40 memristors 40 resistors 120 MOSFETs
MAD	1	2	17 memristors 73 switches 8 drivers

16-bit adder	**Delay-C_{out}**	**Delay-sum**	**Complexity**
Conventional CMOS	10	11	1212 MOSFETs
IMPLY	42	48	192 memristors 192 drivers
Hybrid-CMOS	14	23	426 memristors 180 MOSFETs
Threshold gate—GOTO	4	6	160 memristors 32 GOTO pairs 32 MOSFETs
Threshold gate—CSTG	4	6	160 memristors 160 resistors 480 MOSFETs
MAD	2	4	73 memristors 325 switches 24 drivers

The proposed 16-bit IMPLY implementation has the highest delay of all of the implementations. The design takes about 4× the number of steps to resolve the final sum as compared to CMOS. In comparison to the other memristor-based solutions, the IMPLY design still takes about 2× the next slowest design, hybrid-CMOS. In terms of complexity, the IMPLY implementation reduces the complexity from the traditional CMOS implementation of 1212 MOSFETs to 192 memristors and the accompanying switch and driver circuitry. So, although the delay increases by about 4×, the number of components is reduced to 1/6 of the CMOS design.

The proposed hybrid-CMOS approach suffers in comparison to the traditional CMOS in terms of delay due to signal buffering. However, the CMOS usage vastly improves over the traditional CMOS implementation from 1212 MOSFETs to just 180 MOSFETs plus 426 memristors. This is less than half the number of components. In comparison to the IMPLY approach, the hybrid-CMOS approach is faster, but the differential decreases as the adder width increases because of increased signal buffering. Although the complexities cannot be compared directly, the hybrid-CMOS design requires about 150 percent more components than the IMPLY design.

The delay of the threshold gate designs are lower than the IMPLY, hybrid-CMOS, and CMOS designs by a significant margin. The delay only increases by three when the adder quadruples in size. The 16-bit carry-lookahead adder takes only four cycles for the final carry-out and six cycles for the final sum. The complexity is also very competitive, especially for the GOTO implementation. Both implementations require fewer memristors than the previous three designs, and the GOTO implementation requires the fewest components of all the designs.

The delay of the MAD design is the lowest of all the designs. This is partially due to the inherent characteristics of the MAD approach and partially due to optimizations and parallelizability. For example, by transforming the traditional Boolean logic in the CMOS design and leveraging MAD methodologies, the P and G calculations were eliminated. The delay is less than all of the other designs, beating out the IMPLY design by 20× and 12× for the carry-out and sum signals, respectively. The MAD design also requires the fewest memristors and drivers. The number of memristors in the 16-bit MAD design is reduced to just 73 at the cost of more switches than the other implementations.

6.3 Carry-select adders

This subchapter switches gears to focus on the carry-select adder. Carry-select adders are more complex than ripple-carry adders but are also faster. This section investigates how to implement carry-select adders using memristor methodologies. Memristor-based carry-select adders are presented for the IMPLY, hybrid-CMOS, threshold, and MAD contexts. Each design is optimized for its given constructs and analyzed against the other designs and the traditional CMOS design in terms of delay and component count.

6.3.1 IMPLY carry-select adder

First, the use of the IMPLY operation to implement a carry-select adder is described. As seen for the ripple-carry adder, many times the optimal IMPLY-based design has different Boolean expressions for its outputs than the traditional CMOS design. This is because IMPLY operations have different costs for each Boolean operation. Carry-select adders are another example of a design that exhibits this variation. The carry-select adder can be highly optimized for the IMPLY context through Boolean translation.

The most crucial part to optimize for performance in any design is the critical path. Reduction in the delay of the critical path directly reduces the delay in the overall design. The critical path of the carry-select adder lies along the carry-propagation chain. The carry signals propagate across the *n*-bit ripple-carry adder blocks of the carry-select adders as described in Chapter 3. For each block *X* in the carry-select adder, the carry-out is calculated as

$$C_{out} = (C_{in} \text{ AND } P_x) \text{ OR } G_x$$

where P_x and G_x are the propagate and generate signals from block *X*, and C_{in} is the carry-out signal from the previous block *X*−1. Fortunately, the AND and OR IMPLY operations can be simplified when they are known to occur consecutively. This was seen briefly with the design of the IMPLY full adder. Without simplification, the steps for C_{in} AND P_x followed by (C_{in} AND P_x) OR G_x are shown in Table 6.28.

Steps 4 and 5 copy the value of G so that the value can be overwritten in step 7. The value of G must be retained for later use in the adder. However, in a carry-select adder, G is an intermediate value that is not reused, so a copy is not needed and it can be overwritten. This eliminates steps 4 and 5 and their memristors. Second, two NOT operations essentially cancel themselves out in steps 3 and 6. These two steps can also be removed. In all, the steps are simplified to the steps in Table 6.29.

The simplified C_{out} expression requires only one additional memristor M0 and only three IMPLY steps rather than seven. These savings are significant since they reduce the critical path of the design at each block in half.

Table 6.28 IMPLY execution steps for C_{out} in carry-select adders

Step	Execution	Description
1	P IMP 0 → M0	NOT P
2	C_{in} IMP (P IMP 0) → M0	
3	(C_{in} IMP (P IMP 0)) IMP 0 → M1	C_{in} AND P
4	G IMP 0 → M2	NOT G
5	(G IMP 0) IMP 0 → M3	Copy of G
6	M1 IMP 0 → M4	NOT (C_{in} AND P)
7	M4 IMP M3 → M3	(C_{in} AND P) OR G

Table 6.29 Minimally reduced IMPLY execution steps for C_{out} in carry-select adder

Step	Execution	Description
1	P IMP 0 → M0	NOT P
2	C_{in} IMP (P IMP 0) → M0	NOT (C_{in} AND P)
3	(C_{in} IMP (P IMP 0)) IMP G	(C_{in} AND P) OR G

By inspection, the carry-propagation delay is only two steps instead of three for most blocks. The first step, P IMP 0, does not require the carry-in signal and can execute off the critical path. Recall that the P bits in an N-bit carry-select adder are the carry outputs from the n-bit blocks for which the carry-in is set to "1." Each of these blocks begins an n-bit IMPLY ripple-carry addition in parallel at $t = 0$ and resolves its carry-out signal after $2n + 11$ steps. Thus, all of the P bits are ready at time $t = 2n + 11$. Now, the P IMP 0 operation can perform in each block as early as $t = 2n + 12$. For all the blocks except the first, this is earlier than the C_{in} signal is available. This step is hidden by the critical path of the design.

Consider a 16-bit carry-select adder with 4-bit blocks. The first 4 bits, b0 to b3, are in Block0 and use a traditional ripple-carry adder. The remaining sets of four bits each have two ripple-carry adders in their blocks. Table 6.30 shows the key steps in each block to perform the carry propagation.

Table 6.30 Execution steps for the blocks in a 16-bit carry-select adder

T	Block0	Block1	Block2	Block3
1	Start addition	Start addition	Start addition	Start addition
2				
3				
4				
5				
6				
7				
8				
9				
10				
11				
12				
13				
14				
15				
16				
17				
18				
19	Resolve C_{out}	Resolve P,G	Resolve P,G	Resolve P,G
20		P IMP 0	P IMP 0	P IMP 0
21		NOT (C_{in} AND P)		
22		(C_{in} AND P) OR $G = C_{out}$		
23			NOT (C_{in} AND P)	
24			(C_{in} AND P) OR $G = C_{out}$	
25				NOT (C_{in} AND P)
26				(C_{in} AND P) OR $G = C_{out}$

For all n-bit blocks after the first, only steps 2 and 3 above are needed to propagate carry-in to carry-out. This makes the carry-propagation delay of the IMPLY logic equivalent to CMOS Boolean logic. For Block1, the P bit and the C_{in} bit are ready at the same time $2n + 11$. Thus, it cannot hide the latency of the P IMP 0 step. All three steps are on the critical path in the first block.

6.3.1.1 IMPLY multiplexer design

Carry-select adders also use multiplexers extensively. A straightforward implementation of a 1-bit multiplexer with IMPLY operations implements

$$\text{Out} = (\text{A AND S}) \text{ OR } \left(\text{B AND } \bar{\bar{\text{S}}}\right)$$

where S is the select line. The steps are shown in Table 6.31.

Execution begins by creating the inverse of the select line for the AND operation. Then (B AND $\bar{\text{S}}$) is computed, followed by (A AND S) and last, the OR operation. Notice that the original $\bar{\text{S}}$ produced in step 1 is reused in step 5. The memristor requires eight steps and eight memristors. However, note that steps 6 and 7 are unnecessary as they perform a copy operation. Also, it may not be necessary to copy the value of S before overwriting it in step 3. Instead, the optimized steps shown in Table 6.32 are used.

Table 6.31 Steps for a 1-bit IMPLY multiplexer

Step	Mux functionality	Mux goal
1	S IMP 0 → M0	$\bar{\text{S}}$
2	$\bar{\text{S}}$ IMP 0 → M1	
3	B IMP M1 → M1	B NAND $\bar{\text{S}}$
4	M1 IMP 0 → M2	B AND $\bar{\text{S}}$
5	A IMP M0 → M0	A NAND S
6	M0 IMP 0 → M3	A AND S
7	M3 IMP 0 → M4	
8	M4 IMP M2 → M2	(A AND S) OR (B AND $\bar{\text{S}}$)

Table 6.32 Steps for an optimized 1-bit IMPLY multiplexer implementing $Out = \overline{(A\ NAND\ S)}\ OR\ \overline{\left(B\ NAND\ \bar{\bar{S}}\right)}$

Step	Mux functionality	Mux goal
1	S IMP 0 → M0	$\bar{\text{S}}$
2	B IMP S → M1	B NAND $\bar{\text{S}}$
3	M1 IMP 0 → Out	B AND $\bar{\text{S}}$
4	A IMP M0 → M0	A NAND S
5	M0 IMP Out → Out	(A AND S) OR (B AND $\bar{\text{S}}$)

Note that this series of steps assumes that the select line is not needed in future steps. If it is, an additional step is necessary to copy S as shown in the original multiplexer steps. The equivalent schematic is shown in Figure 6.29.

The multiplexer requires two input memristors, A and B, two memristors, S and its inverse M0 for the select lines, and one output memristor, Out, for a total of five memristors. No internal memristors are necessary.

When used in the context of the carry-select adder, this multiplexer is instantiated numerous times. Each *n*-bit block has *n* instances of the multiplexer, one for each of the sum bits in the block. The A and B memristors in each multiplexer correspond to the sum for that bit in each of the two ripple-carry adders in that block. For example, consider a 16-bit carry-select adder with 4-bit blocks. Block1 resolves the sums of bits 4 to 7. It has two ripple-carry adders, which both compute a sum for bits 4 to 7. Then four instances of the multiplexer are used to resolve the final sum for bits 4 to 7. The four sum bits from the first ripple-carry adder are the A inputs to the four multiplexers, and the sum bits from the second ripple-carry adder are the B inputs to the four multiplexers.

In this context, the A and B memristors in the multiplexer are not necessary. As described, these values lie in the sum memristors in the respective ripple-carry adder blocks. They do not need to be copied into the multiplexer. Instead, they can be read directly from the ripple-carry adders. Each multiplexer can read its unique sum bits from the ripple-carry adders as they are resolved. The complexity of each multiplexer is now only three memristors.

The same optimization cannot be performed for the select lines. The select line is the carry-in signal from the previous block. The select lines are used as inputs to

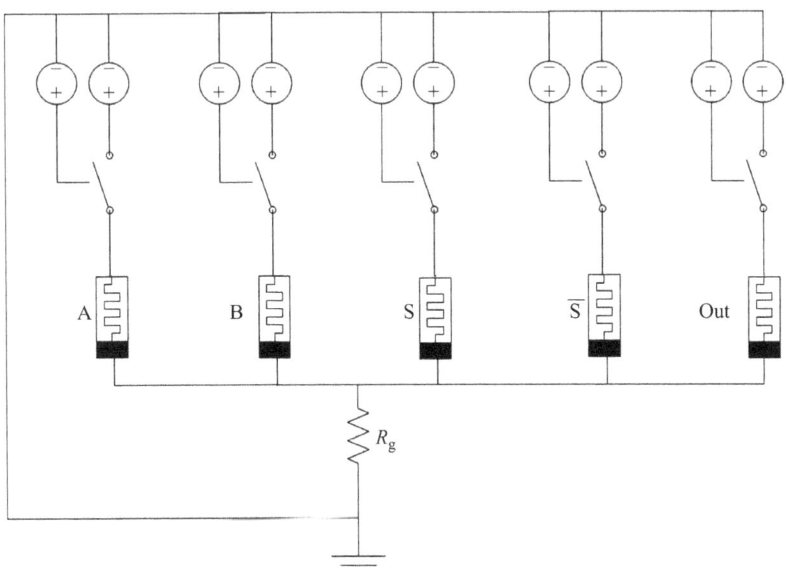

Figure 6.29 IMPLY Multiplexer

all of the 1-bit multiplexers in a block. Their values must be preserved and cannot be overwritten. Also, to allow for parallelism, the values of the select lines are copied to each multiplexer locally so they can execute in parallel.

6.3.1.2 Selecting design parameters

Traditionally, in CMOS implementations, the optimal block width k (in terms of minimizing delay) for a carry-select adder is equivalent to $\sqrt{N}$ for an N-bit adder. However, the IMPLY implementation has different delay characteristics and may have a different optimal block width. If a constant block width of n is used, where N is an integer multiple of n, N/n blocks will exist. An IMPLY-based ripple-carry adder has a delay of $2n + 19$ for the sum, the carry-out is $2n + 11$ steps, and the complexity is $7n + 1$ memristors. So for an N-bit carry-select adder, it takes $2n + 11$ steps to form the carry-out of the first block, three steps for the second block to produce its carry-out, and two steps for each of the $N/n-3$ intermediate blocks. The final block requires six steps for the resolution of the sum multiplexer. Thus, the total delay can be expressed as

$$\text{Delay} = 2n + 14 + \frac{2N}{n}$$

The optimum block size is found by taking the derivative with respect to n and solving for n. The result is

$$n = \sqrt{N}$$
$$\text{Delay} = 4\sqrt{N} + 14$$

This is slightly higher than the traditional CMOS design, which has a delay of $4\sqrt{N} + 2$. However, as the size of the adder increases, the impact of the constant in the delay equation decreases and the two designs converge to an equivalent delay. Also, the IMPLY carry-select adder's delay is significantly less than the optimized IMPLY ripple-carry adder ($2N + 19$). For a 16-bit adder, the comparison in delay is 30 for the carry-select adder vs. 51 for the ripple-carry adder and for a 64-bit adder the difference in delay increases to 46 vs. 147 steps. This is a similar trend to CMOS, in which the carry-select adder is faster than the ripple-carry adder by a significant margin.

The schematic for a 16-bit carry-select adder with 4-bit blocks is shown in Figure 6.30.

Each column represents four bits of the adder. There are two 4-bit ripple-carry adders in each column except for the first block. The first block only requires a single ripple-carry adder since the carry-in signal is known. Notice that the lower 4-bit adder in each column has one fewer memristor than the upper 4-bit adder. This is because the carry-in is tied to 0 in these blocks and no memristor is needed in bit 0. The sets of three memristors along the bottom are the 1-bit multiplexers needed to resolve each of the sum bits for bits 4–15. Each column has four corresponding to the four sum bits in each block. The remaining memristors perform the carry-propagation logic.

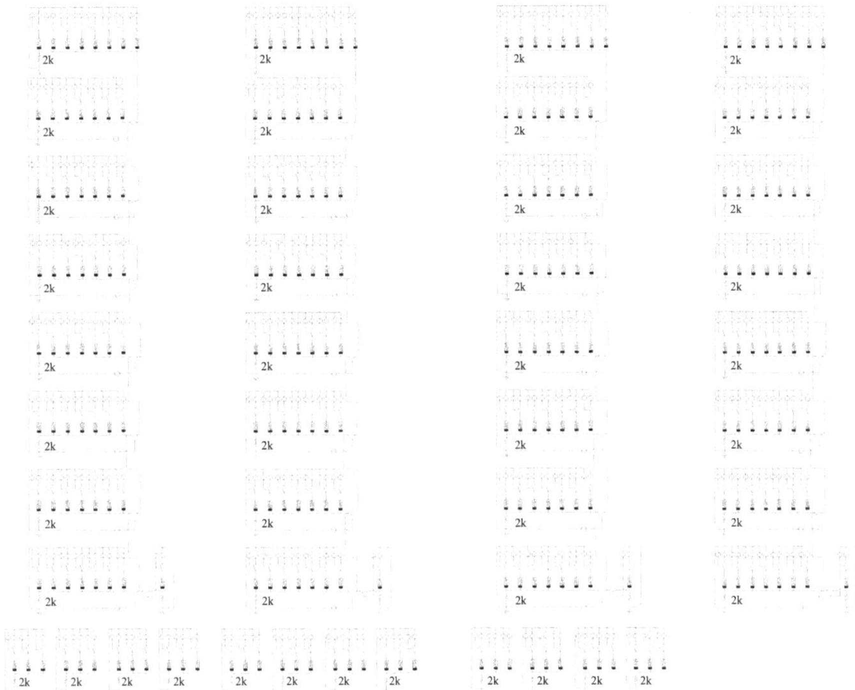

Figure 6.30 16-bit IMPLY carry-select adder schematic

The total complexity of the carry-select adder is calculated as follows:
For an 8-bit multiplication, the total complexity is

Ripple-carry adders:

$$= \left(\frac{2N}{n} - 1\right)(7n + 1) - \left(\frac{N}{n} - 1\right) \text{ memristors}$$
$$= 14N - 7n + \frac{N}{n} \text{ memristors}$$

Propagation logic:

$$= \frac{N}{n} - 2 \text{ memristors}$$

Sum multiplexers:

$$= 3(N - n) \text{ memristors}$$
$$= 3N - 3n \text{ memristors}$$

Total complexity:

$$= 17N - 10n + \frac{2N}{n} - 2 \text{ memristors}$$

This result is around 2.5× the complexity of a ripple-carry adder depending on the selection of n. For $n = \sqrt{N}$, the number of memristors becomes $17N-8\sqrt{N}-2$ or 238 for a 16-bit adder. Note that the carry-propagation logic makes up less than 1 percent of the overall memristor count.

6.3.1.3 Pipelining

The carry-select adder can be pipelined in a similar manner to the IMPLY ripple-carry adder. The granularity for pipelining is left to the designer. Each full adder in the individual n-bit blocks could be pipelined as described for the ripple-carry adders. Alternatively, the full adders in the n-bit blocks are not pipelined and the pipelining occurs across the blocks. In this mode, each time a block propagates its carry-out bit, it begins a new n-bit addition. In other words, it can be pipelined at the bit level, the n-bit block level, or another desired granularity.

To pipeline at the granularity of the n-bit blocks, the circuitry can be reset and reused once the sum and carry signals have been resolved and read by the carry logic. These carry-out values correspond to P and G. Once these values are read by the carry-propagation logic and sum multiplexer, the ripple-carry adder can begin again. However, this uncovers a slight issue with the design. The values of P and G are used in the carry-propagation logic, and G is overwritten to serve as the carry-out memristor. To start a new addition in this block, the value of G needs to be reset, but it cannot be overwritten until it has been used as the carry-out signal.

There are two solutions to overcome the issue of the carry-out signal for pipelining. First, the implementation can take two extra steps to copy the value of the carry-out signal from the G memristor to another memristor before resetting the memristors and beginning a new addition in the block. This costs two steps and two memristors per block. Alternatively, the implementation can wait until the carry-out signal has been used by the next block in which case it is safe to overwrite. The carry-out signal is used twice, once in the first step of the sum multiplexer and once in the second step of carry propagation. This option requires two steps per block but no memristors. Thus, this implementation will use the second approach.

For 4-bit blocks, the carry-out in each ripple-carry adder is produced at $t = 19$. The final carry-out from the block is resolved two or three cycles after the ripple-carry adders complete. During the same cycles, the sum memristors are sensed by the 1-bit multiplexers. At $t = 20$, the A inputs are read by the 1-bit multiplexers and at $t = 21$, the B inputs are read. At the latest, all these operations complete at $t = 22$. Then, the carry-out signal in the G memristor is read twice. In parallel with these read operations, all of the memristors in the other full adders in the n-bit blocks can reset and read the next addition's inputs into the input memristors in the ripple-carry adders. The only remaining step is to clear the final full adder, which holds carry-out memristor G and read its inputs for the next addition. Since this memristor is in the final full adder in the ripple-carry adder, this reset operation can be performed in parallel with the first full adder beginning the new addition. No delay is incurred on the critical path. In all, a new addition can begin every 24 steps. An example of the pipelined execution of four 4-bit blocks in a 16-bit carry-select adder for two additions is shown in Table 6.33.

Block1 determines the critical path. It is the only block, which has a carry-propagation delay of three steps, the rest have two. In the first addition, the adders in Block1 complete in step 19. The carry-out is resolved in step 22. In parallel, the A and B inputs are read. The value of the carry-out is read in steps 23 and 24, as seen by the operation on C_{in} in the subsequent block. In step 25, a new addition begins. Finally, the sum is resolved once the multiplexer completes in step 28. Since the multiplexer is a separate unit from the adder that has already read its inputs, these units can perform in parallel on different additions. Each of the following blocks performs the same steps, two cycles after the previous. Now, an addition completes every 24 cycles.

6.3.2 Hybrid-CMOS carry-select adder

This section presents a carry-select adder implemented with the hybrid-CMOS approach.

First, the implementation of a multiplexer using hybrid-CMOS gates is considered. A schematic for a multiplexer implementing

$$\text{Out} = (\text{A AND S})\ \text{OR}\ \left(\text{B AND}\ \overline{\overline{\text{S}}}\right)$$

in a hybrid-CMOS design is shown in Figure 6.31.

The hybrid-CMOS multiplexer has the same gate delay as the traditional CMOS multiplexer. The complexity is also the same at a gate level: one NOT gate, two AND gates, and an OR gate. This translates to fewer components than in traditional CMOS. The hybrid-CMOS design requires two MOSFETs and six memristors. The CMOS design requires 17 MOSFETs. For an N-bit multiplexer, the select line can be shared across all of the bits so the complexity is 2 MOSFETs $+ 6N$ memristors for the hybrid-CMOS design and $15N + 2$ MOSFETs for the traditional CMOS design.

One consequence of the hybrid-CMOS design is that the outputs must be buffered in order to restore signal strength. Each buffer adds 4 MOSFETs to the design for a total of $4N$ MOSFETs for an N-bit multiplexer.

The carry-propagation logic in the hybrid-CMOS design is the same as the traditional CMOS design. Each block uses an AND gate and an OR gate. In all, there are four memristors. However, because the carry-out signal is used in three places and falls on a long chain of gates, the signals must be buffered. Each carry-propagation block requires four memristors and four MOSFETs. Fortunately, since the buffers are made of two CMOS NOT gates, they also produce the inverse of the carry output. This can remove delays in the sum multiplexers.

The sum multiplexers only incur a delay of two steps rather than the usual three steps. The sum multiplexers use the carry-out signal as its select line. One of

Table 6.33 Pipelined execution steps for the blocks in a 16-bit IMPLY carry-select adder

T	Block0	Block1	Block2	Block3
1	Start addition 0	Start addition 0	Start addition 0	Start addition 0
2				
3				
4				
5				
6				
7				
8				
9				
10				
11				
12				
13				
14				
15				
16				
17				
18				
19	Resolve C_{out}	Resolve P,G	Resolve P,G	Resolve P,G
20		P IMP 0	P IMP 0	P IMP 0
21	C_{out} read by propagation logic	NOT (C_{in} AND P)		
	Reset memristors	A inputs read		
22	C_{out} read by multiplexer	(C_{in} AND P) OR G = C_{out}		
	Load inputs	B inputs read		
23	Start addition 1	C_{out} read by propagation logic	NOT (C_{in} AND P)	
		Reset memristors	A inputs read	
24		C_{out} read by multiplexer	(C_{in} AND P) OR G = C_{out}	
		Load inputs	B inputs read	
25		Start addition 1	C_{out} read by propagation logic	NOT (C_{in} AND P)
			Reset memristors	A inputs read
26		Reset FA3 memristors including C_{out}	C_{out} read by multiplexer	(C_{in} AND P) OR G = C_{out}
			Load inputs	B inputs read
27			Start addition 1	C_{out} read by propagation logic
				Reset memristors
28		Sum multiplexer for addition 0 completes	Reset FA3 memristors including C_{out}	C_{out} read by multiplexer
				Load inputs
29				Start addition 1
30			Sum multiplexer for addition 0 completes	Reset FA3 memristors including C_{out}

Table 6.33 (*Continued*)

T	Block0	Block1	Block2	Block3
31				
32				Sum multiplexer for addition 0 completes
33				
34				
35				
36				
37				
38				
39				
40				
41	Resolve C_{out}			
42				
43		Resolve P,G		
44		P IMP 0		
45	C_{out} read by propagation logic	NOT (C_{in} AND P)	Resolve P,G	
	Reset memristors	A inputs read		
46	C_{out} read by multiplexer	(C_{in} AND P) OR G = C_{out}	P IMP 0	
	Load inputs	B inputs read		
47	Start addition 1	C_{out} read by propagation logic	NOT (C_{in} AND P)	Resolve P,G
		Reset memristors	A inputs read	
48		C_{out} read by multiplexer	(C_{in} AND P) OR G = C_{out}	P IMP 0
		Load inputs	B inputs read	
49		Start addition 2	C_{out} read by propagation logic	NOT (C_{in} AND P)
			Reset memristors	A inputs read
50		Reset FA3 memristors including C_{out}	C_{out} read by multiplexer	(C_{in} AND P) OR G = C_{out}
			Load inputs	B inputs read
51			Start addition 2	C_{out} read by propagation logic
				Reset memristors
52		Sum multiplexer for addition 1 completes	Reset FA3 memristors including C_{out}	C_{out} read by multiplexer
				Load inputs
53				Start addition 2
54			Sum multiplexer for addition 1 completes	Reset FA3 memristors including C_{out}
55				
56				Sum multiplexer for addition 1 completes

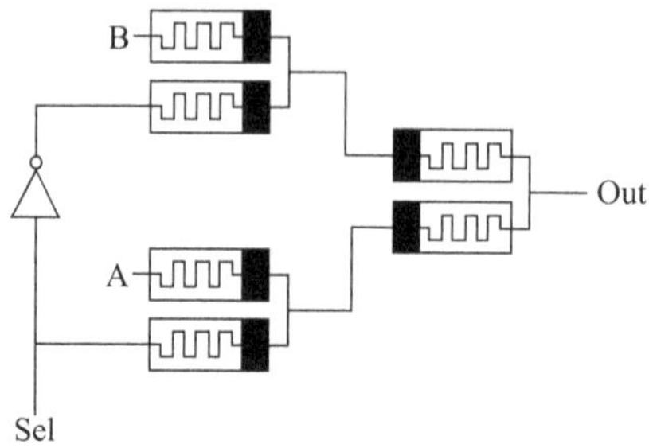

Figure 6.31 Hybrid-CMOS multiplexer

the gate delays in a traditional CMOS multiplexer comes from the NOT gate on the select line. However, the buffered output on each carry-out signal produces the inverted value during the propagation logic. Thus, each multiplexer only has a delay of two steps. For the same reason, the MOSFETs in the multiplexers can be removed entirely.

The delay for an N-bit carry-select adder with block size n consists of the delay through the ripple-carry adders, carry propagation, and the final sum multiplexer. In the hybrid approach, it takes $3n + 3$ steps to form the carry-out of the first block, 3 steps for each of the $(N/n)-2$ intermediate blocks, and 2 steps in the final block. Thus,

$$\text{Delay} = 3n - 1 + \frac{3N}{n}$$

Since the design has the same gate structure as the traditional CMOS carry-select adder, the optimal block size is still $\sqrt{N}$. For $n = \sqrt{N}$, the delay simplifies to $\text{Delay} = 6\sqrt{N}-1$. For a 16-bit adder, this delay is 23 vs. 30 for the IMPLY implementation discussed in the previous section, reaping about a 1/5 improvement.

The components required are $(2N-n)$(14 memristors + 12 MOSFETs) for the ripple-carry adders, $(N/n-1)(6n$ memristors) for the multiplexers, and $(N/n-2)$(4 memristors + 4 MOSFETs) for the carry-propagation logic. The total component count is

The total complexity is computed as follows:

Ripple-carry adders:

$$\begin{aligned} &= (2N - n)(14 \text{ memristors} + 12 \text{ MOSFETs}) \\ &= 28N - 14n \text{ memristors} + 24N - 12N \text{ MOSFETs} \end{aligned}$$

Propagation logic:

$$\begin{aligned} &= \left(\frac{N}{n} - 2\right)(4 \text{ memristors} + 4 \text{ MOSFETs}) \\ &= \frac{4N}{n} - 8 \text{ memristors} + \frac{4N}{n} - 8 \text{ MOSFETs} \end{aligned}$$

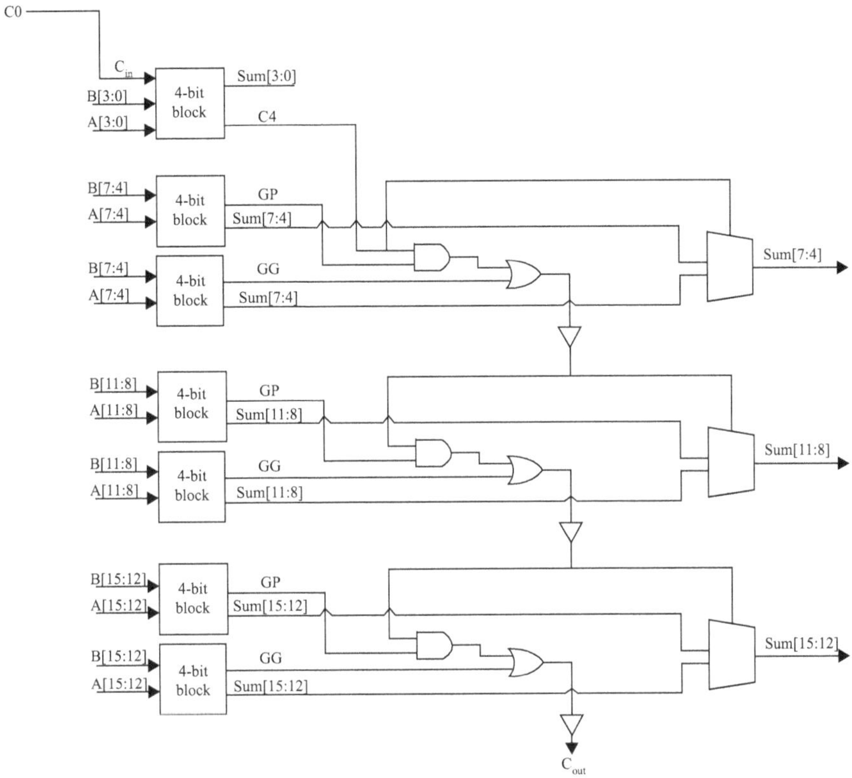

Figure 6.32 Optimized hybrid-CMOS carry-select adder schematic

Sum multiplexers:

$$= \left(\frac{N}{n} - 1\right)(6n) \text{ memristors}$$

$$= 6N - 6n \text{ memristors}$$

Total complexity:

$$= 34N - 20n + \frac{4N}{n} - 8 \text{ memristors} + 24N - 12n + \frac{4N}{n} - 8 \text{ MOSFETs}$$

For a 16-bit adder with 4-bit blocks, this becomes 472 memristors + 344 MOSFETs. The full gate level schematic for this implementation is shown in Figure 6.32.

Each set of four bits has two 4-bit adder blocks, each holding a 4-bit ripple-carry adder. The top block of each pair has the carry-in bit tied to 1 and produces the propagate bit. The lower block of each pair has the carry-in bit tied to 0 and produces the generate bit. Notice that there are buffers on the outputs of each OR

Table 6.34 Truth table for a threshold gate implementation for carry-propagation logic in a carry-select adder

GPC_{in}	Weighted sum	C_{out}
000	0	0
001	1	0
010	1	0
011	2	1
100	2	1
101	3	1
110	3	1
111	4	1

gate, which in turn feed the carry-out bit and its inverse to the subsequent set of blocks.

6.3.3 Threshold gate carry-select adder

This section presents a design for a threshold gate carry-select adder. The design is explained at the gate level and then implemented with two variations. The first is constructed with GOTO-based threshold gates, and the second is constructed with CSTG-based threshold gates. The two designs are each presented and discussed.

The carry-propagation logic equation, $C_{out} = (C_{in} \text{ AND P}) \text{ OR G}$, can be implemented with a single-threshold gate by selecting appropriate weights for the inputs. Consider a 3-input threshold gate with G, P, and C_{in} as inputs and C_{out} as the output. There are two expressions, which will produce a 1 at C_{out}: G and (C_{in} AND P). Each of these expressions must have the same overall weight as each other at the threshold gate to satisfy the OR operation. This way, both are equally capable of producing a "1." Since there are two variables in the second term and it is an AND operation, these must each have half of the weight of the variable in the first term. In this way, the second term will only hit the same threshold as the first term if both of the inputs are 1, satisfying the AND operation. Assign the weights {P:1, C:1, G:2} to the gate inputs. The equation for the output is essentially

$$C_{out} = \begin{cases} 1 \text{ if } (2 * G + 1 * P + 1 * C_{in}) \geq 2 \\ \text{else } 0 \end{cases}$$

which has the same truth table as the logic equation as shown in Table 6.34.

Next, a multiplexer is implemented using threshold gates. A multiplexer can be implemented using two threshold gates and a NOT gate as shown in Figure 6.33.

This multiplexer executes

$$\text{Out} = (\text{A AND S}) \text{ OR } \left(\text{B AND } \overline{\overline{\text{S}}}\right)$$

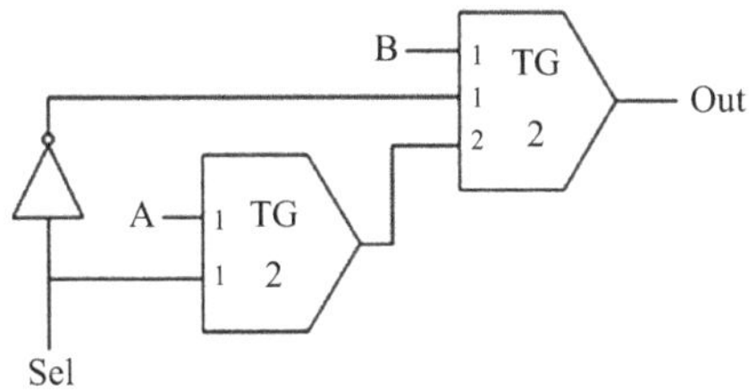

Figure 6.33 Threshold gate multiplexer

Figure 6.33 shows a single bit multiplexer but the same design can be instantiated multiple times to form an n-bit multiplexer. An n-bit multiplexer requires $2n$ threshold gates and 1 CMOS gate since the output of the single select line inverter can be used in each of the n bits.

6.3.3.1 GOTO implementation

Now consider the design of a carry-select adder using GOTO-based threshold gates. For a GOTO implementation of an N-bit carry-select adder with n-bit blocks, there are $[(N/n)-1]$ n-bit multiplexers. For each threshold gate, there is one GOTO pair per threshold gate output and one memristor per threshold gate input. Each n-bit multiplexer has a total complexity of 2 MOSFETs, $5n$ memristors, and $2n$ GOTO pairs with a delay of two. There are $[(N/n)-2]$ instances of carry-propagation logic, each requiring a single-threshold gate. Each of these gates needs three memristors and one GOTO pair. The complexity breaks down as follows:

Ripple-carry adders:

$$= \left(\frac{2N}{n} - 1\right)(7n \text{ memristors} + 2n \text{ GOTO pairs} + 2n \text{ MOSFETs})$$

$$= (14N - 7n \text{ memristors} + 4N - 2n \text{ GOTO pairs} + 4N - 2n \text{ MOSFETs})$$

Multiplexers:

$$= \left(\frac{N}{n} - 1\right)(5n \text{ memristors} + 2n \text{ GOTO pairs} + 2 \text{ MOSFETs})$$

$$= \left(5N - 5n \text{ memristors} + 2N - 2n \text{ GOTO pairs} + \left(\frac{2N}{n} - 2\right) \text{ MOSFETs}\right)$$

Carry-propagation logic:

$$= \left(\frac{N}{n} - 2\right)(3 \text{ memristors} + 1 \text{ GOTO pair})$$

$$= \left(\frac{3N}{n} - 6\right) \text{ memristors} + \left(\frac{N}{n} - 2\right) \text{ GOTO pairs}$$

Total complexity:

$$= \left(19N - 12n + \frac{3N}{n} - 6\right) \text{memristors} + \left(6N - 4n + \frac{N}{n} - 2\right) \text{GOTO pairs}$$

$$+ \left(4N - 2n + \frac{2N}{n} - 2\right) \text{MOSFETs}$$

Now determine the optimal block size n for a GOTO pair threshold gate carry-select adder. The total delay is $(n + 2)$ steps for the first ripple-carry adder block, $[(N/n)-2]$ steps for the carry propagation, and 2 steps for the final multiplexer. The total delay is $n + (N/n) + 2$ GOTO delays. By taking the derivative with respect to n, the optimal block size is still $n = \sqrt{N}$. Thus the total complexity becomes $(19N - 9\sqrt{N} - 6)$ memristors + $(6N - 3\sqrt{N} - 2)$ GOTO pairs + $(4N - 2)$ drivers.

6.3.3.2 CSTG implementation

If CSTG is used for the threshold gates, there is one memristor per threshold gate input, one resistor per input, two MOSFETs per input, and four MOSFETs per threshold gate output. In total, the CSTG threshold implementation has a complexity as follows:

Ripple-carry adders:

$$= \left(\frac{2N}{n} - 1\right)(7n \text{ memristors} + 7n \text{ resistors} + 22n \text{ MOSFETs})$$

$$= (14N - 7n \text{ memristors} + 14N - 7n \text{ resistors} + 44N - 22n \text{ MOSFETs})$$

Multiplexers:

$$= \left(\frac{N}{n} - 1\right)(5n \text{ memristors} + 5n \text{ resistors} + 18n + 2 \text{ MOSFETs})$$

$$= \left(5N - 5n \text{ memristors} + 5N - 5n \text{ resistors} + \left(18N - 18n + \frac{2N}{n} - 2\right) \text{MOSFETs}\right)$$

Carry-propagation logic:

$$= \left(\frac{N}{n} - 2\right)(3 \text{ memristors} + 3 \text{ resistors} + 10 \text{ MOSFETs})$$

$$= \left(\frac{3N}{n} - 6\right) \text{memristors} + \left(\frac{3N}{n} - 6\right) \text{resistors} + \left(\frac{10N}{n} - 20\right) \text{MOSFETs}$$

Total complexity:

$$= \left(19N - 12n + \frac{3N}{n} - 6\right) \text{memristors} + \left(19N - 12n + \frac{3N}{n} - 6\right) \text{resistors}$$

$$+ \left(62N - 40n + \frac{12N}{n} - 22\right) \text{MOSFETs}$$

The delay is 1 more than GOTO implementation due to the extra delay in the ripple-carry adder. The total delay is $n + (N/n) + 1$ gate delays. For block size $n = \sqrt{N}$, the total complexity is $\left(19N - 9\sqrt{N} - 6\right)$ memristors + $\left(19N - 9\sqrt{N} - 6\right)$ resistors + $\left(62N - 28\sqrt{N} - 22\right)$ MOSFETs.

6.3.4 MAD gate carry-select adder

This section presents a carry-select adder using MAD gates. This carry-select adder is the only presented design, which changes the basic structure of the traditional CMOS carry-select adder. The IMPLY, hybrid-CMOS, and threshold gate implementations simply translated and optimized the design into the given context. The MAD design leverages its key properties to remove entire blocks from the design.

The first block in the carry-select adder is the same as the original design, but the remaining n-bit blocks have a slightly modified form. Instead of two n-bit adders per block, there is one. The properties of the MAD full adder can be used to remove the need for the second adder in each adder block. Consider the carry-out logic on the standard MAD full adder. It reads the values of the inputs A and B from the input circuitry and pairs this information with the incoming carry-in signal to determine if it should drive the V_{set} signal to the C_{out} memristor.

The design could instantiate two distinct n-bit adders as done in the traditional CMOS design. Instead, consider a single n-bit ripple-carry adder, which accomplishes both additions. Two entire instantiations of this adder are not needed, only two instantiations of the output logic. On one instantiation, the carry-in signal will be tied to "0" and on the other, the carry-in signal will be tied to "1." Only one instantiation of the input memristors and the voltage division circuitry is necessary. From the work with MAD gates, it is possible to perform multiple operations in parallel that use the same inputs. This modified full adder is an example of this principle. The input memristors A and B will be read and used to resolve the sum and carry outputs for both adder scenarios in parallel. This optimized full adder is shown in Figure 6.34.

The input circuitry is identical to the original full adder. The circuitry for the C_{out} and sum memristors have been duplicated as indicated by memristors C_{out1} and Sum_1. The circuitry for these output memristors has also not changed from the original design; only the signals, which drive the carry-in switches, have changed.

The components with a "0" suffix, the C_{in0} and NC_{in0} signals and memristors C_{out0} and Sum_0, belong to the n-bit adder corresponding to when the carry-in set to "0." The components with a "1" suffix correspond to the n-bit adder corresponding to when the carry-in set to "1." In the first full adder, C_{in0} and NC_{in1} will

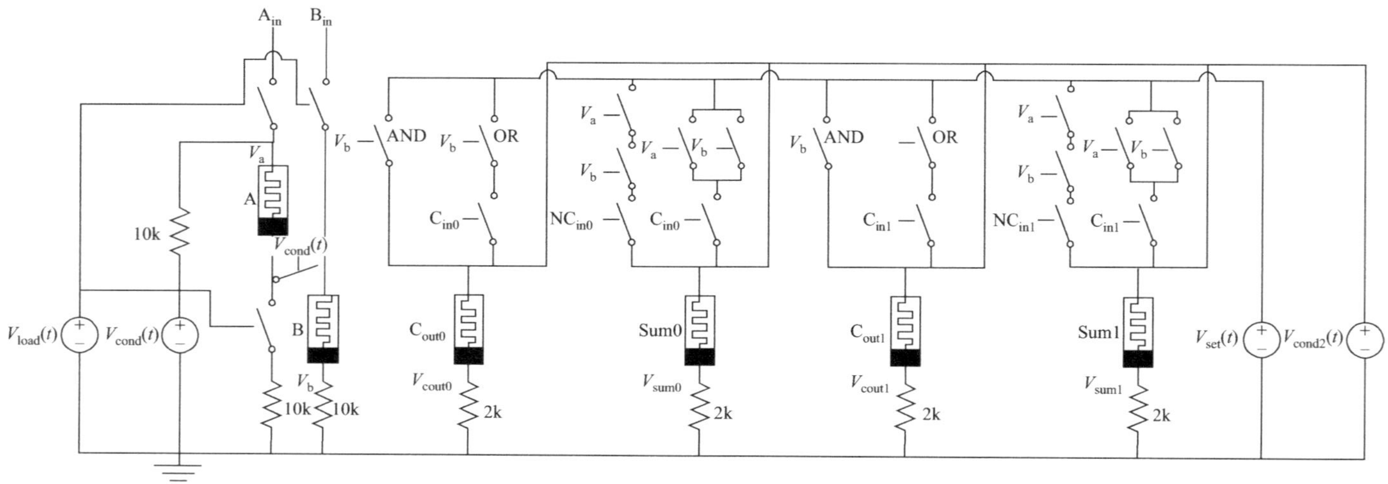

Figure 6.34 Optimized MAD adder for carry-select adders

be tied to ground to represent a carry-in signal of "0" and NC_{in0} and C_{in1} will be tied high to represent a carry-in signal of "1." The remaining full adders in the ripple-carry adder will get these four signals by reading the two carry-out memristors in the previous adder. The p- and n-terminal of the C_{out0} memristor in full adder i will drive the C_{in0} and NC_{in0} input signals to the next full adder $i+1$. The p- and n-terminal of the C_{out1} memristor in full adder i will drive the C_{in1} and NC_{in1} input signals.

By replicating this adder n times and propagating both of the carry signals through the full adders, each n-bit block will require n steps to produce the final two carry-out values. This is the same delay as the standard MAD ripple-carry adder. The total complexity is reduced from two ripple-carry adders ($8n$ memristors, $26n$ switches, and $6n+2$ drivers) to $6n$ memristors, $22n$ switches, and $3n+1$ drivers.

The propagation logic between blocks will use the same logic as the carry logic within the n-bit adders, requiring only a single delay. The circuitry for this is shown in Figure 6.35.

Note that the circuitry takes the same form as the traditional MAD full adder except the drivers on the switches have changed. V_{prop} applies a V_{set} signal to the carry-out memristor that is gated by two paths. The first path has a single switch,

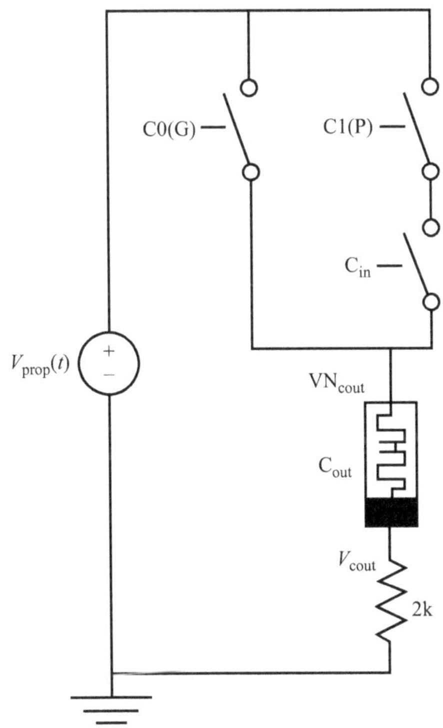

Figure 6.35 Carry-propagation logic for the MAD carry-select adder

driven by the value of G, or the carry-out bit from the C_{out0} memristor in the *n*th full adder of the ripple-carry adder. The second path has two switches, one driven by the carry-in signal from the previous block and the other by P. P is the carry-out bit from the C_{out1} memristor in the *n*th full adder of the ripple-carry adder. Thus, the equation for setting C_{out} is

$$C_{out} = (C_{in} \text{ AND P}) \text{ OR G}$$

The driver to read the value of C_{out} is not shown but is a standard V_{cond} driver connected in parallel with the V_{prop} driver.

Last, a MAD multiplexer is implemented. A 1-bit MAD multiplexer can be implemented with one memristor, one resistor, and four switches. The circuitry for these multiplexers is shown in Figure 6.36.

To execute the multiplexer, the V_{set} signal is applied to the sum memristor. There are two paths from V_{set} to the sum memristor, each gated by two switches. The switches on the first path are gated by input A and the input select line S. The switches on the second path are gated by input B and the inverse of the select line $\overline{S}$. V_{set} will be applied to the sum memristor, setting it to "1," if either of these paths is full satisfied. In other words, if both input A and the select line are "1," the left path will connect V_{set} to the memristor and if both input B and the inverse of the select line are "1," the right path will. This effectively implements the equation for the multiplexer,

$$\text{Out} = (\text{A AND S}) \text{ OR } \left(\text{B AND } \overline{\overline{\text{S}}}\right)$$

For proper functionality, the design must ensure that the values of A, B, and the select line are all read with a V_{cond} voltage during the same step that V_{set} is applied to the sum memristor. Note that the V_{cond} voltage shown in Figure 6.36 is only used

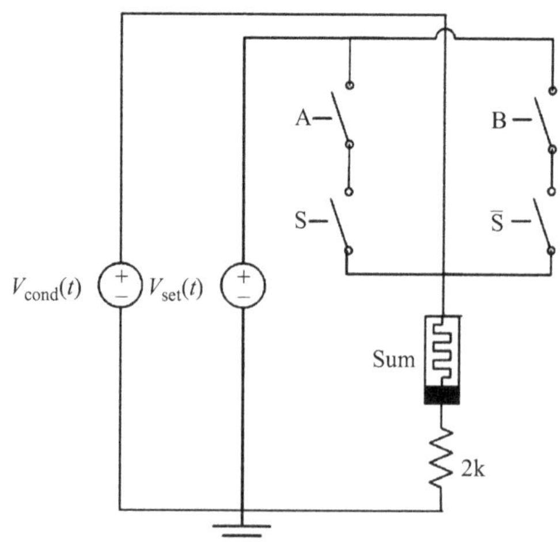

Figure 6.36 MAD multiplexer for a carry-select adder

to read the value of the sum memristor at the end of execution. This multiplexer enables the sum bits to resolve one step after the carry-out bit arrives.

The V_{set} and V_{cond} signals for the multiplexer can be shared across all multiplexers in the same block as long as these values are set and read during the same steps in execution. Thus, only two drivers are needed for an n-bit MAD multiplexer. The total delay for an N-bit MAD carry-select adder is n steps for the ripple-carry adder, $(N/n) - 2$ steps for the carry-propagation logic blocks and a final step for the sum multiplexer. The total delay is $n + (N/n) - 1$ steps. Solving the derivative for n results in the same optimal value as CMOS, $n = \sqrt{N}$. Thus, the total delay of an N-bit MAD carry-select adder is $2\sqrt{N} - 1$, less than half the CMOS delay of $4\sqrt{N} + 2$.

Consider a 16-bit adder. This adder has a delay of seven steps. The first four bits resolve in the first four steps. They resolve one cycle at a time since it is a standard n-bit MAD ripple-carry adder. Then, the n-bit blocks resolve in each subsequent step. Bits 4–7 resolve their carry-out and sum memristor in step 5, bits 8–11 resolve in step 6, and bits 12–15 resolve in step 7.

The area of the adder is also less than alternative memristor-based carry-select adder designs. Since each n-bit block only requires a single n-bit ripple-carry adder, the component area is $6n$ memristors, $22n$ switches, and $3n + 1$ drivers per block. The first n-bit block is also simplified to a standard MAD ripple-carry adder since it has a known carry-in value. The carry-propagation logic requires one memristor, three switches, and one driver in order to select the carry-out signal from the two candidates and store its value. Each block's multiplexer requires $4n$ switches, n memristors, and two drivers.

Therefore, the total component area for an N-bit carry-select adder is

Ripple-carry adders:

$$= (N - n)\left(6 \text{ memristors} + 22 \text{ switches} + \left(3 + \frac{1}{n}\right) \text{drivers}\right)$$

$$+ n\left(4 \text{ memristors} + 13 \text{ switches} + \left(3 + \frac{1}{n}\right) \text{drivers}\right)$$

$$= (6N - 2n) \text{ memristors} + (22N - 9n) \text{ switches} + \left(3N + \frac{N}{n}\right) \text{drivers}$$

Multiplexers:

$$= (N - n)\left(1 \text{ memristors} + 4 \text{ switches} + \frac{2}{n} \text{ drivers}\right)$$

$$= (N - n) \text{ memristors} + (4N - 4n) \text{ switches} + \left(\frac{2N}{n} - 2\right) \text{drivers}$$

Carry-propagation logic:

$$= \left(\frac{N}{n} - 2\right)(1 \text{ memristor} + 3 \text{ switches} + 2 \text{ drivers})$$

$$= \left(\frac{N}{n} - 2\right) \text{memristors} + \left(\frac{3N}{n} - 6\right) \text{switches} + \left(\frac{2N}{n} - 4\right) \text{drivers}$$

Total complexity:

$$= \left(7N - 3n + \frac{N}{n} - 2\right) \text{ memristors} + \left(26N - 13n + \frac{3N}{n} - 6\right) \text{ switches}$$

$$+ \left(3N + \frac{5N}{n} - 6\right) \text{ drivers}$$

For the optimal block size, this becomes

$$= \left(7N - 2\sqrt{N} - 2\right) \text{ memristors} + \left(26N - 10\sqrt{N} - 6\right) \text{ switches} + 3N$$

$$+ 5\sqrt{N} - 6 \text{ drivers}$$

or 102 memristors + 370 switches + 62 drivers for a 16-bit carry-select adder. A schematic for a 16-bit MAD carry-select adder is shown in Figure 6.37. Redundant drivers are replicated in the schematic for readability, but many can be shared and optimized out as described.

The right-most column is the ripple-carry adder for the least significant three bits. The remaining columns are the optimized n-bit blocks to compute both additions at once. The carry-out logic can be seen between each of the n-bit blocks. Last, the 12 multiplexers for selecting the sum outputs are along the bottom row of the schematic.

6.3.4.1 Pipelining

The carry-select adder can also benefit from pipelining. Similar to the IMPLY approach, the pipelining can occur at any granularity, from the individual full adders to the n-bit blocks and so on. If additions are pipelined at the n-bit block granularity, then a new addition can begin once the block has propagated its P and G bits and resolved its sum bits. Each n-bit block can resolve its P and G bits at $t = n$. It takes one more delay step to resolve its carry-out and sum output into the result memristors. Then the bits in the block need to all be reset and the next inputs loaded. This means that a new addition can begin every $n + 3$ cycles. For the 16-bit adder with 4-nit bit blocks that have been discussed, this is a new addition every seven cycles. Table 6.35 shows the key execution stages of the four 4-bit blocks in a 16-bit carry-select adder.

The delay of the n-bit blocks determines the throughput of the design. An addition begins every seven cycles. Each block performs the same steps, each beginning a single cycle after the previous block.

6.3.5 Carry-select adder analysis and comparison

A comparison of the delay for the various memristor-based carry-select adder designs as compared to CMOS is shown in Table 6.36. The GOTO-based implementation is used for the threshold gate design comparisons.

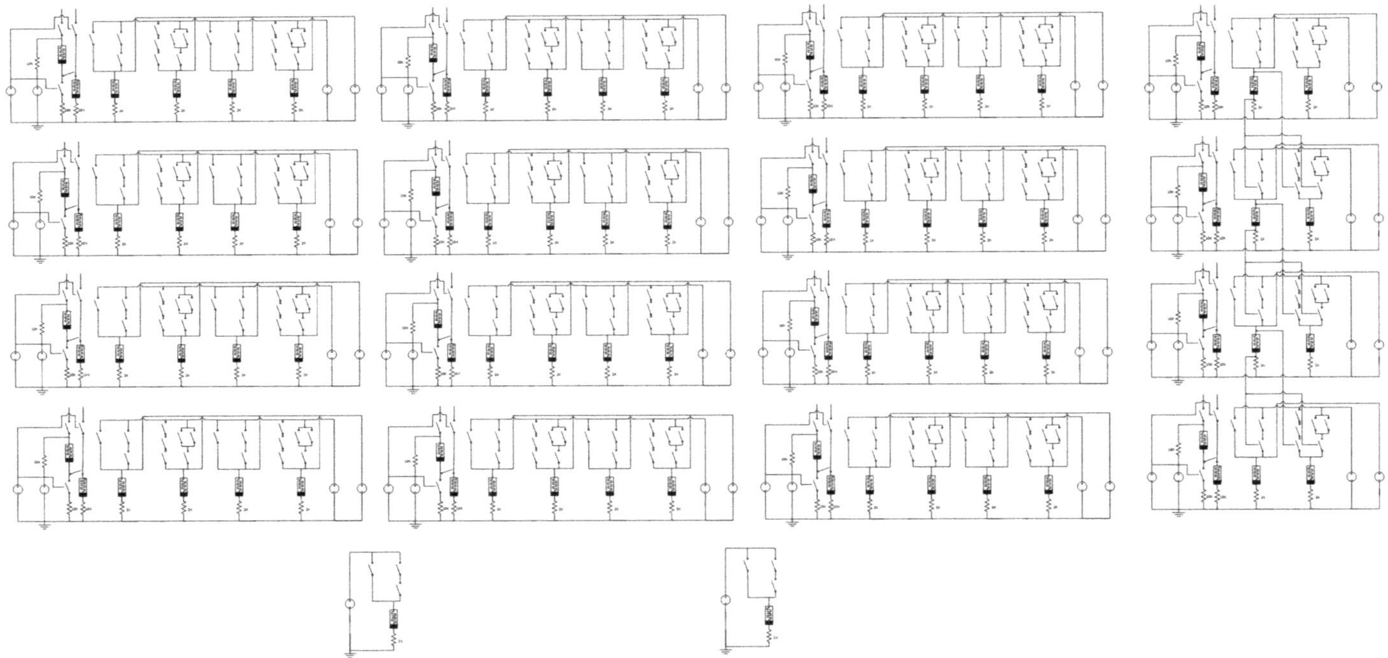

Figure 6.37   16-bit MAD carry-select adder

Table 6.35 Pipelined execution steps for the blocks in a 16-bit MAD carry-select adder

T	Block0	Block1	Block2	Block3
1	Start addition 0	Start addition 0	Start addition 0	Start addition 0
2				
3				
4	Resolve C_{out}	Resolve P,G	Resolve P,G	Resolve P,G
5	Reset memristors	Resolve sum, C_{out}		
6	Load inputs	Reset memristors	Resolve sum, C_{out}	
7	Start addition 1	Load inputs	Reset memristors	Resolve sum, C_{out}
8		Start addition 1	Load inputs	Reset memristors
9			Start addition 1	Load inputs
10	Resolve C_{out}			Start addition 1
11		Resolve P,G		
12	Reset memristors	Resolve sum, C_{out}	Resolve P,G	
13	Load inputs	Reset memristors	Resolve sum, C_{out}	Resolve P,G
14	Start addition 2	Load inputs	Reset memristors	Resolve sum, C_{out}
15		Start addition 2	Load inputs	Reset memristors
16			Start addition 2	Load inputs
17				Start addition 2

Table 6.36 Delay comparisons for the carry-select adder designs

Approach	Delay	Delay for $n = \sqrt{N}$
IMPLY	$2n + 14 + (2N/n)$	$4\sqrt{N} + 14$
Hybrid-CMOS	$3n - 1 + (3N/n)$	$6\sqrt{N} - 1$
Threshold gates	$n + 2 + (N/n)$	$2\sqrt{N} + 2$
MAD	$2n - 1$	$2\sqrt{N} - 1$
CMOS	$2n + 2 + (2N/n)$	$4\sqrt{N} + 2$

In comparison to the traditional CMOS approach, the IMPLY approach has nearly an identical delay. Regardless of the size of the adder, the delay for the IMPLY design is 12 more cycles. As the size N increases, this constant becomes negligible and the delays of the two designs converge to be equivalent.

The hybrid-CMOS design has an increased delay due to signal buffering on the carry-propagation signals and the ripple-carry adders. The threshold gate and proposed MAD designs exhibit half the delay of traditional CMOS with the MAD design edging out the threshold design by a three delays. The delay is also less than half of the delay of the IMPLY design and about 1/3 of the Hybrid-CMOS design's delay.

Recall that both the IMPLY and MAD designs can also be pipelined to allow for higher throughput. Table 6.37 shows the latency for each of the proposed approaches to complete X consecutive additions.

Table 6.37 Latency for X additions on the proposed carry-select adders

Approach	Pipelined delay for $n = \sqrt{N}$
IMPLY	$4\sqrt{N} + 24X - 10$
Hybrid-CMOS	$6X\sqrt{N} - X$
Threshold gates	$2X\sqrt{N} + 2X$
MAD	$2\sqrt{N} + 7X - 8$
CMOS	$4X\sqrt{N} + 2X$

Table 6.38 Component counts for carry-select adders

Approach	Memristors	MOSFETs	GOTOs	Drivers	Switches
IMPLY	$17N - 8\sqrt{N} - 2$			$17N - 8\sqrt{N} - 2$	$22N - 10\sqrt{N} - 4$
Hybrid-CMOS	$34N - 16\sqrt{N} - 8$	$24N - 8\sqrt{N} - 8$			
GOTO Threshold Gates	$19N - 9\sqrt{N} - 6$	$4N - 2$	$6N - 3\sqrt{N} - 2$		
MAD	$7N - 2\sqrt{N} - 2$			$3N + 5\sqrt{N} - 6$	$26N - 10\sqrt{N} - 6$
CMOS		$93N - 42\sqrt{N} - 22$			

The hybrid-CMOS, threshold gate, and CMOS designs do not benefit from pipelining, and all additions must be serialized. These designs see no improvement in throughput in Table 6.37. However, the IMPLY and MAD designs do. The MAD approach now has the lowest delay and the highest throughput. The delay reduction as compared to the CMOS and hybrid-CMOS approaches is now multiplicative. The threshold approach, which appeared comparable to the MAD design for a single addition, is less competitive since no pipelining can be performed. As the number of additions increases, the threshold approach requires $(2\sqrt{N})+2$ steps for each addition, whereas the MAD approach requires a constant 7. Thus, for all adder widths greater than four, the MAD approach has a higher throughput than the threshold gate approach. The IMPLY approach also sees improved throughput from pipelining a new addition every 24 steps. Although this is not as performant as the MAD approach, it is still a viable alternative over the other three approaches.

A general comparison of the complexity of the memristor-based designs in comparison to traditional CMOS is shown in Table 6.38.

In comparison to the traditional CMOS design, the alternative memristor-based designs all require fewer components. The number of MOSFETs is reduced by about 75 percent for the hybrid-CMOS approach and to nearly none in the threshold gate implementation. Recall that the threshold gate implementation only needs inverters for the ripple-carry adders and the select line on each n-bit multiplexer. The carry-propagation logic requires no CMOS. The remaining designs do not use any MOSFETs. Overall, the threshold gate design requires the fewest components. However, the IMPLY and MAD designs benefit from the fact that they do not use any MOSFETs in their designs.

As expected, the IMPLY and MAD designs require the fewest memristors. The presented MAD design uses significantly fewer components than the CMOS and hybrid-CMOS designs. It also requires less than half the number of memristors as the IMPLY design. Although the number of switches is roughly equivalent between the two designs, the IMPLY approach uses about 5 times the number of drivers and 2.5 times the number of memristors. However, it uses fewer switches.

6.4 Conditional-sum adders

Memristor-based conditional-sum adders are presented for the IMPLY, hybrid-CMOS, threshold, and MAD contexts. Each adder is explained, optimized, and analyzed in for its given approach. Last, the designs are compared against each other and the traditional CMOS design of a conditional-sum adder in terms of delay and component count.

6.4.1 IMPLY conditional-sum adder

There has been previous literature that explored a naive IMPLY-based conditional-sum adder implementation [16]. The work presented an 8-bit conditional-sum adder, which requires 454 memristors and a has a delay of 39 steps. The work also presented a 16-bit conditional-sum adder, which requires 975 memristors and has a delay of 46 steps. In comparison with the designs for IMPLY ripple-carry adders, it is evident that the conditional-sum adder is more complex but has a faster delay.

This section presents a memristor-based conditional-sum adder designed from IMPLY-based modified half adders and multiplexers. Recall that a modified half adder takes two inputs and produces four outputs corresponding to the carry-out and sum bits when the carry-in is “0” and when the carry-in is “1.”

In the design of a modified half adder, the first steps from an IMPLY full adder are not used. There are two limitations. First, the full adder computes the term A XOR B rather than A OR B, which is not useful in a modified adder. Second, the ripple-carry adder execution step count is based on the assumption that the carry-out is the only output on the critical path. For a modified half adder, both the carry and sum outputs are on the critical path.

Instead, the modified half adder steps are selected to equally prioritize the carry outputs and sum outputs. A breakdown of the execution steps to implement a 1-bit modified half adder with IMPLY operations is given in Table 6.39.

The IMPLY modified half adder is one of the few modules that uses the same Boolean equations for its outputs as the traditional CMOS design. The optimal

Table 6.39 Steps for a 1-bit IMPLY modified half adder

Step	Functionality	Description
1	A IMP 0 → M0	NOT A
2	B IMP 0 → M1	NOT B
3	M1 IMP 0 → M3	Copy of B
4	A IMP M1 → M1	A IMP (B IMP 0) = A NAND B
5	M0 IMP M3 → C_1	(A IMP 0) IMP B = A OR B = C_1
6	M1 IMP M2 → C_0	A AND B = C_0
7	Reset M1,M0	
8	M3 IMP M0 → M0	NOT C_1
9	M2 IMP M1 → M1	NOT C_0
10	M1 IMP M0 → S_1	(NOT C0) IMP (NOT C1) = (NOT C0) NAND C1 = S_1
11	M0 IMP M4 → S_4	(NOT C0) AND C1 = S_0

expressions for the outputs in the CMOS domain are also the optimal expressions in the IMPLY domain.

The execution takes a total of 11 total steps, 5 and 6 for the carry signals and 10 and 11 for the sum bits. This design intentionally resolves C_1 and S_1 before the C_0 and S_0 counterparts to hide latency in the adder design. In a conditional-sum adder, the outputs of the modified adders either serve as select lines to a multiplexer or as inputs to a multiplexer. In the case that they serve as inputs, they can either be input A or B. In the multiplexer, A is selected when the select line is "1" and B is selected when the inverse of the select line is "1." When the select line enters the multiplexer, it takes one step to be inverted. In this IMPLY design, outputs C_1 and S_1 serve as the "A" inputs and C_0 and S_0 serve as the "B" inputs. In this way, the extra cycle delay to obtain C_0 and S_0 is hidden by the extra cycle delay to obtain the inverted select line. Thus, the critical path does not see a penalty in latency.

The execution steps assume that none of the outputs are read during execution. Recall that the IMPLY full adder design has two reserved steps in which the subsequent bit in the chain read the carry signal. This improved the critical path of the carry-propagation chain and moved the sum outputs to complete later. In many cases, the presented modified half adder will have similar reserved steps to read the carry-out signals. This will increase the total delay for the sum outputs by two steps but will not add any components.

The modified half adder also has a step for resetting two of the memristors for use as the sum outputs. Alternatively, this step can be removed and two additional memristors can be used. Depending on the optimization trade-offs, this alternative may be preferred. However, it is not advised to insert any reset steps along the critical path of the adder unless necessary.

A single bit modified half adder requires two input memristors, one intermediate memristor and four output memristors for C_0, C_1, S_0, and S_1 for a total of seven memristors. The schematic is shown in Figure 6.38.

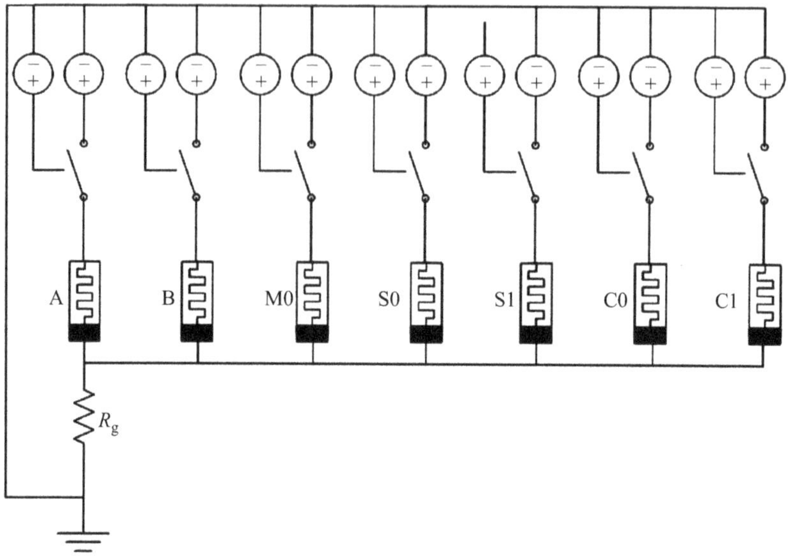

Figure 6.38 IMPLY modified half adder

The schematic looks nearly identical to a full adder. The key differences lie in the driver logic on the memristors rather than the circuitry. This modified half adder would exist in a single row in the crossbar structure.

For a conditional-sum adder, many of the outputs from the modified half adders have a fan-out greater than one. An output with fan-out $F > 1$ needs to be copied F times so it can be used in parallel across the F units. In IMPLY steps, this requires 2*F steps and 2*F extra memristors. However, aspects of the modified half adder and overall design can reduce this overhead.

6.4.1.1 Optimizations to reduce copy operations

In a modified half adder, the two sum outputs S_0 and S_1 are inverses of each other. Referring back to Table 6.39, step 11 performs NOT S_1. This step can be changed to perform S1 IMP 0 $\rightarrow X_0, X_1$, which stores the result into two memristors X_0 and X_1. By selecting X_0 and X_1 as input memristors in the two multiplexers that use S_0, both units obtain this value in parallel in step 11. This single-read operation is able to set multiple memristors in a single step by applying V_{cond} to S_1 and V_{set} to both X_0 and X_1. This removes the need for memristor M4 in the modified half adder. However, it increases the size of the multiplexers to five memristors each in order to hold their inputs locally.

Similarly, S_1 can be provided to its destinations. Step 11 can be seen as the first step in a copy operation for S_1. The step from S1 IMP 0 $\rightarrow X_0, X_1$ to S1 is extended to IMP 0 $\rightarrow X_0, X_1, X_2, X_3$, which also stores the result into two memristors X_2 and X_3. By selecting X_2 and X_3 as memristors in the two multiplexers that use S_1, both units perform the first step in their copy operation in parallel during step 11. Then, these two units must just perform a single additional step for X_i IMP 0 → Y_i to obtain S_1. Memristors X_2 and X_3 can be reset for later use.

If this 12th step is added to the execution of the modified half adders, the design can compute all four outputs and copy the sum outputs into their destinations. Since the sum outputs are not on the critical path, these copy steps do not impact the final delay of the adder. Also, since both the X_i and Y_i memristors are inputs to the next units in the adder, there are no additional memristors introduced in these steps.

The C_0 and C_1 outputs are not inverses of each other so their copy operations cannot be overlapped. However, they can still perform the multidestination set operations for both of their copy operations. Note that the carry outputs are used in two ways in a conditional-sum adder: as inputs to multiplexers and as select lines to multiplexers. In either case, the carry-out signal usually has a fan-out greater than one and requires a copy operation. For the instances in which the carry-out signal is on the critical path, this adds two steps to the overall delay.

It was shown that the delay through a multiplexer once the select line and input memristors are resolved is four delays. These multiplexers are used in a hierarchical fashion with the modified half adders to propagate the carry signal efficiently.

Consider a 4-bit conditional-sum adder. This adder requires three modified half adders, one full adder, three 2-bit multiplexers, and one 3-bit multiplexer. The full adder requires seven memristors, and the modified half adders require six memristors each. All the inputs for the multiplexer for bit 1 have a fan-out of 1. Thus, this multiplexer only requires three memristors per bit. The other two 2-bit multiplexers have inputs with fan-out greater than 1 and each require 5 memristors per bit. The final 3-bit multiplexer's inputs all have a fan-out of 1 so the multiplexer requires 3 memristors per bit. In total this is 60 memristors.

The full adder takes 13 cycles to resolve its carry-out and two more steps to invert and copy it for use as the select line. This addresses the fan-out issue. The modified half adder takes 12 steps to resolve and prepare its outputs for the multiplexer. The multiplexer completes in four cycles, completing at $t = 19$. In parallel, the 2-bit multiplexers for bits 2 and 3 have also resolved. Now, the final 3-bit multiplexer can complete. First, two steps copy the select line to be used across the three bits. Then, the four steps of the multiplexer execute. The total execution time for the 4-bit conditional-sum adder is 25 steps.

To expand to an 8-bit conditional-sum adder, four modified half adders, four 2-bit multiplexers, two 3-bit multiplexers, and one 5-bit multiplexer are added. The modified half adders add 24 memristors. The multiplexers all require at least 3 memristors per bit for a total of 57 memristors. Additionally, 12 of the 1-bit multiplexers must hold their inputs from the previous stage to handle fan-out. This adds 24 memristors. The total complexity of an 8-bit conditional-sum adder is 165 memristors. A complete breakdown of the complexity of the conditional-sum adder for various adder widths is given in Table 6.40.

The delay of the conditional-sum adder grows logarithmically with the width of the adder. Each time the adder doubles in width, an additional level with a final

Table 6.40 Complexity for IMPLY conditional-sum adders

Width	2	4	8
Proposed	19 memristors	60 memristors	165 memristors

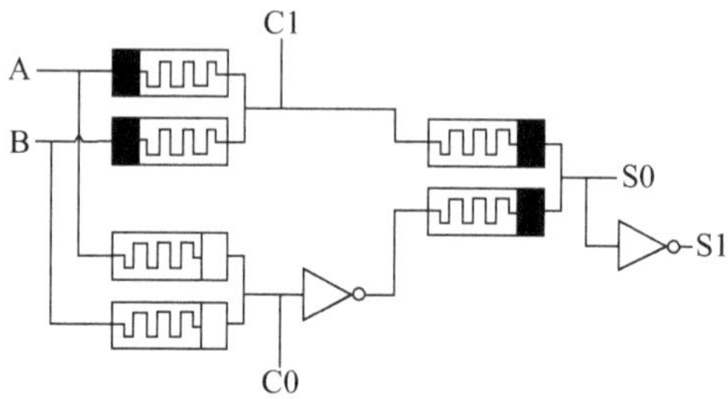

Figure 6.39 Hybrid-CMOS modified half adder

multiplexer is added to the hierarchy. Thus, the delay grows by the delay of a multiplexer each time the width doubles. For a 2-bit adder, the outputs are ready after the initial full adder and modified half adder complete and the multiplexer resolves. The full adder resolves its carry-out signals after 13 steps. Six more steps are required to copy the values of the carry-out signals and resolve the multiplexer. Thus, the 2-bit adder requires 19 steps. A 4-bit adder requires 25 steps, an 8-bit adder requires 31, and a 16-bit adder requires 37.

6.4.1.2 Pipelining

The IMPLY conditional-sum adder can be pipelined to perform a series of additions with a higher throughput. The full adder for bit 0 has a higher delay than the modified half adders. Thus, this unit determines the maximum throughput of the design. Once the full adder has propagated its carry and sum outputs to the next multiplexer, the memristors can be reset and a new addition can begin. This is complete at $t = 15$. After a reset step, the next addition begins. Thus, every 16 cycles a new addition can begin. The description of this process is identical to the explanation for the ripple-carry adder or carry-select adder.

6.4.2 Hybrid-CMOS conditional-sum adder

In this section, a hybrid-CMOS memristor-based conditional-sum adder is presented and analyzed. In the hybrid-CMOS approach, a modified half adder consists of six memristors and four MOSFETs as shown in Figure 6.39.

The carry-out bits are ready after a single-gate delay, the S_0 sum bit is ready after three, and the S_1 sum bit is ready after four. At a gate level, this design is identical to the traditional CMOS modified half adder. Because of the use of inverters and low fan-out, signal restoration buffers are not needed.

Table 6.41 Complexity for hybrid-CMOS conditional-sum adders in comparison to CMOS

Adder width	**2**	**4**	**8**
CMOS	90 MOSFETs	239 MOSFETs	614 MOSFETs
Hybrid-CMOS	32 memristors	86 memristors	224 memristors
	18 MOSFETs	52 MOSFETs	142 MOSFETs

The modified half adder can be used in the conditional-sum adder. For the multiplexers, the design uses the hybrid-CMOS multiplexer design described for the carry-select adder. These multiplexers have the same delay and complexity as traditional CMOS multiplexers. The component count for a hybrid-CMOS multiplexer is two MOSFETs and six memristors. However, for N-bit multiplexers such that $N > 1$, the select line is shared across the bits so the total complexity is $6N$ memristors + 2 MOSFETs. Coupling the modified half adders with the multiplexers produces a complete conditional-sum adder.

Since this design is constructed from hybrid-CMOS gates, some signal restoration is necessary. Specifically, the outputs of the multiplexers, which feed into other multiplexers, are buffered at their outputs. These signals have high fan-out within the multiplexers and across multiplexers. Many also exist on long gate chains as the signal traverses the hierarchy of multiplexers. For each of these reasons, the outputs must be buffered to maintain their signal strength. For the 8-bit implementation, this adds 80 MOSFETs to the design.

At the gate level, the complexity of the hybrid-CMOS conditional-sum adder is nearly identical to the traditional CMOS counterpart. There is minimal additional buffering required on the multiplexers for the hybrid-CMOS implementation.

Consider an 8-bit conditional-sum adder. The complexity breaks down as follows:

Modified half adders and full adder

$$= 7(6 \text{ memristors} + 4 \text{ MOSFETs}) + (14 \text{ memristors} + 12 \text{ MOSFETs})$$

Multiplexers:

$$\begin{aligned} &= 7(12 \text{ memristors} + 2 \text{ MOSFETs}) + 3(18 \text{ memristors} + 2 \text{ MOSFETs}) \\ &\quad + (30 \text{ memristors} + 2 \text{ MOSFETs}) + 80 \text{ MOSFETs} \\ &= 168 \text{ memristors} + 102 \text{ MOSFETs} \end{aligned}$$

Total complexity:

$$= 224 \text{ memristors} + 142 \text{ MOSFETs}$$

The specific component requirements for various adder sizes are given in Table 6.41.

For all adder widths, the hybrid-CMOS designs require fewer components and reduces the number of MOSFETs to less than 1/4 of the CMOS design. This is due to the fact that the memristor-based gates require fewer components than CMOS gates.

In the traditional CMOS design, the delay of the adder increases logarithmically with the size of the adder by the delay of a multiplexer. This trend is similar in the hybrid-CMOS designs. The only difference is that a buffer is added on the carry-out propagation path at the output of each multiplexer. Thus, the delay increases by four each time the adder width is doubled rather than three. A 2-bit hybrid-CMOS conditional-sum adder has a delay of eight cycles. The initial full adder takes five cycles to produce its carry-out signal and the multiplexer takes three more steps to resolve the sum multiplexers. Given this information, a 4-bit adder requires 12 steps and an 8-bit adder requires 16.

6.4.3 Threshold gate conditional-sum adder

Conditional-sum adders can also be implemented using threshold gates. For the purposes of conditional-sum adders, it is preferential to use the CSTG implementation because it produces the output of the threshold gate and its inverse in the last buffer. The GOTO implementation only produces the threshold gate output. This is especially useful for producing the sum outputs. This fact allows a modified half adder to be created from threshold gates as shown in Figure 6.40.

This modified half adder has a complexity of three threshold gates. These gates require 6 memristors, 6 resistors, and 24 MOSFETs. Note that both NOT gate have been removed from the critical path of the design. This is because the threshold gate produces both outputs as a by-product of the threshold detection. Second, the sum outputs are both produced by the final threshold gate. This reduces the delay from four to two for the sum bits and removes four MOSFETs from the design. The delay is one for the carry output bits C_0 and C_1 and two for the sum output bits S_0 and S_1.

Now construct an 8-bit conditional-sum adder from the multiplexers and the modified half adders. Recall that an N-bit multiplexer with CSTG threshold gates has a delay of two and $2N+1$ gates for a total of $18N+2$ MOSFETs, $5N$ memristors, and $5N$ resistors. In this design, the total cost is

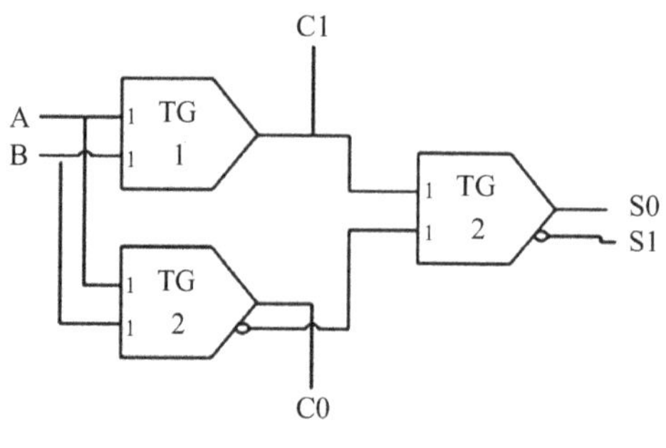

Figure 6.40 CSTG threshold gates modified half adder

Full adder:

$$= 7 \text{ memristors} + 7 \text{ resistors} + 22 \text{ MOSFETs}$$

Modified half adders:

$$\begin{aligned} &= 7(6 \text{ memristors} + 6 \text{ resistors} + 24 \text{ MOSFETs}) \\ &= 42 \text{ memristors} + 42 \text{ resistors} + 168 \text{ MOSFETs} \end{aligned}$$

Multiplexers:

$$\begin{aligned} &= 7(10 \text{ memristors} + 10 \text{ resistors} + 38 \text{ MOSFETs}) \\ &\quad + 3(15 \text{ memristors} + 15 \text{ resistors} + 56 \text{ MOSFETs}) \\ &\quad + 1(25 \text{ memristors} + 25 \text{ resistors} + 92 \text{ MOSFETs}) \\ &= 140 \text{ memristors} + 140 \text{ resistors} + 526 \text{ MOSFETs} \end{aligned}$$

Total complexity:

$$= 189 \text{ memristors} + 189 \text{ resistors} + 716 \text{ MOSFETs}$$

The delay is one step for the carry-out from the first full adder, two steps for each level of multiplexers in the hierarchy, and a final step to wait for the sum inputs to arrive to the final multiplexer and compute its result. For an N-bit conditional-sum adder, the delay is $2k$ where $k = 1 + \log_2(N)$. For an 8-bit adder, the addition completes in just eight steps. This is an improvement over the standard CMOS implementation.

6.4.4 MAD gate conditional-sum adder

This section presents the design of an 8-bit MAD-based conditional-sum adder architecture. The modified adder structure presented for the carry-select adder is reused. As described for the MAD carry-select adders, each of the modified half adders can produce its four outputs, $C0_i$, $C1_i$, $S0_i$, and $S1_i$ in a single step. The circuit for this modified half adder is shown in Figure 6.41.

The switches for the carry-in signal have been removed since they are no longer used. The switches corresponding to a carry-in of 0 have been replaced with open circuits and the switches corresponding to a carry-in of 1 have been replaced with wire. The block requires six memristors, ten switches, and four drivers. Note that all of the drivers can be shared across the modified half adders in the design.

A naïve approach would use this modified half adder design for every bit of the adder. Then, the first level of 2-bit multiplexers would execute to resolve the outputs to the next level in the hierarchy. Each 2-bit multiplexer requires an additional eight switches and two memristors. However, this incurs a cost in both area and delay that is unnecessary if the design is optimized for the adder's context. Instead, the multiplexers can be merged into the modified half adders themselves.

Let the even bits' modified half adders resolve the four outputs using the circuit in Figure 6.41. Let the odd bits' modified half adders use the adder presented for the MAD carry-select adder as shown in Figure 6.42.

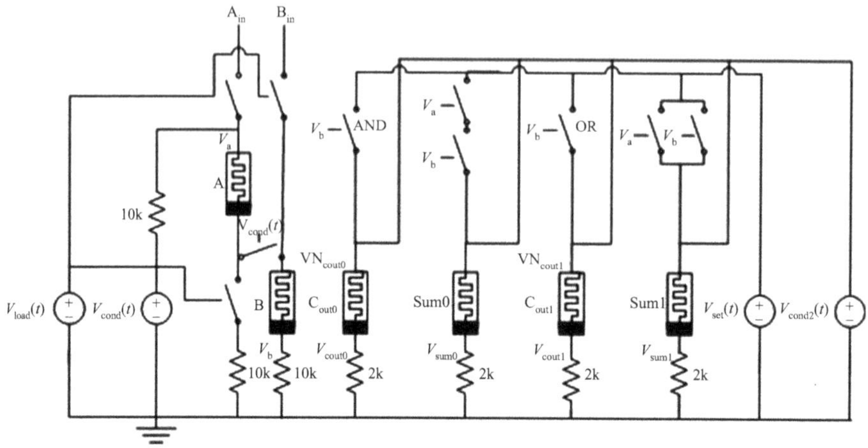

Figure 6.41 MAD modified half adder for even bits in a conditional-sum adder

To discuss the execution of the conditional-sum adder, focus on each pair of bits. Bits 0 and 1 are a pair, bits 2 and 3 are another pair, and so on. Each of these pairs has an odd bit and an even bit. Execution begins as normal for the even bits and they resolve their outputs at $t = 1$. At $t = 2$, the modified half adders for the even bits apply the read voltage V_{cond2} to the carry-out memristors. The voltages at V_{cout0} and V_{cout1} are then sensed for their values and used in conjunction with the logic in the odd bit's modified half adder. During the same step, the odd bit's adder applies V_{cond} is to read the input memristors A and B. V_{set} is also applied to set the four output memristors. Based on the sensed values of A, B, and the incoming carry signals from the even bit's modified half adder, all four outputs can be resolved to their final values.

The first level of multiplexer logic in the conditional-sum adder can now be removed. The values of the output memristors in the odd bits can now be read and propagated to multiplexers in later levels of the addition. All of the remaining multiplexers require one memristor, four switches, and two drivers per output. The drivers can be shared across all multiplexers on the same level in the adder design as long as these values are set and read at the same steps in execution.

Each hierarchical level of the conditional-sum adder incurs a single bit delay. In other words, the delay only increases by a single step each time the width of the adder doubles. A 2-bit adder requires a single step for the full adder and a single step for the modified half adder and combined multiplexer. Thus, a 2-bit conditional-sum adder requires two steps, a 4-bit conditional-sum adder requires three steps, an 8-bit adder requires four steps, and so on. In general, the MAD design requires $\log(N)+1$ steps to complete an addition.

The total complexity consists of the area of the modified half adders and the multiplexers. A schematic for an 8-bit MAD conditional-sum adder is shown in Figure 6.43.

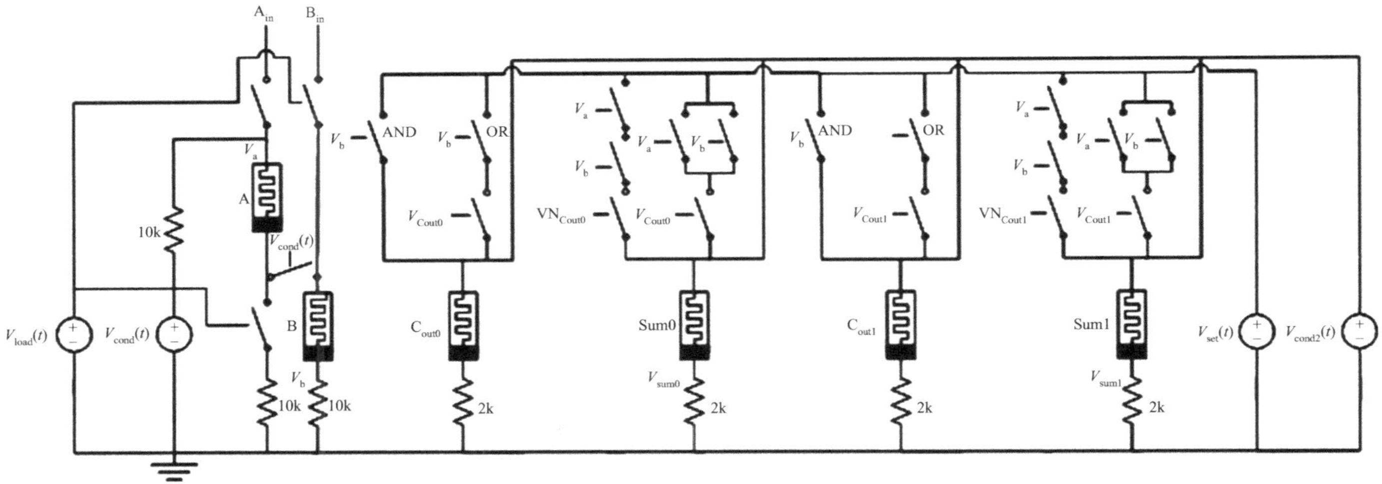

Figure 6.42   MAD modified half adder for odd bits in a conditional-sum adder

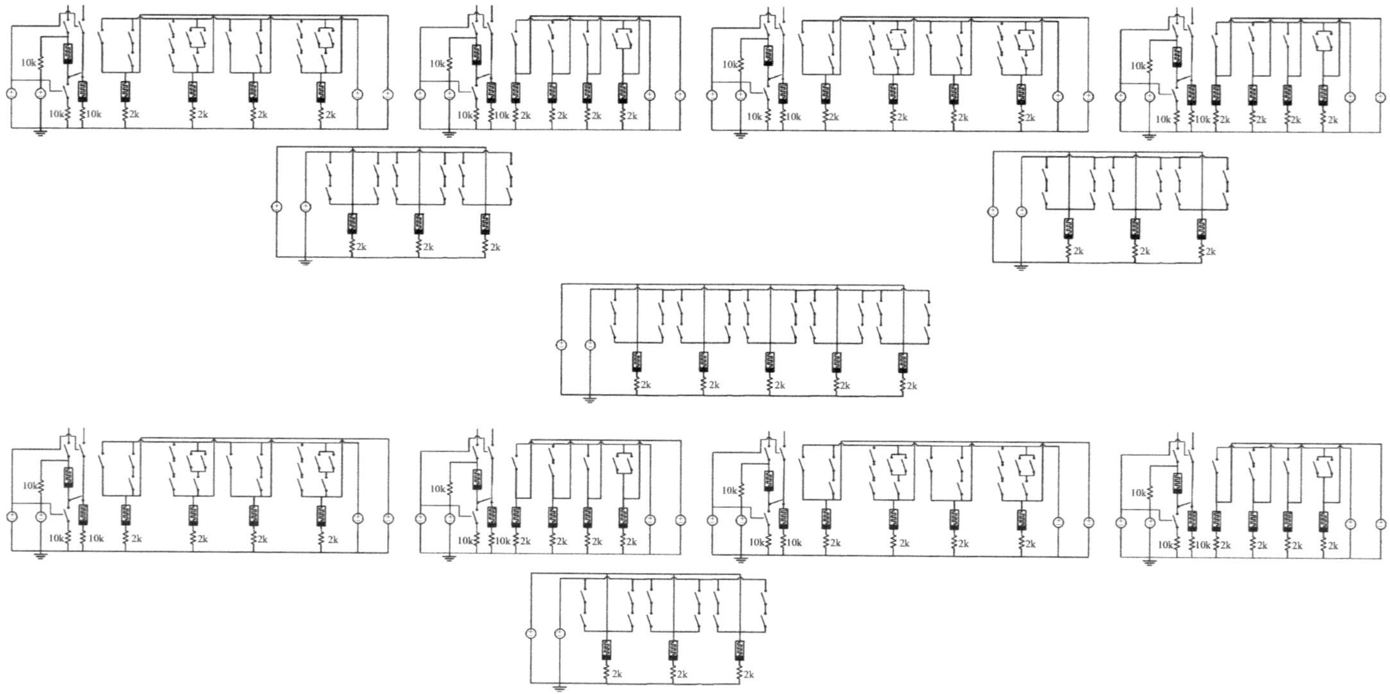

Figure 6.43 8-bit MAD conditional-sum adder

Notice that there is one fewer level of multiplexers in this design as compared to the traditional CMOS design. The design only has 3- and 5-bit multiplexers now. Redundant drivers are shown to make functionality clear; however, many can be optimized out as described. For example, all of the multiplexers on the same level in the design can share the V_{cond} and V_{set} drivers since execution traverses through the hierarchy in lockstep. By the same logic, all of the even-bit adders can share their drivers and all of the odd-bit adders can as well. This implies that the number of drivers only increases by 2 each time the adder width doubles in order to read and write the outputs of the multiplexers in the additional stage.

For the 8-bit MAD conditional-sum adder design, the total cost is

Even bit adders:

$$= \frac{N}{2}(6 \text{ memristors} + 10 \text{ switches}) + 4 \text{ drivers}$$
$$= 24 \text{ memristors} + 40 \text{ switches} + 4 \text{ drivers}$$

Odd bit adders:

$$= \frac{N}{2}(6 \text{ memristors} + 22 \text{ switches}) + 4 \text{ drivers}$$
$$= 24 \text{ memristors} + 88 \text{ switches} + 4 \text{ drivers}$$

Multiplexers:

$$= 14(1 \text{ memristor} + 4 \text{ switches}) + 4 \text{ drivers}$$
$$= 14 \text{ memristors} + 56 \text{ switches} + 4 \text{ drivers}$$

Total complexity:

$$= 62 \text{ memristors} + 184 \text{ switches} + 12 \text{ drivers}$$

The conditional-sum adder is also able to be pipelined. The first level of the design, which holds the adders, takes two steps to finish. The remaining levels in the design take a single step to resolve. Thus, the first level is the limiting factor on the throughput of the design. The first level of the adder is complete after two steps. Then, the adders reset their memristors in step 3, load the next inputs in step 4, and begin execution again in step 5. A new addition can begin every four steps.

6.4.5 Conditional-sum adder analysis and comparison

A full component comparison for the four presented conditional-sum adders and the traditional CMOS design for various adder widths is shown in Table 6.42.

The IMPLY implementation has consistently fewer components than the CMOS for the conditional-sum adder. It requires about 66 percent the number of components that the CMOS design requires. The hybrid-CMOS approach uses more memristors than the IMPLY approach and requires additional MOSFETs for the NOT gates. This complexity is still less than that of traditional CMOS by a significant margin. The reduction in components in the hybrid-CMOS implementation increases as the adder size increases. For the 8-bit adder, the

Table 6.42 Complexity for conditional-sum adders

Adder width	**2**	**4**	**8**
CMOS	90 MOSFETs	239 MOSFETs	614 MOSFETs
IMPLY	19 memristors 19 drivers 19 switches	60 memristors 60 drivers 60 switches	165 memristors 165 drivers 165 switches
Hybrid-CMOS	32 memristors 18 MOSFETs	86 memristors 52 MOSFETs	224 memristors 142 MOSFETs
Threshold CSTG	84 MOSFETs 23 memristors 23 resistors	264 MOSFETs 70 memristors 70 resistors	716 MOSFETs 189 memristors 189 resistors
MAD	12 memristors 8 drivers 32 switches	27 memristors 10 drivers 76 switches	62 memristors 12 drivers 184 switches

Table 6.43 Delay for conditional-sum adders, $k = 1 + \log_2(N)$

Adder width	**2**	**4**	**8**	**16**	**General**
CMOS	8	11	14	17	$3k+2$
IMPLY	19	25	31	37	$6k+7$
Hybrid	8	12	16	20	$4k$
Threshold CSTG	4	6	8	10	$2k$
MAD	2	3	4	5	k

hybrid-CMOS implementation uses less than 1/4 the number of MOSFETs as the CMOS implementation.

The threshold gate implementation requires slightly more components than the traditional CMOS implementation. This trend is expected since the design used CSTG rather than GOTO pairs. CSTG are higher complexity at the benefit of faster delay.

The MAD design uses fewer components than all alternative designs as the adder width increases. Also, the MAD design uses about 1/3 the number of memristors as compared to the threshold and hybrid-CMOS designs for an adder width of eight. The IMPLY approach, known for its low area, requires about three times as many memristors and almost 10 times as many drivers for an adder width of eight.

Delay comparisons for the proposed conditional-sum adders are shown in Table 6.43.

The hybrid-CMOS approach has an extra delay at each hierarchical level compared to CMOS due to the multiplexer output buffering. The IMPLY design has the highest delay increase at each level, adding six steps each time the adder width doubles. Essentially, the number of steps in the IMPLY design is double the CMOS design.

The remaining two designs all improve in terms of delay over the CMOS design. The threshold gate implementation only adds two cycles each time the adder width doubles and the MAD design only adds one. Overall, the MAD design has the lowest delay, with a 16-bit addition taking only five cycles. This is twice as fast as the next best alternative, threshold gates, which is also a competitive delay.

The IMPLY and MAD approaches can again be pipelined, allowing a new addition to begin every 16 and 4 steps, respectively. Thus, for large numbers of consecutive additions, the IMPLY approach becomes a better alternative than the threshold, hybrid-CMOS, or CMOS approaches and the MAD approach gains an even larger delay advantage over all other approaches.

Chapter 7
Memristor-based multiplier designs[1]

This chapter presents contributions in the design and implementation of memristor-based multipliers. Minimal prior work exists for memristor-based designs for multipliers. The complexity of these designs coupled with the complexity of memristor models and their programing leads to a high design overhead. However, there have been a few prior works that have explored multipliers in the context of memristors for the IMPLY approach.

Similar to Chapter 6 on adders, the focus is on four different multipliers: shift-and-add, Booth, array, and Dadda multipliers. For each multiplier, implementations using IMPLY, hybrid-CMOS, threshold gate, and MAD approaches are examined. Each implementation is explained and analyzed in terms of complexity and delay.

Due to the increasing complexity of these designs, the CSTG threshold gate implementations are not considered. The component area of CSTG implementations precludes them from being desirable for these units. Recall that a single three-input CSTG threshold gate requires three memristors, three resistors, and ten MOSFETs. Thus, for some of the more complex designs, only the GOTO pair implementations are presented. Complete schematics and simulations are also given.

7.1 Shift-and-add multipliers

First, shift-and-add multipliers are considered. For the purposes of this text, the focus is on shift-and-add multipliers that have equal-width inputs. The practices, principles, and optimizations discussed can be extended to any width multiplier.

Each section presents a memristor-based shift-and-add multiplier constructed from either IMPLY, hybrid-CMOS, threshold, or MAD constructs. The section concludes with a discussion on the cost and performance trade-offs of each design with respect to each other and the traditional CMOS design.

7.1.1 IMPLY shift-and-add multiplier

This section presents an optimized IMPLY-based shift-and-add multiplier. In general, shift-and-add multipliers are lower in area with higher delay than other

[1]This chapter is based on [60,61].

multipliers. IMPLY-based memristor designs make the same trade-off, favoring a low area design over a fast design. Thus, for applications for which shift-and-add multipliers are appropriate, IMPLY is a reasonable approach to maintain the same priorities.

Shift-and-add multipliers perform a multiplication by executing a series of serialized additions. For a multiplication A*B, the multiplier will add A to a running total B times to reach the product. Because the shift-and-add multiplier's main computational unit is an adder, it takes a similar form to a standard ripple-carry adder. It requires additional control logic including a multiplexer and shift registers.

As described in the introduction to shift-and-add multipliers, the multiplier and the product are stored in shift registers so that they can be shifted after each iteration. The multiplexer is used to select either the A operand or the value "0" in each iteration, depending on the value of the multiplier. The full execution of the multiplier is reviewed as the IMPLY design is introduced.

First, a shift register is implemented using the IMPLY operation. Then the design is completed, and optimizations for the IMPLY context are considered to drastically reduce the design overhead.

7.1.1.1 Shift-register implementation

First, an N-bit shift-register implementation for the multiplier is presented. This register will hold the values of multiplier B and the running total of the product register. Figure 7.1 shows the schematic for a 4-bit shift register.

A right-shift operation essentially moves each bit i into position $i-1$. A "move" operation can be achieved using IMPLY operations by performing a COPY operation from each bit i to bit $i-1$.

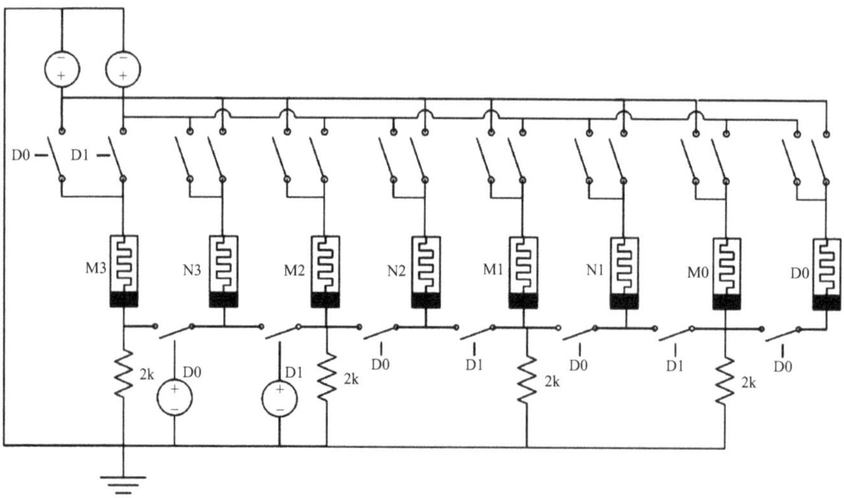

Figure 7.1 4-bit IMPLY shift-register schematic

A COPY operation is achieved using IMPLY operations by using two cleared memristors and performing two consecutive NOT operations. For the first NOT, V_{cond} is applied to the memristor of interest, and V_{set} is applied to a cleared memristor N_a. The result ends up in N_a. For the second NOT, V_{cond} is applied to N_a, and V_{set} is applied to a second cleared memristor M_b. Now a copied value of the input lies in M_b.

Note, a left-shift operation is nearly identical except the copy operation that applies the V_{set} voltage to the memristor for bit $i + 1$ rather than $i - 1$ in the second step of the copy. Thus, in the example, memristor M_b should be selected as the memristor for bit $i + 1$. Thus, the principles of this shift-register design can be used to design either a left- or right-shift register.

To perform a shift on N bits, the COPY process is performed for each bit i of the register in parallel. Let the shift register be called M, and each bit of the shift register be denoted as M_i. Based on the description of a COPY operation, another memristor is required per bit to serve as the cleared memristor, which will hold the result of the first NOT operation (N_a in the earlier example). Call each of these memristors N_i. In the first step of the shift, V_{cond} is driven on all of the M_i memristors, and V_{set} is driven on all of the N_i memristors. Now, the N_i memristors hold the results of the N-bit NOT operation on the M_i memristors.

In the second step, a second NOT operation is performed that applies V_{cond} to the N_i memristors and V_{set} to N cleared memristors. For efficiency in memristor reuse, the cleared memristors will be the same M memristors. Each N_i will set memristor M_{i-1} for bit $i - 1$. Thus, the destination of the COPY operation is the "next" bit, and no more memristors are required.

One challenge with this design is that all of the M memristors need to be reset before the second NOT operation in the shift occurs. This reset step cannot happen until the value of the memristor has been read in the first NOT operation. The reset step can be added between the first and second steps of the COPY operations as shown in Table 7.1.

Initially, all of the N memristors are reset and the M memristors hold the value 4'b1100. This value is shown in bold in Table 7.1, so its value can be tracked through execution.

Table 7.1 IMPLY shift-register steps

T	M_3	N_3	M_2	N_2	M_1	N_1	M_0	N_0
0	**1**	0	**1**	0	**0**	0	**0**	0
1	1	0	1	0	0	1	0	1
2	0	0	0	0	0	1	0	1
3	0	0	1	0	1	1	0	1
4	**0**	0	**1**	0	**1**	0	**0**	0
5	0	1	1	0	1	0	0	1
6	0	1	0	0	0	0	0	1
7	0	1	0	0	1	0	1	1
8	**0**	0	**0**	0	**1**	0	**1**	0

In step 1, a NOT operation is performed on each M and N memristor pair in parallel, with the results in the N memristors. In step 2, all of the M memristors are reset so they can be destinations in the next NOT operation. In step 3, the second NOT operation is performed, and V_{cond} is applied to all of the N memristors, while V_{set} is applied to all of the M memristors to the right. Since M_3 is the most significant bit, a 0 is shifted into its position. After step 3, the shifted value 4'b0110 is in memristors M_3 to M_0. In step 4, the N memristors are reset. Now, another shift operation can execute in the same manner. After two shift operations and eight steps, the value 4'b0011 sits in memristors M_3 to M_0 as expected.

Since all of the bits are performing the same operations at the same time, only two drivers D_0 and D_1 are necessary for the entire register, one for all of the M memristors and one for the N memristors. When the first NOT operation is performed, V_{set} drives the N memristors and V_{cond} drives the M memristors. When the second NOT operation is performed, V_{set} drives the M memristors and V_{cond} drives the N memristors. Switches are placed between each memristor to logically separate each bit, enabling parallel operations.

Note that the shift register produces the inverse of the register as a "free" by-product of the shift. This value lies in the N memristors after the first step. Thus, for applications, which require the inverse of the operand, this shift register is especially beneficial. The inverse of the operand is resolved after the first IMPLY step of every shift operation.

In the context of an N-bit shift-and-add multiplier, the product requires a shift register of size $2N$, and the multiplier requires a shift register of size N. By using the presented design, each bit in the shift register requires two memristors and two switches. Thus, the product register requires $4N$ memristors and $4N$ switches. The multiplier shift register requires $2N$ memristors and $2N$ switches. Each register needs two drivers regardless of N, one for operating the M memristors and one for operating the N memristors. In all, the shift registers in the shift-and-add multiplier require $6N$ memristors, $6N$ switches, and four drivers.

7.1.1.2 Baseline multiplier implementation

In addition to the shift registers, the shift-and-add multiplier requires an N-bit multiplicand register, a ripple-carry adder, and a multiplexer. The multiplicand register does not need to be a shift register since its value is constant across execution.

The design uses the presented optimized N-bit ripple-carry adder that requires $7N+1$ memristors, $7N$ drivers, and $8N-1$ switches. It also uses the 2-to-1 N-bit multiplexer described for the carry-select adder that requires $3N$ memristors, three drivers, and $3N$ switches. An N-bit register requires one memristor and switch per bit, for a total of N each, plus two drivers for loading and reading the N-bit value. Coupling these complexities with the product and multiplier shift registers, the shift-and-add multiplier has the following complexity:

Product shift register:

$= 4N$ memristors $+ 4N$ switches $+ 2$ drivers

Multiplier shift register:

$$= 2N \text{ memristors} + 2N \text{ switches} + 2 \text{ drivers}$$

Multiplicand register:

$$= N \text{ memristors} + N \text{ switches} + 2 \text{ drivers}$$

Ripple-carry adder:

$$= 7N + 1 \text{ memristors} + 8N - 1 \text{ switches} + 7N \text{ drivers}$$

8-bit multiplexer:

$$= 3N \text{ memristors} + 3N \text{ switches} + 3 \text{ drivers}$$

Total complexity:

$$= 17N + 1 \text{ memristors} + 18N - 1 \text{ switches} + 7N + 9 \text{ drivers}$$

The entire design only requires $17N+1$ memristors. This is an example of how minimal the complexity of shift-and-add multipliers is. The ripple-carry adder itself accounts for almost 50 percent of the design's area. In fact, this multiplier requires fewer components than both the IMPLY carry-select adder and the carry-lookahead adder.

One iteration of the multiplication performs a shift, a multiplexer selection, an N-bit addition, and a store operation back into the shift register. As described, the shift operation requires four steps—the first to reset the N_i memristors, the second to produce the inverse results (M_i IMP N_i), the third to clear the initial memristors (M_i), and the fourth to write the final values (N_i IMP M_{i-1}). The N-bit ripple-carry adder requires $2N+19$ steps, and the multiplexer requires six steps. Thus, the initial design of the shift-and-add multiplier requires $2N+29$ steps per iteration, or $2N^2+29N$ steps for the full N iterations of the multiplication.

7.1.1.3 General optimizations

The shift-and-add multiplier can be significantly optimized to reduce its delay and complexity. The shift-and-add multiplier is already very low in complexity, but small simplifications and reduction in idle components can produce an even more lightweight design.

First, the multiplier shift register is incorporated into the product shift register to eliminate the N-bit multiplier register entirely. This is a standard optimization that can be performed for nearly any implementation of a shift-and-add register.

The product shift register is sized to $2N$ to fit the maximum possible product of the two N-bit inputs. The multiplier shift register is of size N. However, the width of each of these values changes across the iterations of the multiplication. At initialization, the product register has a width of 0, and the multiplier register has a width of N. After the first N-bit addition, the product register holds an N-bit value. Then, each of the values is shifted right. This removes the least significant bit of the multiplier and adds a most significant bit to the product register. Now, the multiplier has a width of $N-1$, and the product has a width of $N+1$. This process repeats for each iteration of the multiplication. Thus, during every iteration, the sum of the widths of the two registers is exactly $2N$.

This realization allows the product and multiplier to share the $2N$-bit shift register. Initially, the multiplier occupies bits $[N-1{:}0]$ in the shift register and the product (initialized to 0) occupies bits $[2N-1{:}N]$. Each iteration, the least significant bit of the shift register, also the least significant bit of the multiplier, is used and shifted out of the register. The product bits are also shifted right. After N iterations, all N bits of the multiplier have shifted out of the register and the full $2N$ bits hold the final product. This maximally utilizes the hardware in the $2N$-bit shift register and eliminates the idle redundancy in the shift registers.

Second, steps in the various logic blocks can perform in parallel because the blocks are logically separated. There is no need to require only a single logic block to execute an IMPLY operation in a single step. For example, consider the shift register and the ripple-carry adder. In the baseline design, the shift registers sit idly while the ripple-carry addition executes. Instead, the step that resets the N memristors of the shift register to prepare for a shift operation can be performed during the ripple-carry addition. This hides the delay of this step from the critical path. It also reduces the effective latency of the shift from four steps to three.

Last, recall that in each iteration, the least significant bit of the multiplier determines whether the multiplicand or the value "0" is added to the running product. Since one of the inputs to the multiplexer is known to be 0, its corresponding memristors in the multiplexer can be removed. Instead, the signal will just be tied off. Given this knowledge, the multiplexer is reduced to a single AND operation. Since input B is 0, the second term in the multiplexer equation $\text{Out} = (\text{A AND S})\ \text{OR}\ \left(\text{B AND}\ \overline{\overline{\text{S}}}\right)$ is 0 too. The equation simplifies to $\text{Out} = \text{A AND S}$, which is a simple AND operation. The IMPLY steps can be seen in Table 7.2, with the input B to the multiplexer assumed to be 0.

The new multiplexer only requires two memristors and three steps. Additionally, the design logically separates each bit of the N-bit multiplexer so that the bits can perform in parallel.

The steps for the multiplexer can be incorporated into the first steps of the ripple-carry adder. By logically combining the steps of all these modules, Boolean simplifications can be found that cross module boundaries. For example, the first few steps of a standard full adder or ripple-carry adder perform basic

Table 7.2 Simplified IMPLY shift-and-add multiplexer

Step	**Original**	**Optimized**
1	S IMP 0 → M0	S IMP 0 → M0
2	B IMP S → M1	A IMP M0
3	M1 IMP 0 → Out	M0 IMP Out
4	A IMP M0 → M0	
5	M0 IMP Out → Out	

transformations on the inputs. The first four steps of the IMPLY ripple-carry adder are shown in Table 7.3.

The first two steps are used to find the inverses of the inputs, and the third step is used to copy the B input. The actual execution of the adder does not begin until step 4. However, let input A for the ripple-carry adder be the output of the *N*-bit multiplexer. Thus, the values in the "out" memristors in the multiplexer are the A memristors in the adder. Step 1 finds the inverse of the multiplexer output. This was already found in step 2 of the multiplexer. Thus, step 1 is redundant. The combined multiplexer and adder steps are shown in Table 7.4 for the first bit of the multiplication.

Table 7.3 First four steps of an IMPLY ripple-carry adder

Step	Functionality	Goal
1	A IMP 0	A→M2
2	B IMP 0	B→M3
3	B	M3→M1
4	(A IMP 0) IMP (B IMP 0)	M2→M3

Table 7.4 Steps for one iteration of an IMPLY shift-and-add multiplier for bit 0

	Execution	Description
1	NOT S	S → M0
2	A NAND S	A → M0
3	A (mux result)	M0 → M1
4	B IMP 0	Product Reg → M2
5	B	M2 → M3
6	(A IMP 0) IMP (B IMP 0)	M0 → M2
7	A IMP B	M1 → M3
8	((A IMP 0) IMP (B IMP 0)) IMP 0	M2 → M4
9	B IMP (A IMP 0)	Product Reg → M0
10	(B IMP (A IMP 0)) IMP 0	M0 → M5
11	A XOR B	M3 → M4
12		False (M0,M2,M3)
13	(A XOR B) IMP 0	M4 → M0
14	C_{in} IMP ((A XOR B) IMP 0)	C_{in} → M0
15	C_{in} AND (A XOR B) + AB → C_{out}	M0 → M5
16	Next bit is reading C_{in}	
17	C_{in} IMP 0	C_{in} → M3
18	C_{in} + (A XOR B)	M3 → M4
19	Next bit is reading C_{in}	
20	(C_{in} + (A XOR B)) IMP 0	M4 → M2
21	(C_{in} NAND (A XOR B)) IMP ((C_{in} + (A XOR B)) IMP 0)	M0 → M2
22	(C_{in} NAND (A XOR B)) IMP ((C_{in} + (A XOR B)) IMP 0) IMP 0 = SUM	M2 → Zeroed out Product Reg
23		False all except input A

The first three steps are the multiplexer. The remaining steps are identical to the original ripple-carry adder but with the first step removed. Instead, the value in M0 is used. The number of steps for this module is 23 steps instead of $21 + 3 = 24$ steps for the separate modules.

There are nine memristors in this adder module: three input memristors S, A, and B and six internal memristors M0 to M5. Before combining the multiplexer and adder blocks, there were seven memristors per adder and three memristors per bit in the multiplexer for a total of ten memristors per bit. Thus, one memristor has been removed from each bit of the multiplexer for a total reduction of N fewer memristors.

Steps for the shift register can also be incorporated into the adder. At the end of the addition, the result is stored into the product register as denoted in step 22. After this step, the memristors in the adder and the memristors holding the inverse operand in the shift registers (memristors N_i) are reset in step 23 in order to prepare for the next iteration of the multiplication. This is necessary for proper IMPLY functionality.

Instead, each bit can reset its intermediate memristors as soon as its computation is complete. For example, after bit b propagates its carry-out value and resolves its sum, it can reset its intermediate memristors while bit b+1 performs its remaining computation. Bit 0 can reset its memristors in step 23, bit 1 can reset in step 25, and the rest of the bits are executing the adder IMPLY steps. Similarly, once the sum value is sensed into the shift register in step 22, all of the sum memristors can be cleared while the shift steps occur. Thus, none of the reset steps contribute to the critical path.

7.1.1.4 Optimizations for the memristor context

So far, optimizations have been made to the circuit implementation at a low level, but the high-level design itself has not been changed. However, the shift-and-add multiplier can be significantly optimized in the context of IMPLY operations. The multiplier can use the key properties of the IMPLY operation to reduce the delay and complexity drastically.

At a high level, the IMPLY operations differ from traditional CMOS operations in the way they hold their results. In traditional CMOS, the result is represented as a voltage on a wire and is not stored. The computations performed on the result are dictated by the gates and logic blocks that this voltage traverses through. The IMPLY operation is nearly the opposite. The IMPLY operation represents the results in a memristor storage cell. The computations are performed using voltage signals driven on the memristors. In broad terms, traditional CMOS brings data to the computation, and IMPLY operations bring computation to the data. This property is heavily leveraged in the design of a shift-and-add multiplier.

For example, the multiplicand register can be eliminated. In the traditional CMOS implementation, the register is necessary to hold the value of the multiplicand throughout execution. The CMOS adder does not have the ability to store the multiplicand so it must be stored elsewhere. In an IMPLY memristor-based design, the adder does have the capability to store the multiplicand itself. In fact, it stores all its values, inputs, outputs, and intermediate values. Inputs A and B are already included as memristors in the design of the standard IMPLY ripple-carry adder. Since the multiplicand is one of the operands in every addition, it can permanently lie in the A operands of the adder.

When the multiplicand register is removed, N memristors, drivers, and switches are eliminated from the design. This also saves two steps for performing a copy from the multiplicand register to the adder/multiplexer during each iteration of the multiplication. This is an example of how the unique capabilities of memristors can transform traditional CMOS logic designs.

Similarly, the shift registers can be optimized. In traditional CMOS, the shift register is used to physically move the bit values to their appropriate positions before performing the next iteration. In this way, the bits are "lined up" to the correct combinational circuitry. This is necessary because the circuitry is static and performs "blindly." The circuitry cannot change and acts identically across iterations of the multiplication. The circuitry necessarily always reads the exact same register bits for every input and stores into the exact same register bits for every output. Only values in the wires and registers that the circuitry operates on can change. Thus, the control logic (i.e., shifts) must ensure the intended values sit in these bits before each iteration begins.

The IMPLY designs do not have this limitation. Recall that the computations in IMPLY are performed with the drivers. The drivers are not static and can apply V_{cond} and V_{set} signals to different memristors across iterations. Thus, a shift operation can be achieved by virtually shifting the driver circuitry across the bits. As a simple example, consider an 8-bit IMPLY multiplication.

Let the 2N-bit shift register P_i begin such that $P_i[15:8] = 0$ (for the initial product) and $P_i[7:0]$ = multiplier B[7:0]. The first addition would take $P_i[15:8]$ and add either the multiplicand or 0 based on $P_i[0]$, store the result back into $P_i[15:8]$.

If the design performs a shift, then it shifts the register once to the right. It performs this with two N-bit NOT operations performed in parallel across the bits in the register. This is done by applying the V_{cond} and V_{set} drivers as described earlier. After the shift, the least significant bit of the multiplier is shifted out, and the multiplier is in bits $P_i[6:0]$. The product now lies in bits $P_i[14:7]$. When the second iteration begins, the multiplier's least significant bit is read into $P_i[0]$ and read in bits $P_i[15:8]$ as input to the adder. Again, this is done by applying V_{cond} drivers to the memristors for bits 8 to 15 in the register.

Instead, consider if this physical shift is not performed. A couple of modifications will be made to the format of the shift register. Rather than storing $P_i[7:0]$ = multiplier B[7:0], $P_i[15:8]$ = multiplier B[7:0] is stored. The bits are stored in the most significant N bits. Now, bit 0 of the multiplier will be sensed from $P_i[8]$ in the first iteration. Second, as soon as the least significant bit of the multiplier is read in the first step of an addition, its memristor is reset. This is safe because this bit would otherwise be shifted out in the original design, it is no longer needed. Third, rather than storing the sum into $P_i[15:8]$ in the final step of the addition, the drivers will store the result into $P_i[7:0]$. At the beginning of the next addition iteration, the product input will be read from $P_i[8:1]$. This effectively achieves the shift. To make this change in the design, the driver logic is changed to apply V_{set} to the M memristors in the shift register for bits 8 to 1 instead of bits 15–8.

Note, when this change is made, the memristor that held the least significant bit of the multiplier B at the beginning of the iteration holds the most significant bit of the product for the next iteration. This value is always 0 since it is normally a shifted-in value in the register implementation. The rest of the multiplier, B[7:1], is preserved in bits P_i[15:9]. This achieves the same outcome as the original shift operation.

Thus, there is no need for the shift register at all. Instead, a simple $2N$-bit register is necessary for holding the intermediate multiplier and product values at each iteration. This removes three steps from each iteration (for the shift operation) and $2N$ memristors (for the inverse shift values).

To accommodate this optimization, some changes need to be made to the drivers. Specifically, in step 1 of each iteration, the drivers will read a different bit as the select line. In iteration i, the adder will apply the V_{cond} driver to memristor M_{8+i} to read the least significant bit of the multiplier for the given iteration. Second, the drivers must change to read the running product from a different set of N memristors in each iteration. Specifically, in step 4, the adder will apply the V_{cond} driver to memristors M_{7+i} to M_i. Last, the drivers must change to store the running product into the same set of N memristors that it read from. Specifically, in step 22, the adder will apply the V_{set} driver to memristors M_{7+i} to M_i.

These three modifications achieve the shift functionality in a simple $2N$-bit register without any additional overhead for memristors or steps. Now, there are no steps required for the shift, and the total delay of an addition is 23 steps. A visual representation of the updated product register functionality is shown in Figure 7.2.

By the same reasoning, the product register can be removed entirely. Consider two consecutive iterations of an 8-bit multiplication, i and $i+1$. Let the sum in the first iteration be S_i[7:0]. S_i[7:0] is computed and stored into the 16-bit product register P_i in bits $P_i[Y{:}X]$. Then, iteration $i+1$ begins and reads bits $P_i[Y+1{:}X+1]$ (essentially {0, S_i[7:1]}) as an input. Note that all these steps are unnecessary—the result of one iteration of the addition is essentially inserted directly back into the adder as an operand in the subsequent iteration.

Due to the driver-based nature of IMPLY logic, this data movement can be achieved directly in the final stage of iteration i. Rather than copying S_i[7:0] from the adder into P_i, the sum results can be directly computed into their operand location for iteration $i+1$. In other words, for a bit b in the ripple-carry adder, when S_b is being computed, rather than applying V_{set} to the register memristor for bit b, V_{set} is applied to the B operand in the adder for bit b – 1. Now, at the end of iteration i, {0,S_i[7:1]} lies in the 8-bit B operand across the full adders. Iteration $i+1$ can begin as soon as the intermediate memristors are reinitialized to 0.

Only S_i[0] is not moved directly into the input operand memristors. This bit needs to be logically shifted out of the computation. This value will be stored in the memristor that held the least significant bit of the multiplier in that iteration since that value is no longer needed. Now the design only needs a single N-bit register. This register will hold both the multiplier and the product. At the beginning of an

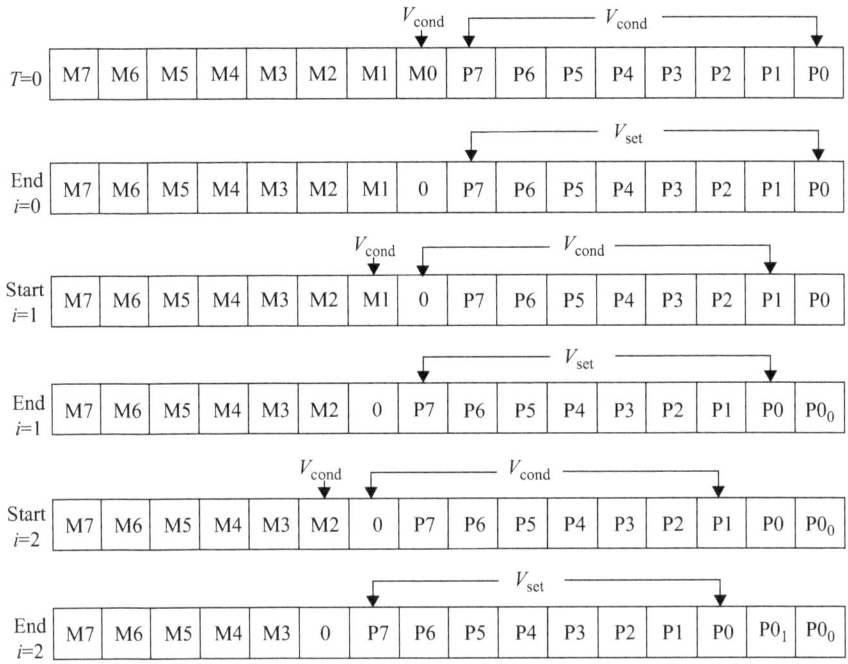

Figure 7.2 Visualization of driver optimizations to achieve shift functionality

Table 7.5 Optimized steps for the IMPLY shift-and-add multiplier

Step	Functionality	Goal
22	(C NAND (A XOR B)) IMP ((C + (A XOR B)) IMP 0) IMP 0 = SUM	M2 → M2($i-1$)
23		False all except A

iteration i, i memristors will hold the computed bits of the product and $N - i$ will hold the multiplier bits.

This optimization is only possible due to the benefits of the IMPLY operation. The delay is maintained at $2N^2 + 21N$ but the modified steps of execution are shown in Table 7.5.

The B memristor has been removed from the adder, and the value of B only lies in the memristor M3. In step 22, the sum is stored directly into the memristor M2 in bit $i - 1$.

With all these optimizations, the only required circuitry is the modified N-bit ripple-carry adder and the N-bit multiplier register. This is a total of $8N$ memristors, $8N$ drivers, and $9N - 1$ switches. The fully optimized schematic for an 8-bit multiplier is shown in Figure 7.3.

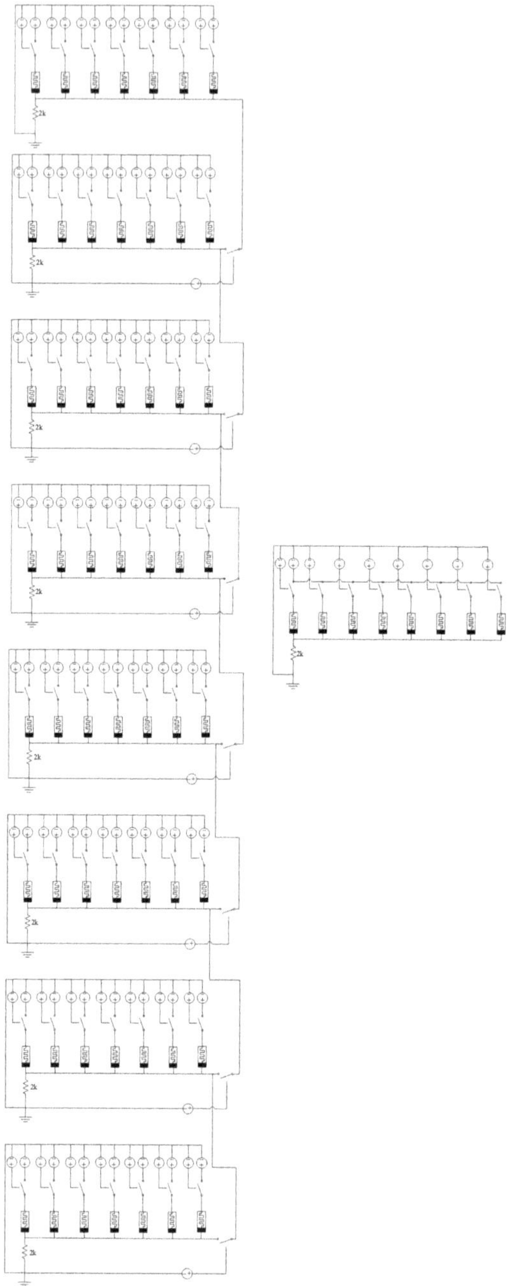

Figure 7.3 IMPLY shift-and-add multiplier schematic

The top two columns of the schematic are the ripple-carry adder with the multiplexer logic incorporated. The bottom row holds the multiplier register. Each memristor in the register is logically disjoint so that it can be sensed in its particular iteration.

As a final optimization, the shift-and-add multiplier overlaps individual iterations of the multiplication. Again, due to the fact that IMPLY operations essentially bring the computation to the data rather than vice versa, the bits of the adder can be individually driven. Since the multiplier is essentially a modified ripple-carry adder, the design can be pipelined in a similar manner as the IMPLY ripple-carry adder.

As soon as bit b completes its first addition iteration, it propagates its carryout, stores the sum result in bit b – 1 in the B operand memristor as discussed, and resets its memristors. At this point, the full adder is able to begin a second iteration of the multiplication by sensing the next bit of the multiplier register. It can do this independent of the other bits in the adder, which are still performing the first iteration.

With this functionality, it is possible to begin a new iteration of the multiplication every 24 cycles. A single full adder stores its sum in step 22 and resets its memristors in step 23. Then, since each full adder computes its sum two steps after the previous full adder, the sum from the next full adder can be shifted into the B input in step 24. In step 25, the full adder will perform S IMP 0, and then the remainder of the adder steps will execute. The rearranged initial steps are shown in Table 7.6.

The B memristor has been removed from the adder, and the value of B only lies in the memristor M3 now. Its inverse still lies in M2 and is now computed in step 5. Since all these memristor mappings have remained unchanged, the remainder of the steps can as well. In step 22, the sum is stored directly into memristor M2 in bit $i-1$. This can be seen in step 1 for bit $i-1$ and step 22 for bit i. Table 7.7 shows the parallel execution of three bits of this ripple-carry adder during the steps in which the shift occurs.

Table 7.6 Optimized steps for the IMPLY shift-and-add multiplier

Step	Functionality	Goal
1	B	(From bit $i+1$) → M3
2	NOT S	S → M0
3	A NAND S	A → M0
4	A (mux result)	M0 → M1
5	B IMP 0	M3 → M2
...	...	...
22	(C NAND (A XOR B)) IMP ((C + (A XOR B)) IMP 0) IMP 0 = SUM	M2 → M2($i-1$)
23		False all except A

Table 7.7 Sum and shift operation in the modified IMPLY shift-and-add multiplier

	Bit 2	**Bit 1**	**Bit 0**
22	M3 → M4	M4 → M2	M2 → Multiplier reg.
23		M0 → M2	False all except A
24	M4 → M2	M2 → $M2_0$	(From bit 1) → M2
25	M0 → M2	False all except A	S → M0
26	M2 → $M2_1$	(From bit 1) → M2	A → M0
27	False all except A	S → M0	M0 → M1
28	(From bit 1) → M2	A → M0	M2 → M3

The bold font indicates the shift operations. In these steps, the sum is being read from bit i and stored into bit $i-1$. The steps must be strategically ordered so that they align across bits.

To perform pipelining correctly, the multiplier register must be modified. In the nonpipelined implementation, bit b_i in the multiplier register is read by each consecutive bit in the ripple-carry adder during iteration i, so V_{cond} is only applied to one memristor in the register at a time. However, if the design is changed to pipeline additions, different bits in the ripple-carry adder will be on different iterations of the adder and will potentially need access to different bits in the multiplier register during the same iteration. Thus, the drivers on the multiplier register must be strategically timed so that no two bits of the multiplier register are needed during the same step. In this way, the circuitry can remain identical, and only the driver logic needs to change to facilitate pipelining.

For a full utilized pipeline, a new addition can begin every 24 steps and a new multiplication can begin every $24N$ steps. This reduces the delay complexity from $O(N^2)$ to $O(N)$ for the entire multiplication.

7.1.2 Hybrid-CMOS shift-and-add multiplier

A hybrid-CMOS design for a shift-and-add multiplier is presented in this section. The design takes a similar gate-level form as the traditional CMOS adder. The main components are a multiplexer, ripple-carry adder, register, and shift registers.

First, the design of a shift register using hybrid-CMOS gates is presented. The shift register needs to be capable of loading an input in parallel (for the product of the adder and the initialization of the multiplier) and shifting a value serially. A schematic for a 4-bit shift register is shown in Figure 7.4.

The shift register has three NAND gates and one D flip-flop per bit, similar to the traditional CMOS design. The write/shift line determines if a value is loaded during the cycle or the current value is shifted. The output of each flip-flop is one bit of the result. The clock feeds into each flip-flop to synchronize the flip-flops. Data are shifted to the right from Q1 to Q4. Each flip-flop has a pair of latches to synchronously hold the data as shown in Figure 7.5.

At a gate-level, the CMOS and hybrid-CMOS designs are identical. The only difference is that now the AND and OR gates are constructed from memristors.

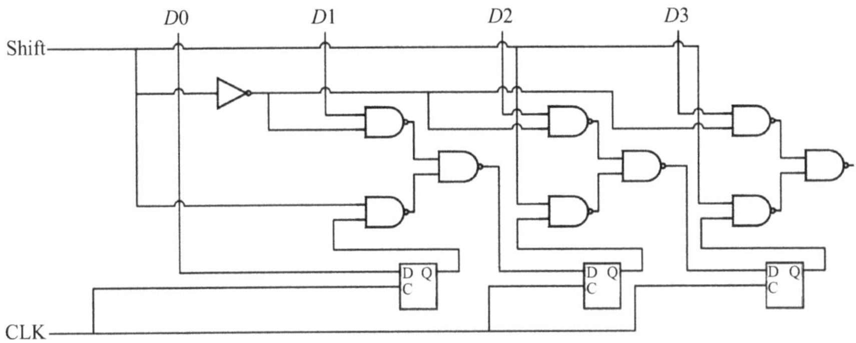

Figure 7.4 4-bit shift-register schematic

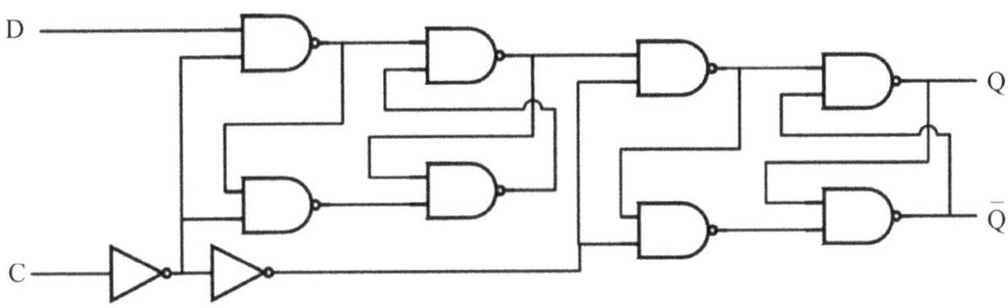

Figure 7.5 D flip-flop for a shift register

This reduces the complexity of each memristor gate to two memristors. Unfortunately, latches and flip-flops make heavy use of NOT gates, which require MOSFETs in hybrid-CMOS designs. Thus, the total complexity for the D flip-flop in hybrid-CMOS is 16 memristors and 20 MOSFETs. This is a significant amount of MOSFETs for a memristor-based design. The surrounding logic for the shift register requires six memristors and six MOSFETs for all of the bits except the first and a single NOT gate with two MOSFETs. Thus, the total complexity for the shift register is $(22N - 6)$ memristors + $(26N - 4)$ MOSFETs.

In general, flip-flops have feedback loops in their circuitry to produce their outputs. The outputs of both latches are fed back into the inputs along with the new inputs. Depending on the clock, the old value will remain the output, or the new value will become the output. This feedback loop is essential to maintain the latch functionality to store the data.

Unfortunately, in the context of hybrid-CMOS gates, this feedback loop causes challenges for signal integrity. Since the output of the single is continuously fed in as an input, the strength of the signal continues to weaken across time. Second, for an N-bit shift register, each flip-flop serves as an input to the next bit in the register. This creates an N-bit chain of flip-flops. As N increases, the strengths of the signals weaken as they propagate through the flip-flops.

Normally, signal restoration buffers would be used to prevent these two sources of signal degradation. However, the shift register does not need any buffering. The presented D flip-flop uses NAND gates for the entire design, essentially buffering every output with the use of a NOT gate. Thus, the number of MOSFETs, which appeared excessive initially, is now more appropriate since the MOSFETs serve a functional and computational purpose. The total cost of an N-bit shift register is still $(22N - 6)$ memristors + $(26N - 4)$ MOSFETs.

A standard register is similar to a shift register except that each flip-flop is not connected to the other flip-flops in the register. Instead, the inputs and outputs are connected to other modules. A register has the same complexity as a set of D flip-flops, $16N$ memristors, and $20N$ MOSFETs.

The registers are used in conjunction with the previous chapter's optimized hybrid-CMOS ripple-carry adders and multiplexers to produce the complete shift-and-add multiplier. Apart from signal restoration, the gate-level circuitry is identical to the traditional CMOS design. The ripple-carry adder requires $14N$ memristors and $12N$ MOSFETs. The multiplexer requires $6N$ memristors two MOSFETs. Since the multiplexer output feeds the ripple-carry adder, a buffer is needed on each output, or $4N$ MOSFETs. The total complexity is defined as follows:

$2N$-bit shift register:

$$= (44N - 6) \text{ memristors} + (52N - 4) \text{ MOSFETs}$$

N-bit register:

$$= 16N \text{ memristors} + 20N \text{ MOSFETs}$$

Ripple-carry adder:

$$= 14N \text{ memristors} + 12N \text{ MOSFETs}$$

Multiplexer:

$$= 6N \text{ memristors} + (4N + 2) \text{ MOSFETs}$$

Total complexity:

$$= (80N - 6) \text{ memristors} + (88N - 2) \text{ MOSFETs}$$

The majority of the complexity comes from the registers in the design. In fact, about 75 percent of the total number of components is from the registers. The hybrid-CMOS design cannot perform the extensive optimizations to remove the registers as was done for the IMPLY shift-and-add multiplier. The hybrid-CMOS approach is entirely combinational like the traditional CMOS design. Hybrid-CMOS gates constructed from memristors do not store the results of their computations the way result memristors do for IMPLY operations. The shift registers and registers are still required in the hybrid-CMOS context.

Recall that the only purpose of these registers is to store the multiplicand input and the adder output during each iteration of the multiplication. This is an example of how storage can be a large contributor to the complexity of modern arithmetic units and speaks to the motivation for using memristors as both computational and storage elements. Although hybrid-CMOS designs do not, many memristor-based methods, i.e., IMPLY and MAD operations, remove the need for these structures by leveraging memristor storage capabilities.

The delay of the shift-and-add multiplier is the number of iterations times the gate delay of each iteration. Each iteration consists of an addition, multiplexer, and shift operation. The addition takes $3N + 4$ steps and the multiplexer takes three steps plus an additional step to buffer its output. The shift operation takes two steps to load and four steps to shift for a total of six steps. Thus, each iteration takes $3N + 14$ steps. The entire multiplication takes $3N^2 + 14N$ steps.

7.1.3 Threshold-gate shift-and-add multiplier

Threshold gates are also a viable design approach for shift-and-add multipliers. Threshold gates can implement all logical structures, including registers, shift registers, and multiplexers and adders as seen previously. This section discusses the implementation of a register and shift register using threshold gates and connects these logic blocks with threshold-based multiplexer and ripple-carry adder to complete a shift-and-add multiplier. For this design, the threshold gates are implemented with GOTO pairs since they minimize the use of CMOS.

First, a $2N$-bit shift register is constructed from threshold gates. The simple shift register consists of a series of $2N$ flip-flops, each of which consists of two latches. The shift register for the shift-and-add multiplier also needs the ability to load a value across the bits in parallel. The diagram for the implementation of a flip-flop with both load and shift functionality is given in Figure 7.6.

D is the value being shifted in from the previous bit, NEW is a value being loaded, and LD is a signal that is driven high when NEW should be accepted and driven low when the value is being shifted. The top threshold gate performs the load logic. The next two threshold gates serve as one latch, while the bottom two threshold gates serve as another. The threshold gates are driven by the CLK and $\overline{\text{CLK}}$ signals according to the shift functionality.

The first threshold gate for the load logic, call it T0, is effectively an AND gate that executes CLK AND LD AND NEW. The output of the threshold gate is only "1" if all of the inputs are a "1" corresponding to the load signal being high, the input being a "1," and the clock being high. If any of these three signals is low, or "0," the output will be 0. This is correct functionality because it is the only circumstance for which the register should load a "1" into its flip-flop.

Now consider the two threshold gates for the first latch in the flip-flop. The first threshold gate, call is T1, is an AND gate that executes CLK AND $\overline{\text{LD}}$ AND D. The output of the first threshold gate is D if CLK is high, and the load signal is low. If the load signal is high, the shifted value of D from the previous bit is not accepted. In this way, if a shift operation is performed during the clock signal, the output of the first threshold gate will be D. The threshold gates

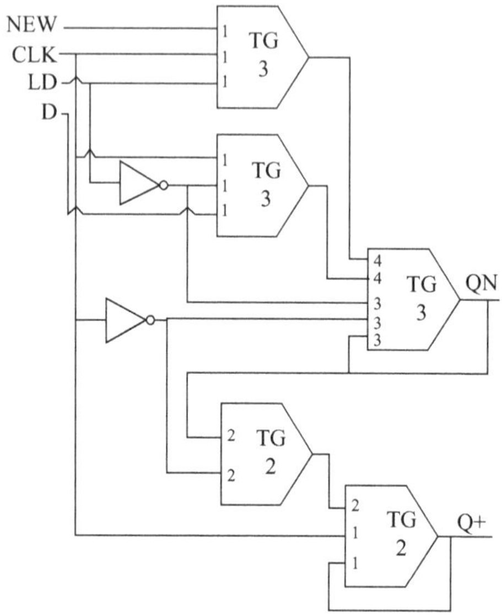

Figure 7.6 Diagram for a flip-flop with load functionality designed from threshold gates

T0 and T1 are mutually exclusive because of their LD terms so it is guaranteed that at most one of them has an output of "1" at any given time.

The third threshold gate T2 executes

$$\text{Out} = \text{T0}_{\text{out}} \text{ OR } \text{T1}_{\text{out}} \text{ OR } \left(\overline{\text{CLK}} \text{ AND } \overline{\text{LD}} \text{ AND QN}\right).$$

The threshold of T2 is 3 and the weights for the inputs from T0 and T1 are 4. If either of the outputs from T0 or T1 is a "1," the output of T2 will be a "1." The only other satisfactory input to produce a "1" is the third Boolean expression, $\left(\overline{\text{CLK}} \text{ AND } \overline{\text{LD}} \text{ AND QN}\right)$. This term is mutually exclusive with T0_{out} and T1_{out} due to the CLK and LD signals. Thus, only one of these three terms can be "1" in any given step.

Together, the three threshold gates produce the output QN as the following:

$$\text{Out} = (\text{CLK AND LD AND NEW}) \text{ OR}$$
$$\left(\text{CLK AND } \overline{\text{LD}} \text{ AND D}\right) \text{ OR } \left(\overline{\text{CLK}} \text{ AND } \overline{\text{LD}} \text{ AND QN}\right).$$

Essentially, if the clock is high, either load or shift a value. If load is a "1," the new value is taken, if load is a "0," the shifted value is taken. If the clock is low without a load operation, the current latched value is maintained. Thus, when the CLK is low, QN maintains its value, and when CLK is high, QN accepts a new value.

The lower most two threshold gates, T3 and T4, execute in the identical manner as the first latch except the load signal has been removed. It operates on the opposite clock edge to properly latch the value. The output equation becomes

$$\text{Out} = (\text{CLK AND Q}) \text{ OR } \left(\overline{\text{CLK}} \text{ AND QN}\right).$$

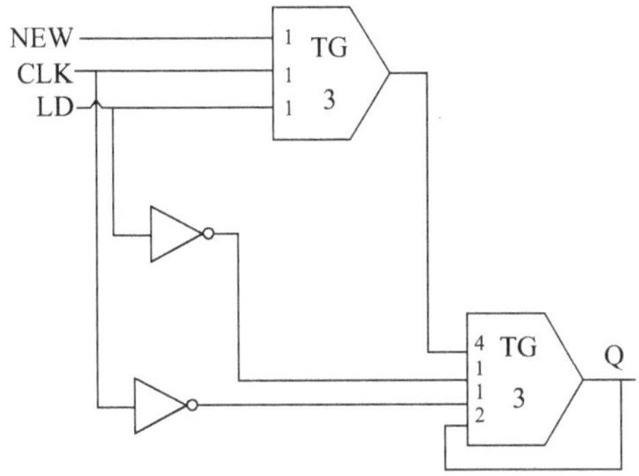

Figure 7.7 Simple threshold-gate latch

Essentially, Q+ becomes a "1" if the incoming value QN from the previous latch is a "1" and CLK is low or if the current value of Q is a "1" and the CLK is high. Thus, when the CLK is high, Q+ maintains its value, and when CLK is low, Q+ accepts the value of QN. This is the correct behavior for a latch. The delay of this flip-flop is six gate delays to perform a shift and two gate delays to perform a load.

When these flip-flops are incorporated into an N-bit shift register, the NOT gates for the LD and CLK signals in each flip-flop do not need to be repeated in each cell. Only one NOT gate is required per signal for a total of four MOSFETs for the shift register. The rest of the components in the shift register require no CMOS when a GOTO implementation is chosen. Each flip-flop uses 16 memristors and 5 GOTO pairs. A $2N$-bit shift register uses $32N$ memristors, $10N$ GOTO pairs, and 4 MOSFETs.

There is also an N-bit register to hold the multiplicand A. Again, since the threshold gates implement combinational logic and are not storage elements, this register cannot be eliminated as it was in the IMPLY design. Each flip-flop in this register has the ability to load and store a value using the circuitry shown in Figure 7.7.

This latch executes the equation

$$\text{Out} = (\text{CLK AND LD AND NEW}) \text{ OR } \left(\left(\overline{\text{CLK}} \text{ OR } \overline{\text{LD}}\right) \text{ AND Q}\right).$$

In other words, the output takes the new value if CLK and the load signal are high; otherwise, it maintains its current value. This correctly functions as a latch: only a load operation will change the value. The latch requires seven memristors and two GOTO pairs per bit. For the shift-and-adder multiplicand register, this is a total of $7N$ memristors and $2N$ GOTO pairs. The inverters for CLK and LD are not necessary since they already exist for the shift register.

The shift register is used in conjunction with a threshold-based ripple-carry adder and multiplexer to complete the shift-and-add multiplier design. The previous chapter presented a ripple-carry adder, which requires $N + 2$ gate delays and $2N$ gates. It also presented a multiplexer that requires two threshold gates per bit and a CMOS NOT gate for the select line. This work proposes a design, which implements both the ripple-carry adder and the multiplexer into a single logic unit with a total of $N+2$ delays and $2N$ gates. The modified ripple-carry adder uses a full adder as shown in Figure 7.8.

This adder implements both the multiplexing of the multiplicand A as well as the addition of this operand with the running product PROD and carry-in C_{in}. The first gate in a standard threshold gate full adder has three inputs, A, B, and C_{in}, and performs a majority operation. In the modified full adder, input A must first execute through the multiplexer.

Recall that the simplified multiplexer logic is essentially an AND operation, A AND Sel. Thus, the original design for a full adder is simply enhanced to weight the A and select line inputs each with half the weight of the A input in the original design. Each of these inputs now has half the weight of the other two inputs, PROD and C_{in}. In this way, the inputs A and Sel will only achieve the equivalent weight of the other inputs if both are a "1," essentially performing an "AND" within the threshold gate itself. For example, the threshold gate for C_{out} is still a majority gate for the modified inputs (A AND S), PROD, and C_{in}.

There are three scenarios that will meet the threshold of the first gate and produce a "1" output for C_{out}. First, if the product bit is a "1" and the carry-in signal is a "1," the output will be "1." Second, if the product bit is "1" and the output of the multiplexer is "1," the output will be a "1." Finally, if the carry-in signal is a "1" and the output of the multiplexer is "1," the output will be a "1."

These scenarios can be translated one-to-one to the threshold-gate design. First, if the product bit is a "1" and the carry-in signal is a "1," the output will be "1." Second, if the product bit is "1," A is "1," and the select line is "1," the output will be a "1." Finally, if the carry-in signal is a "1," A is "1," and the select line is "1," the output will be a "1."

The same logic is used to determine the weights on the sum threshold gate. The sum output will only be a "1" if the output of the multiplexer, "B" and C_{in} are all "1" or if exactly one of them is a "1."

The adder requires the same number of gates and has the same gate delay as the original full adder. The delay for an N-bit addition is also maintained at $N+2$ gate delays. However, the number of inputs per full adder increased such that the

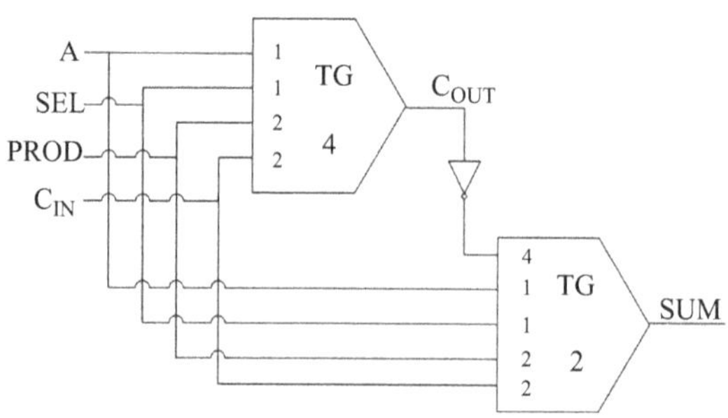

Figure 7.8 Threshold-gate 1-bit adder design optimized for use in a shift-and-add multiplier

cost of each full adder is now nine memristors, two MOSFETs, and two GOTO pairs. The total cost for the N-bit adder in the shift-and-add multiplier is $9N$ memristors + $2N$ MOSFETs + $2N$ GOTO pairs.

The total complexity of the threshold-based design is as follows:

Ripple-carry adder + multiplexer:

= $9N$ memristors + $2N$ MOSFETs + $2N$ GOTO pairs

$2N$-bit shift register:

= $32N$ memristors + 4 MOSFETs + $10N$ GOTO pairs

N-bit register:

= $7N$ memristors + $2N$ GOTO pairs

Total complexity:

= $48N$ memristors + $2N$ + 4 MOSFETs + $14N$ GOTO pairs

The delay of each iteration is $N + 2$ gate delays through the adder/multiplexer, two gate delays to load the sum into the product shift register, and six gate delays to perform the shift operation. For an N-bit multiplication, the total delay is $N^2 + 10N$ gate delays.

7.1.4 MAD gate shift-and-add multiplier

A MAD gate design for a shift-and-add memristor-based multiplier is presented in this section. The design is optimized at a unit-level and at the macro level. First, the design minimizes the delay internal to each full adder in the ripple-carry adder, then it focuses on optimizations across the logic block.

The MAD full adder presented for a ripple-carry adder is used as a baseline block for the adder in the shift-and-add multiplier. This adder requires 4 memristors, 13 switches, and 4 drivers and only one step to compute the sum and carryout. The adder can be simplified by leveraging information about the shift-and-add multiplier context. The resultant full adder is shown in Figure 7.9.

The P memristor represents the bit operand of the running product register while A and B represent the multiplicand and multiplier. The output circuitry remains unchanged.

A memristor, labeled "0," has been added in parallel with the A input memristor. In a shift-and-add multiplier, both the "0" and "A" inputs to the multiplexer are known and constant throughout the iterations. Thus, rather than having a single-input A memristor like the baseline full adder, this adder has two memristors to hold the value of the multiplicand (A) and the value 0. These represent the two inputs to the multiplexer in the multiplier.

The value of A can be loaded once at the beginning of the multiplication and held resident in the full adders for the entire multiplication. This is possible because memristors in the MAD context perform logic and storage operations.

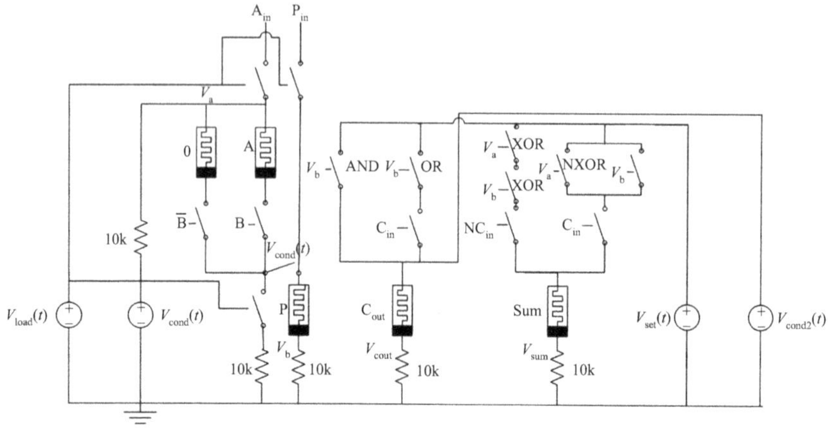

Figure 7.9 MAD full adder for a shift-and-add multiplier

This optimization removes the need for the multiplicand register entirely since the multiplicand now lies in the full adders permanently. This also eliminates the overhead of copying the multiplicand operand into the ripple-carry adder during each iteration. Thus, both area and latency are improved. It is also an option to replace the "0" memristor with a resistor R_m such that $R_m = R_{high}$. This achieves the same behavior and is left as a design-time decision to optimize complexity. For the purposes of these text, the "0" memristor is kept for clarity.

At initialization, the V_{load} signal is driven high and the value of the multiplicand operand A is set by A_{in}. The multiplier is also set into a separate N-bit register of N memristors and switches. Then, at the beginning of each iteration i, the value of the product register is set by P_{in} into the P memristor. This happens once per iteration since the value of P changes across iterations.

In the next step, the multiplier and carry-in bit are both sensed simultaneously to resolve the full adder in a single step. The multiplier bit B_i and its inverse are sensed by applying the read voltage V_{cond} to it in the multiplier register. The voltages are used as the drivers on the switches labeled "NOT B" and "B." This is identical to the sensing behavior each full adder in a MAD ripple-carry adder does to read its carry-in signal and compute its outputs.

The p terminal of the multiplier memristor will be read for B, and the n terminal will be read for NOT B. Depending on the voltages read at the terminals of the multiplier memristor, the gate will close for either the multiplicand, A, or the "0" memristor for the addition.

If B is a "1," the voltage at the p terminal will be large enough to satisfy the threshold to close the B gates. The voltage at the n terminal will not be large enough to satisfy the threshold to close the "NOT B" gates. If B is a "0," the opposite is true. If B is a "0," the voltage at the p terminal will not be large enough to satisfy the threshold to close the B gates. The voltage at the n terminal will be large enough to satisfy the threshold to close the "NOT B" gates.

Exactly one of the switches will close, and the standard MAD voltage division circuit for the inputs will be created with the P memristor as the second input operand. This incorporates the multiplexer functionality into the adder without any delay. It comes at the cost of two switches in the full adder. At this time, V_{cond} is applied to connect the selected input A and the product memristor in series to ground. V_{cond} is also applied to the carry-in bit in the previous adder (indicated by the value of C_{in}).

The voltages at V_a and V_b resolve the final carryout and sum memristors as normal. The gates are labeled "AND" and "OR" to indicate the Boolean operation achieved by the given threshold voltage of that gate. For example, the AND switch only closes when the voltage sensed at node V_b is greater than the threshold voltage denoting that both inputs are "1." Similarly, the OR switch only closes when the voltage sensed at V_b is greater than the threshold voltage denoting that at least one input is a "1." These voltages are described in the introduction to MAD gates.

7.1.4.1 Removal of the shift registers

This full adder can now be connected in the same way the full adders for the standard MAD ripple-carry adder were. A similar optimization as was done for the IMPLY design is performed by leveraging the driver nature of MAD gates. Recall that the IMPLY design was optimized to remove the need for a shift register. Instead, the result of the addition was stored directly back into the input operand in the adder, performing the shift at the same time. The same optimization will be performed here.

If the design used a ripple-carry adder and a shift register, there would be a large overhead associated with data movement between each iteration. The ripple carry would complete and resolve the sum in the sum memristors. A MAD copy operation would store the sum into the shift register. Then, another copy operation would shift each value into the next bit in the register. Finally, another copy operation would store the shifted product into the P bits in the MAD. It is clear that these steps are redundant and necessary.

Instead, the design foregoes all of the copy operation, and each full adder i stores its sum directly into the previous full adder $i+1$ in the chain. The design will largely remain unchanged except that the logic shown in Figure 7.9 for the sum memristor will now be producing sum_{i-1}. Imagine cutting out the block of logic that is producing the sum in a full adder and moving it in its entirety to the previous full adder. Nothing else needs to change.

The same signals are sensed, the same drivers are used, the same Boolean expressions are operated, and the same threshold voltages are used. The only difference is that it occurs to a memristor in a different location in the design. To perform this, rather than applying V_{sct} to the sum memristor in full adder i, V_{set} is applied to the sum memristor in full adder $i-1$. This achieves the shift functionality as part of the addition steps. This is possible because the sum memristor and its drivers are completely independent from the rest of the circuit. The design no longer needs the product shift register or the accompanying shift delay.

7.1.4.2 Removal of the sum memristors

There is one challenge for the updated design. At the end of an addition, the design has successfully performed a shift operation. The sum from bits [$N-1$:0] essentially lies in bits [$N-2$:0] now. However, before the next addition can begin, the new sum needs to be in the P input memristors. This requires a step to copy the value from the sum memristor in each full adder to the P memristor.

This copy operation is again unnecessary. To perform the shift, the design only has to change to perform the logic on a different memristor. Originally, the sum memristor was selected in the previous bit. Instead, the P memristor is selected in the previous bit. In other words, the driver and switch logic shown for the sum memristors will actually be performed on the P memristor. Cumulatively, the full adder is modified to store into the previous bit's P memristor rather than its own sum memristor.

The underlying timing and logic do not change, but their targets do. This removes another N memristors by eliminating the sum memristors. The modified MAD full adder is shown in Figure 7.10.

Now, the P memristor has the ability to be loaded from outside of the adder or set by the sum operation. The only time the P memristor is loaded from outside of the adder is at the very beginning of the multiplication. In each subsequent intermediate addition iteration, the value of the P memristor is set by the sum logic. The P memristor is still connected to the "0" and "A" memristors to perform in the input voltage division circuitry.

By similar logic, the multiplier bits do not need to be in a shift register. Each memristor will be sensed one iteration at a time by the drivers to logically perform a shift without physically performing one. In iteration i, bit i is read as the B input.

In total, the optimized design requires N replications of the modified full adder and an N-bit register. The total complexity of the design is $5N$ memristors, $3N+2$

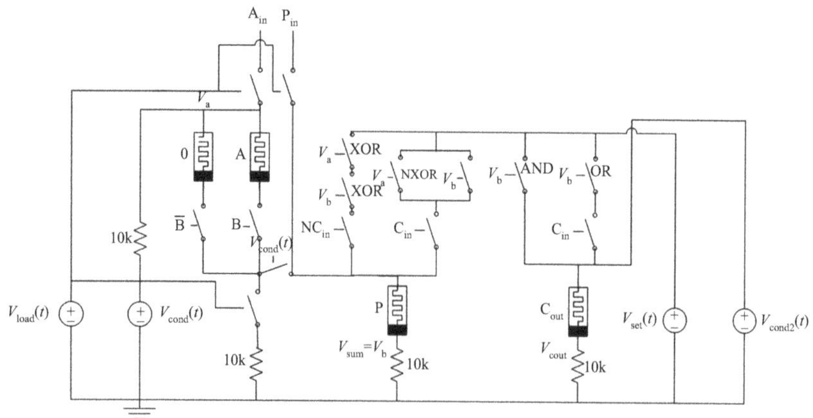

Figure 7.10 Optimized MAD full adder for a shift-and-add multiplier

drivers, and 16N switches. The full schematic for an 8-bit implementation can be seen in Figure 7.11.

The delay of a single iteration consists of the delay of the first full adder + one step per consecutive bit for carry propagation. Thus, the total delay of an addition is $N+1$ steps (the same as the standard MAD ripple-carry adder), and the total delay for the multiplication is N^2+N steps.

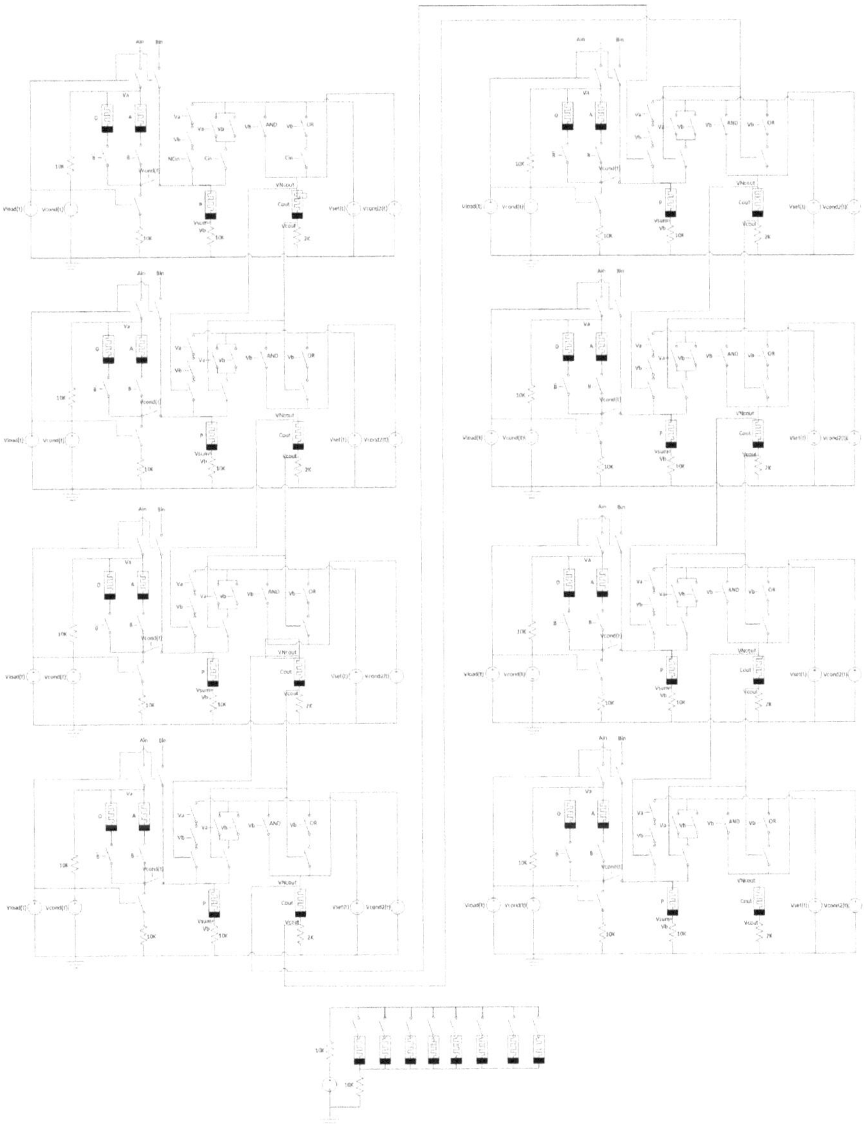

Figure 7.11 8-bit MAD shift-and-add multiplier

7.1.4.3 Pipelining

The MAD implementation can also be modified to accommodate pipelined additions. In the same manner as explained for the MAD ripple-carry adder, each bit can begin the next iteration of the addition as soon as it propagates its carryout and sum results to the next bit. In the next cycle, it receives the sum result from the next bit into its P memristor and it resets its carryout and product memristors. Now it can start the next iteration. This implies that a new addition can occur every four cycles.

Consider bit b. When the carry-in is ready, the full adder is able to set its carry-out memristor and the previous full adder's sum memristor. In the second cycle, the full adder resets the product register. In the third cycle, bit b+1 reads bit b's carry-out memristor and resolves its sum into the P memristor in bit b. In the fourth step, bit b resets the carry-out memristor.

Note that this optimization poses some challenges to the current design. During each iteration of the standard shift-and-add multiplier, the least significant bit of the multiplier is used as the select line for the input A in the addition. However, with pipelining enabled, different bits of the full adder can execute different iterations during the same time step. This means that it is possible that different bits of the full adder are using different bits of the multiplier register at the same time. In order to enable this capability, the multiplier register will have N distinct drivers, one per bit. In this way, the bits of the multiplier are disjoint and can be read independently of each other and in parallel if needed.

The total throughput of the MAD shift-and-add multiplier is now every $4N$ steps rather than N^2+N.

7.1.5 Shift-and-add multiplier analysis and comparison

The proposed designs significantly decrease the delay and complexity of the multipliers as compared to the traditional CMOS designs. A full comparison of the delay and complexity of each design are given in Table 7.8.

The hybrid-CMOS design is the only design that has a greater delay than the CMOS design for large N. This is explained by the buffering necessary in the ripple-carry adder.

Table 7.8 Delay and area comparisons of the proposed shift-and-add multipliers

	Traditional CMOS	**Proposed IMPLY**	**Proposed hybrid-CMOS**	**Threshold gates**	**Proposed MAD**
Delay	$2N^2+23N$	$2N^2+21N$	$3N^2+14N$	N^2+10N	N^2+N
Pipelined delay	N/A	$24N$	N/A	N/A	$4N$
Complexity	$132N+6$ MOSFETs	$8N$ memristors	$80N-6$ memristors	$48N$ memristors	$5N$ memristors
		$8N$ drivers	$88N-2$ MOSFETs	$2N+4$ MOSFETs	$3N+2$ drivers
		$9N-1$ switches		$14N$ GOTO pairs	$16N$ switches

The proposed IMPLY design follows closely with traditional CMOS in terms of delay, but it improves by a few cycles each iteration. The IMPLY operation is $2N$ fewer steps, a competitive delay given the serialized nature of IMPLY steps. This improvement is largely attributed to the optimizations to remove the shift functionality entirely from the design.

The threshold gate reduces the delay by over $2X$ from $2N^2 + 23N$ to $N^2 + 10N$. The proposed MAD design further reduces the delay to only $N^2 + N$ steps, a significant saving, especially as N increases. These designs require less than half the number of steps in the optimized IMPLY, hybrid-CMOS, and CMOS designs.

When the designs are fully utilized, both the IMPLY and MAD designs benefit from the ability to pipeline consecutive additions. For both proposed designs, the latency reduces from $O(N^2)$ to $O(N)$. For the IMPLY implementation, the throughput of an N-bit multiplication reduces from one multiplication every $2N^2 + 21N$ steps to $24N$. For the MAD implementation, the throughput is $4N$ steps per multiplication—only 64 steps for an entire 16-bit multiplication. This is a hugely significant improvement over traditional CMOS designs, which requires 880 steps. This comes at the cost of just $N - 1$ more drivers to the design. The hybrid-CMOS, threshold, and traditional CMOS design cannot perform pipelining, so their throughputs remain the same.

In terms of complexity, the presented IMPLY and MAD designs require the fewest components. This is expected since both of these designs are able to remove the shift registers from the design. The MAD design requires fewer memristors and drivers, while the IMPLY design requires fewer switches. Both of the designs significantly reduce the number of components to about 1/6 the number of components in the CMOS design.

The threshold-gate implementation requires more complexity but reduces the amount of CMOS to a negligible amount. The number of components is about half of the CMOS design. The hybrid-CMOS implementation is the only design, which requires more components than the traditional CMOS design. This is due to the high component counts in the shift registers. These contribute to the majority of the cost of the design.

7.1.6 IMPLY and MAD Booth multipliers

In this section, both the IMPLY and MAD designs are extended to implement a Booth multiplier. These two approaches are selected due to the optimizations their dual logic and storage capabilities were able to achieve in the shift-and-add multiplier. The Booth multipliers presented are radix-2 multipliers, operating on two bits of the multiplier at a time. The multipliers can be extended to higher radices by performing the same modifications for the given radix.

Recall that Booth multipliers are similar to shift-and-add multipliers except they support 2's complement and operate on multiple bits of the multiplier at once. Depending on the bit(s) of the multiplier, either the multiplicand A, 0, or the 2's complement of the multiplicand, $\overline{A}$ are added to the running product register. These designs assume that the value for the 2's complement of the multiplicand is calculated and initialized into the design before execution.

This can be done using a standard ripple-carry adder, and the details will not be described.

7.1.6.1 IMPLY Booth multiplier implementation

The only change necessary to enhance the IMPLY shift-and-add multiplier design to a Booth multiplier is the multiplexer. The multiplexer requires a 2-bit select line for the two bits of the multiplier that are read during each iteration. The 2-bit multiplexer can be simplified since two of the inputs are always "0." In the Booth design, when the select lines are 2'b00 or 2'b11, the input selected is "0," so the equation for the multiplexer simplifies to $\text{Out}_{\text{mux}} = \left(A_i\overline{M_i}M_{i-1}\right)\text{OR}\left(\overline{A_i}M_i\overline{M_{i-1}}\right)$ where $\overline{A_i}$ represents the 2's complement of A for bit i and M_i and M_{i-1} represent the select lines from the two bits from the multiplier. The multiplexer executes the steps in Table 7.9.

This multiplexer has been simplified and optimized by rearranging the IMPLY operations to minimize the number of steps. First, the three-input AND operations are performed. They operate on disjoint values. Last, the OR operation is performed.

Note that the first two steps use two memristors T0 and T1 to perform a COPY operation. In the context of this multiplier, it may be possible to eliminate these steps if it is found that the multiplier is not needed again during execution. This would not be true of M_i since it is necessarily needed in the next iteration as a select line again. The steps purposefully selected M_{i-1} for the copy operations so they can be removed.

In iteration k, the steps of the multiplexer that only rely on values from the multiplier register (and not on the multiplicand) can execute during iteration $k-1$ while the addition is completing. This hides the effective latency of the multiplexer in each iteration. For example, as the first addition is executing, the multiplexer would normally be sitting idle. Instead, the multiplexer can begin executing to resolve the input to the adder for the iteration two. The multiplexer logic knows which multiplier bits are needed and no other external information is required to begin operation.

Table 7.9 Modified steps for a 2-bit IMPLY multiplexer for Booth multipliers

Step	Execution	Description
1	$M_{(i-1)}$ IMP T0	NOT $M_{(i-1)}$
2	T0 IMP T1	Copy $M_{(i-1)}$
3	M_i IMP T1	M_i NAND $\overline{M_{i-1}}$
4	M_i IMP T2	NOT(M_i)
5	T2 IMP T0	M_{i-1} NAND $\overline{M_i}$
6	$\overline{A}$ IMP T1	$\overline{A}$ NAND $M_i\overline{M_{i-1}}$
7	A IMP T0	A NAND $\overline{M_i}M_{i-1}$
8	T0 IMP T3	A AND $\overline{M_i}M_{i-1}$
9	T1 IMP T3	$\left(\overline{A}M_i\overline{M_{i-1}}\right)$ OR $\left(A\overline{M_i}M_{i-1}\right)$

Note that steps 1–5 in Table 7.9 can all occur during the execution of the previous addition iteration, so only steps 6–9 are on the critical path. In the original IMPLY shift-and-add multiplier, the first three steps of each addition performed the multiplexer logic. Steps 6–9 for the updated multiplexer logic replace these three steps. As a result, the Booth multiplier only takes a single step more per iteration than the shift-and-add multiplier. Memristors T2 and T3 are equivalent to memristors M0 and M1 in the original shift-and-add multiplier's execution steps. For the Booth implementation, only two additional memristors, T0 and T1, are required for the multiplexer.

The only other modification to the multiplier design is that the drivers on the multiplier register must change to sense two bits in each iteration i, M_i and $M_{(i-1)}$, rather than one. This is similar to the enhancement to the shift-and-add multiplier to enable pipelining because both require the ability to read multiple bits of the multiplier in a single time unit. Again, this change does not affect the circuitry, but it does change the application of the voltage signals for each bit in the multiplier. Now, in each iteration, the multiplexer functionality will apply V_{cond} to two memristors rather than one.

7.1.6.2 MAD Booth multiplier implementation

Similar to the IMPLY design, the MAD shift-and-add multiplier design must be slightly altered to become a Booth multiplier.

First, the multiplexer logic will be modified from a 2-to-1 multiplexer to a 4-to-1 multiplexer. In the MAD design, the multiplexer logic is incorporated into each full adder itself. To transform this logic into a 4-to-1 multiplexer, a third memristor that holds the 2's complement of the multiplicand will be added in series with the multiplicand and "0" memristor. This memristor will serve as a third operand to the multiplexer.

The logic that selects between these potential operands also changes. Let M_i be the bit in the multiplier that selects the input operands in the current iteration. Originally, the value of M_i was used to select either "0" (if $M_i = 0$) or the multiplicand (if $M_i = 1$). This was done by placing switches on each of the memristor input operands, gated by M_i and $\overline{M}_i$ such that only one switch would close on a given iteration. Now the circuitry will use the values of two multiplier bits, M_i and M_{i-1}. Figure 7.12 shows the modified adder for the Booth multiplier.

The third memristor for the multiplexer is now in parallel with the original two multiplexer memristor inputs, "0" and A. The pairs of switches on each memristor correspond to the multiplier bit values that will select that input: M_iM_{i-1} = 2'b11 and M_iM_{i-1} = 2'b00 will select 0, M_iM_{i-1} = 2'b01 will select A, and M_iM_{i-1} = 2'b10 will select the 2's complement of A. The remainder of the full adder remains unchanged.

The enhancement for Booth's multiplier adds one memristor and six switches to each full adder. The total complexity of the adder and multiplier register with N distinct drivers is $6N$ memristors, $22N$ switches, and $4N + 1$ drivers.

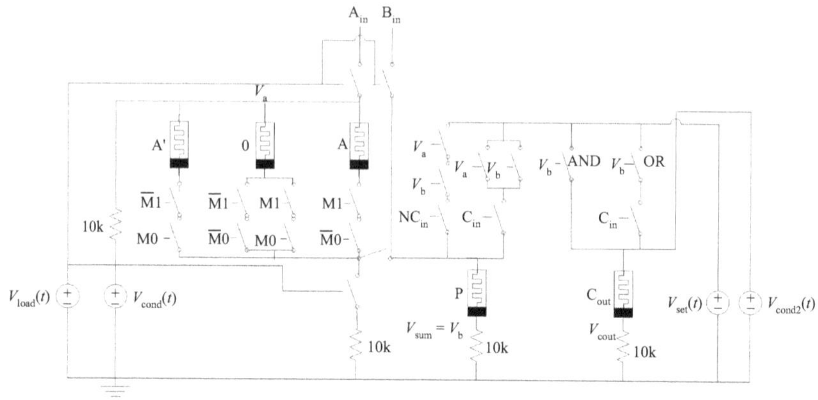

Figure 7.12 MAD full adder for Booth multiplier

Table 7.10 Delay and complexity comparisons of the proposed Booth multipliers

	CMOS	IMPLY	MAD
Delay	$2N^2 + 26N$	$2N^2 + 22N$	$N^2 + N$
Amortized delay	$2N^2 + 26N$	$25N$	$4N$
Complexity	$166N + 8$ MOSFETs	$10N$ memristors $10N$ drivers $11N - 1$ switches	$6N$ memristors $4N + 1$ drivers $22N$ switches

7.1.6.3 Booth multiplier comparisons

A comparison of the proposed Booth multipliers against the traditional CMOS design is given in Table 7.10.

The IMPLY design has been optimized to favor latency over area. As a result, the IMPLY-based design requires four fewer steps than the traditional CMOS design per iteration. Each bit of the adder needs two more memristors than the shift-and-add multiplier to implement the four-input multiplexer. It is possible to lower the area by reusing memristors from the original design rather than introducing the new memristors T0 and T1. However, this will increase the latency to add reset steps. Alternatively, the copy operation using T0 and T1 can be removed entirely if the multiplier bit is not needed again in the IMPLY steps.

The total complexity for an N-bit IMPLY Booth multiplier is $10N$ memristors, $10N$ drivers, and $11N - 1$ switches. This is less than 20 percent the number of components that CMOS requires which is $166N + 8$ MOSFETs. It can also be pipelined to require a total of only $25N$ steps. This is only one step more per iteration than the IMPLY shift-and-add multiplier.

The N-bit MAD Booth multiplier requires $6N$ memristors, $4N + 1$ drivers, and $22N$ switches. This is a nearly identical component count to the IMPLY design and also less than 20 percent of the number of components for the CMOS design.

The number of steps for the MAD Booth multiplier is the same as the shift-and-add multiplier, $N^2 + N$ steps. This is less than half the number of steps for the CMOS and IMPLY designs. It can also be pipelined by the same logic as the shift-and-add multiplier to have a throughput of $4N$ steps per multiplication. This is significantly faster than CMOS and IMPLY, reducing the delay from 928 steps in CMOS and 400 in IMPLY to only 64 steps in MAD.

7.2 Array multipliers

This chapter presents memristor-based array multipliers for the IMPLY, hybrid-CMOS, threshold, and MAD contexts. Each design is explained, optimized, and analyzed in terms of delay and cost. The designs are compared against each other and the traditional CMOS design.

7.2.1 IMPLY array multiplier

One prior work [15] designed a 4-bit by 4-bit IMPLY-based array multiplier, which requires 73 steps to complete the multiplication. This work improves upon this design to decrease the number of steps to just 58 for a 4-bit multiplication. The optimized IMPLY memristor-based array multiplier is presented in this section, and the steps to obtain the design are outlined. The multiplier uses multiple instantiations of the presented optimized ripple-carry adder in conjunction with AND gates to create the array structure.

Recall the optimized N-bit IMPLY ripple-carry adder discussed in the earlier chapter on adder arithmetic units. The ripple-carry adder has a delay of $2N+19$ steps and uses $7N + 1$ memristors. In the array multiplier, the N-bit array multiplier requires $N - 1$ instances of the N-bit ripple-carry adders. Each row in the array multiplier is a ripple-carry adder. The number of memristors in this set of ripple-carry adders is

$$\begin{aligned} \text{Memristors} &= (N-1)(7N+1) \\ &= 7N^2 - 6N - 1 \end{aligned}$$

Additionally, the design requires N^2 AND gates to compute the initial inputs to the ripple-carry adder array. Each AND gate needs four memristors, contributing a total of $4N^2$ memristors to the design. Last, since the inputs to the individual full adders are used more than once inside the adders, they must be copied from the output of the AND gates before use, requiring two memristors per full adder input. This adds $2N^2$ memristors to the design. Similarly, every sum output must be copied to its next destination, requiring one memristor per output. In all, the copy operations introduce $2N^2 - 4N$ memristors to the design. In total, the baseline array multiplier design requires $15N^2 - 10N - 1$ memristors.

The critical path of the multiplier traverses through an AND gate and then through both dimensions of the array. Within a row, the critical path traverses through the carry propagation path of the ripple-carry adder. Each full adder propagates its carry bit to the subsequent adder in two cycles.

Across rows, each full adder requires the entire full adder delay of 21 steps to produce its sum bit to pass as an input to the subsequent row. For example, consider the first and second row in an array multiplier. The first row executes its ripple-carry adder and completes in $2N + 19$ steps. Then, it copies its sum outputs to the next row. These sum outputs serve as the A inputs to the second row. Based on the execution steps for an IMPLY full adder, the addition does not begin until these inputs arrive. It will take 13 steps for the first bit of the second row to propagate its carry signal and a total of $2N + 19$ cycles for the second row to execute. Thus, the propagation time within a row is only two cycles, but the propagation time across rows is 13 cycles. The propagation time across rows for the sum output is the full 21 steps.

The total delay for the baseline N-bit array multiplier breaks down as follows:

$$\begin{aligned}\text{Steps} = {} & 3 \text{ steps for the initial AND gate} \\ & + 2 \text{ steps for the copy operation} \\ & + 2(N-1) + 19 \text{ steps for the first row ripple-carry add} \\ & + 2(N-2) \text{ steps for the copy operations across rows} \\ & + 21(N-3) \text{ steps to traverse vertically through the} \\ & \qquad \text{intermediate rows} \\ & + 2N + 19 \text{ for the final row ripple-carry add} \\ & = 27N - 26 \text{ steps}\end{aligned}$$

7.2.1.1 Optimized implementation

This work performs a series of optimizations to reduce the component count to $7N^2 - 8N + 9$ memristors and delay to $25N - 42$ steps. As the size of N increases, these optimizations show increased benefits.

First, the full adders with only two inputs are converted to half adders. This eliminates the need for the initial carry-in memristors in the ripple-carry adders. Second, the design leverages the fact that each AND gate result is used by exactly one adder. Thus, the AND operations can perform within the full adders themselves, enabling memristor reuse and removing the need to copy the outputs of the AND operation into the full adder before execution. This saves both memristors and delay steps.

Together, these two optimizations reduce the half adder modules to require eight IMPLY steps and seven memristors. The steps for a half adder with the AND operation incorporated are given in Table 7.11.

A and B represent the inputs to the AND gate. P and Q represent the inputs to the adder. In an array multiplier, some adders use the outputs of two AND gates as adder inputs while others use the output of only one AND gate as an adder input. To accommodate this variability, steps 1–3 (as marked with *) of the half adder can perform multiple instances in parallel. If the adder only uses one AND gate as an input, steps 1–3 will be instantiated once and will require two memristors. If the adder uses two AND gates as inputs, steps 1–3 will be instantiated twice and require four memristors. Steps 1–3 will perform (A_i AND B_i) and (A_j AND B_j) in parallel.

Table 7.11 Execution steps for an IMPLY half adder

Step	Functionality	Description
1*	B IMP 0 → M0	NOT B
2*	A IMP (B IMP 0)	A NAND B = NP
3*	(A IMP (B IMP 0)) IMP 0 → M1	A AND B = P
4	P IMP $\overline{Q}$ → $\overline{Q}$	P NAND Q
5	(P IMP $\overline{Q}$) IMP 0 → C_{out}, M2	P AND Q = C_{out}
6	$\overline{P}$ IMP Q → Q	P OR Q
7	Q IMP M2 → M2	P XNOR Q
8	M2 IMP 0 → Sum	P XOR Q = Sum

*The results of the AND and NAND operations are now both available in P, $\overline{P}$, Q, and $\overline{Q}$

The full adders in the multiplier will also incorporate the first three steps into the beginning of their execution to compute their inputs from an AND gate. Steps 4–8 will not be used. Instead, the normal execution of the full adder will be performed to produce the sum and carryout.

Now, the first adder in each ripple-carry adder completes its C_{out} in five steps instead of 13. The remaining full adders still take two steps to compute the carry-out and nine steps to compute the sum once the carry-in signal arrives. The total delay of a ripple-carry adder reduces to $2N + 12$ steps.

At the end of these steps, the input P and its inverse are available to the adder. If the adder has both of its inputs from AND gates, then Q and its inverse are also available now. If the adder only has one of its inputs from AND gates, then the inverse of Q is not yet available. To obtain the inverse of Q, the operation Q IMP 0 is performed in parallel with step 3 on an additional memristor. Steps 4–8 are identical across all of the adders, regardless of the inputs.

Table 7.11 is optimized for latency. However, note that in an array multiplier, many of the adders receive one of more of their inputs prior to their carry-in signals. For adders that are not in the first row, one of their inputs comes from an AND gate. All these adders can determine the output of this AND gate and corresponding input before their other inputs arrive from the first row. Similarly, the full adders receive their input from the previous row before the carry-in signal on their row has propagated and arrived.

Thus, steps 1–3 can perform preprocessing on the inputs that do not fall on the critical path. The delay for these steps is effectively hidden by the latency of other operations. At the end of steps 1–3, memristors can be reset during idle steps in preparation for the arrival of the remaining input(s). In this way, the adders can reuse memristors and reduce the effective latency without impacting the critical path of the multiplier.

By strategically timing the driver circuitry to accommodate these changes, the full and half adders are able to reduce the number of memristors to seven and six, respectively. All full and half adders are modified in this way except for the first half adder and first full adder. These adders do not use this technique because the resolution of their inputs is along the critical path, and thus the delay is optimized at the cost of increased memristors.

Third, optimizations are performed across the adders that share inputs and outputs to remove the need to copy operands across rows. Rather than storing the result of the sum into a resident memristor in the final step of the addition, each adder directly stores its sum result into a memristor in the next adder(s) that needs it. This is done with the same protocol as described for the shift operation in the shift-and-add multiplier. V_{set} is applied to the input memristors in the subsequent adder(s) rather than the local output sum memristor. This removes the $2N^2-4N$ memristors introduced for copy operations from the design.

Last, a few small optimizations can be performed on various adders to remove memristors. For example, bit 0 of the sum is not on the critical path. It is the first bit to be computed in the adder and sits idle while the remainder of the array executes. Only the carryout of bit 0 is on the critical path. This enables more memristor reuse at the cost of higher delay. The updated steps for the first half adder in the multiplier are shown in Table 7.12.

Rather than using seven memristors and seven IMPLY steps, the design uses five memristors and eight steps. The delay of the carry-out signal remains unchanged at five steps so the critical path is unaffected.

Similar logic is used to minimize area elsewhere off the critical path. This is left as an exercise to the reader. The final schematic for a 4-bit array multiplier is shown in Figure 7.13.

Each row corresponds to a ripple-carry adder with the integrated AND gates except for those on the first row. Each adder is driven by inputs from the adder above it in the array and to the right of it in the array. This is denoted by the driver and switch placement between each bit. Note that the first row has more operands

Table 7.12 Optimized execution steps for bit 0 of an IMPLY array multiplier

Step	Functionality	Description
1*	B IMP 0 → M0	NOT B
2*	A IMP (B IMP 0)	A NAND B = NP
3*	(A IMP (B IMP 0)) IMP 0 → M1	A AND B = P
4	P IMP $\overline{Q}$ → $\overline{Q}$	P NAND Q
5	(P IMP $\overline{Q}$) IMP 0 → C_{out}	P AND Q = C_{out}
6	$\overline{P}$ IMP Q → Q	P OR Q
7	Reset M0, M1, M2 (P, $\overline{P}$, $\overline{Q}$)	Reset
7	Q IMP M0 → M0	P XNOR Q
8	M0 IMP M1 → Sum	P XOR Q = Sum

*The results of the AND and NAND operations are now both available in P, $\overline{P}$, Q, and $\overline{Q}$

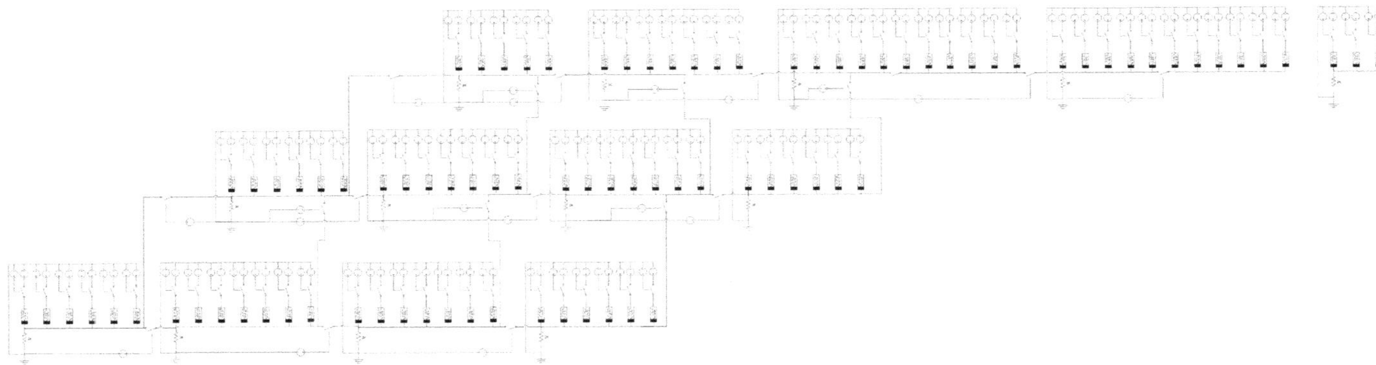

Figure 7.13 4-bit IMPLY array multiplier schematic

in the first two full adders. As noted, this first full adder and half adder are on the critical path, so delay is optimized at the expense of additional memristors.

For an N-bit array multiplier, there are $N^2 - 2N$ full adders and N half adders. All but one of the full adders requires seven memristors each and all but one of the half adders require six memristors each. The final design requires an additional nine memristors for the first full adder and half adder, resulting in a total of $7N^2 - 8N + 9$ memristors for an N-bit multiplier. This is almost half of the baseline requirement, $15N^2 - 10N - 1$ memristors. Each memristor also requires a switch and a driver to apply the IMPLY voltages. Last, there are an addition $N(N - 1)$ drivers and switches to accommodate the transfer of outputs and inputs across the adders.

In terms of delay, the proposed design reduces the delay from $27N - 26$ to $25N - 42$. Delay steps are reduced during the vertical traversal of the array multiplier by removing copy operations. The delay along the horizontal traversals is reduced by strategic IMPLY operation ordering. The total number of steps breaks down as follows:

$$
\begin{aligned}
\text{Steps} = &+2(N-1) + 12 \text{ steps for the first row ripple-carry add} \\
&+21(N-3) \text{ steps to traverse vertically through the} \\
&\quad \text{intermediate rows} \\
&+2N + 9 \text{ for the final row ripple-carry add.} \\
&= 25N - 42 \text{ steps}
\end{aligned}
$$

The first row's ripple-carry adder begins with a half adder that incorporates its AND gate into the first three steps. As shown, this half adder produces its carry-out signal in five steps and reduces the delay of the ripple-carry adder to $2N + 12$ steps. The traversal through the full adders in the rows still requires the full 21 steps. The final row's adder has already completed its AND operation by the time its A and B inputs arrived. Thus, steps 1–3 in the half adder have completed, and it only takes two steps to compute its carryout. This reduces the delay for the final ripple-carry addition to $2N+9$ steps.

7.2.1.2 Pipelining

Similar to other IMPLY designs, it is possible to pipeline multiplications through the array multiplier. Recall that an IMPLY full adder completes in 21 steps and an IMPLY half adder completes in eight steps. In this case, the delay of the full adder limits the pipeline granularity of the design. It is not possible to begin a new addition faster than an individual full adder can. The throughput for a full adder is equivalent to the throughput of the multiplier as a whole.

In the optimized array multiplier, the sum outputs are stored directly into their destination adders. The C_{out} signal is read as part of the 21 steps of the adder. Thus, the only necessary step for the next addition is to reset the memristors. Thus, a new multiplication can begin every 22 steps.

Alternatively, the array multiplier can be pipelined for a single multiplication. This is done by pipelining the components of the array multiplier themselves.

Array multipliers are constructed from several instantiations of the same ripple-carry adder circuitry. However, consider an array multiplier that has a single instantiation of the ripple-carry adder that is reused $N-1$ times for a single N-bit multiplication. In this way, an N-bit IMPLY-based multiplier can be reduced to just one row of the array if area is the primary concern.

For a pipelined design with a single row of adders, the driver logic must change. The AND operations and the first ripple-carry addition execute normally. However, instead of storing the results of the sum and carryout into the destinations in the next ripple-carry adder row, they are stored directly back into the adders themselves. Before the final store operation, the input memristors in each full adder are reset. Then, each full adder logically maps itself to be the same indexed full adder in the next row. Full adder 0 is the first full adder for the first row, second row, and $N-1$th row. The V_{cond} and V_{set} signals are driven accordingly to ensure that each full adder's outputs are driven to the correct full adders for this mapping.

Execution continues in this feedback manner, requiring one more delay step per adder row than the original design to reset the input memristors. Thus, a signal multiplication requires $22N$ steps rather than 22 steps in the first pipelined design. However, this alternative reduces the complexity to only one row of the design, or $7N$ memristors.

7.2.2 Hybrid-CMOS array multiplier

This section presents a hybrid-CMOS implementation for an array multiplier. The proposed design uses the same array structure as a traditional CMOS design with $N-1$ N-bit ripple-carry adders. There is also an array of $N \times N$ AND gates at the beginning of the multiplication to produce the dot products.

A schematic for the hybrid-CMOS based 4-bit array multiplier is shown in Figure 7.14.

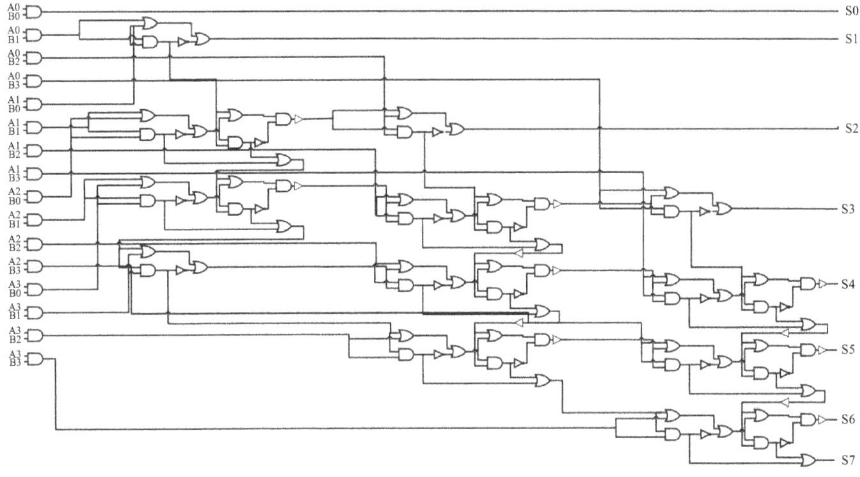

Figure 7.14 4-bit hybrid-CMOS array multiplier schematic

The left column of the schematic shows the 16 AND gates to compute the initial dot products for input to the array. The remaining three blocks correspond to the $N-1$ rows of modified ripple-carry adders. The first column has a half adder, followed by two full adders and finally a second half adder. The second column, corresponding to the second row of the array, has a half adder followed by three full adders. The final column has a half adder followed by three full adders. The distinction between half adders and full adders is made with the complexity of the block.

The array multiplier lends itself to the hybrid-CMOS approach because it does not use any NOT gates outside of the adders themselves. The initial array of AND gates does not use any NOT operations and the propagation of signals does not have any gates either. Thus, the use of NOT gates is very minimal in the design. Also, the fan-out of each AND gate is one, so no signal buffering is necessary on these gates either. Thus, the use of CMOS is kept to a minimum in the hybrid-CMOS implementation.

In this design, the AND operation can be performed in a single step. This computes the inputs to the first adders after less delay than the IMPLY design. Also, recall the optimized XOR gate used in the hybrid-CMOS adders. This XOR gate performs both the A AND B and A XOR B operations as a by-product of the XOR operation. The XOR gate is essentially a half adder because A AND B is equivalent to the carryout and A XOR B is equivalent to the sum. Each half adder in this design is implemented using a single XOR gate. This gate requires six memristors and one NOT gate.

Through simulation, signal restoration was found to be necessary on all adder outputs due to gate concatenation. Although the NOT gates inside the adders help mediate the need for signal restoration, it does not completely remove it. The critical path contains a chain of $3N-4$ full adders. This length is too great to maintain signal integrity, even with the NOT gates in the full adders.

Each row's ripple-carry adder expects the inputs to be fully buffered to correctly function. Since there are N^2-N total adders, each with two outputs, the multiplier requires $2N^2-2N$ CMOS buffers, or $8N^2-8N$ MOSFETs.

The AND gates require $2N^2$ memristors as expected, but the complexity of the ripple-carry adders needs to be recalculated. Since some of the full adders have been replaced with half adders, the complexity of the ripple-carry adders has decreased. It is simpler to express the complexity in terms of the individual full adders and half adders rather than the full ripple-carry adders. The design has N^2-2N full adders and N half adders. The total complexity of the hybrid-CMOS array multiplier becomes

Full adders:

$$= (N^2 - 2N)(14 \text{ memristors} + 4 \text{ MOSFETs})$$

Half adders:

$$= N(6 \text{ memristors} + 2 \text{ MOSFETs})$$

Adder buffers:

$$= (N^2 - N)(8 \text{ MOSFETs})$$

AND gates:

$$= N^2(2 \text{ memristors})$$

Total complexity:

$$= 16N^2 - 22N \text{ memristors} + 12N^2 - 14N \text{ MOSFETs}$$

The standard hybrid-CMOS ripple-carry adder requires $14N$ memristors and $12N$ MOSFETs. If simply $N - 1$ of the ripple-carry adders were implemented, the total complexity would be $14N^2 - 14N$ memristors and $12N^2 - 12N$ MOSFETs. Adding this to the N^2 AND gates, the design would use $16N^2 - 14N$ memristors and $12N^2 - 12N$ MOSFETs. Thus, this design has reduced both the number of memristors and MOSFETs.

One drawback of the design is that the signal restoration lies on the critical path. Each propagated carry signal and sum signal must be buffered. Thus, the delay of the ripple-carry adders becomes $3N + 4$ rather than the traditional $2N + 4$, and the delay through a full adder is seven rather than six. The total delay for an N-bit multiplication is one step for the initial AND operation, $3(N - 1) + 4$ steps through the first ripple-carry adder, $7(N - 3)$ steps to traverse the array vertically, and another $3N + 4$ steps through the final ripple-carry adder, for a total of $13N - 15$ steps.

7.2.3 Threshold-gate array multiplier

This section presents two threshold gate–based designs for array multiplier. The first uses GOTO pairs and the second uses CSTG gates. The GOTO pair implementation has the benefit of fewer components per gate, while the CSTG implementation has the benefit of less inverters in the overall design.

Independent of the underlying gate structure, both designs use multiple instantiations of their respective threshold-gate ripple-carry adder designs from the previous chapter coupled with an array of AND gates. The AND gates are also implemented with either GOTO pairs or CSTG, respectively.

7.2.3.1 GOTO implementation

As shown in the chapter on arithmetic units, the threshold N-bit ripple-carry adder implemented with GOTO pairs and memristors requires $N + 2$ gate delays and $7N$ memristors, $2N$ MOSFETs, and $2N$ GOTO pairs. Additionally, a threshold gate consisting of two memristors and one GOTO pair can be used to implement an AND gate. For a refresher on this, refer to the description of a memristor-based latch in the shift register for the shift-and-add multiplier.

Since there are a total of $N - 1$ N-bit ripple-carry adders, the total cost of the adder array is $(N - 1)(7N \text{ memristors} + 2N \text{ MOSFETs} + 2N \text{ GOTO pairs})$. There are a total of N^2 AND gates that require $2N^2$ memristors and N^2 GOTO pairs. Together, the complexity for the array multiplier is $9N^2 - 7N$ memristors, $2N^2 - 2N$ MOSFETs, and $3N^2 - 2N$ GOTO pairs.

The critical path delay through the adder traverses an AND gate, performs an $N - 1$ bit addition, traverses $N - 3$ rows vertically in the array, and completes a final N-bit addition in the bottom row of the adder. The initial AND gate takes one step, an $N - 1$ bit addition requires $N + 1$ step, three steps for each full adder traversed vertically, and a final $N + 2$ steps. The total delay is $5N - 5$ steps.

The design can be optimized by transforming the full adders into half adders wherever applicable. This will help reduce the area and delay of the design. The full adders for the least significant bit in each row of the array multiplier can be simplified to half adders. An implementation of a half adder with threshold gates is shown in Figure 7.15.

P_i and Q_i are the inputs to the adder. The half adder looks nearly identical to the full adder except it does not require the carry-in input. This reduces the number of inputs on each of the gates by one memristor. The first threshold gate is an AND gate, executing P_i AND Q_i. The threshold of the gate, two, will only be met if both P_i and Q_i are "1."

The second gate executes P_i XOR Q_i. Consider each input combination of P_i and Q_i. If both inputs are a "0," then the output of the C_{out} gate is a "0," which is inverted to a "1." Each of the inputs P_i and Q_i are "0," so the total sum of the inputs to the XOR gate is two. The threshold of the gate is three, so the output of the XOR gate for the sum will be "0." If one of the inputs is "0" and one of the inputs is "1," the output of the C_{out} gate is a "0," which is inverted to a "1." The sum of the inputs to the XOR gate is three. This satisfies the threshold of the XOR gate and the sum output is "1." Finally, if the inputs P_i and Q_i are both "1," the C_{out} gate is a "1," which is inverted to a "0." The total sum of the inputs to the XOR gate is two. This does not satisfy the threshold and the sum output is "0." This correctly implements an XOR gate.

Similar to the IMPLY approach, the AND operation can also be incorporated into the adders. Note that a threshold gate to perform an AND operation is roughly equivalent to a traditional CMOS AND gate. It does not take advantage of any of the characteristics of threshold gates that make them more efficient. If the AND gate is incorporated into the adder, some of the benefits of these gates begin to have

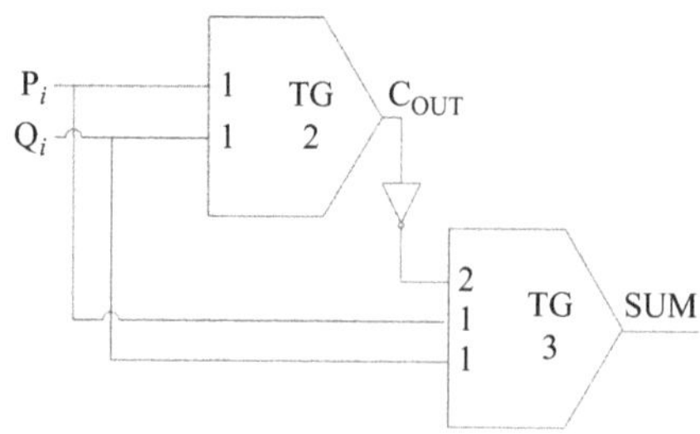

Figure 7.15 Threshold-based half adder

an impact on the design. An implementation of a half adder that includes the AND functionality is shown in Figure 7.16.

A_i and B_i are the two inputs that require an AND to produce the first operand to the half adder. Q_i is the second operand, which does not require a Boolean operation before feeding into the adder. Essentially, A_i AND $B_i = P_i$ from the original half adder design. In the context of an array multiplier, Q_i is one of the outputs from an earlier adder.

The AND operation is consumed into the threshold gates in the half adder. The first threshold gate computes C_{out} that is computed as $C_{out} = (P_i$ AND $Q_i)$ in a half adder. Since P_i is now equivalent to (A_i AND B_i), this becomes $C_{out} = ((A_i$ AND $B_i)$ AND $Q_i)$ that is simply a three-input AND operation. Recall that a two-input AND operation is achieved with a threshold gate by giving the inputs equal weight w and assigning the threshold of the gate to $2w$. In a three-input AND operation, the threshold of the gate is $3w$.

The second threshold gate is more complex to implement the XOR operation. It requires performing two transformation steps. First, double the weights of all of the inputs to the original XOR gate and adjust the threshold gate accordingly. This does not change the functionality of the gate since all of the inputs and thresholds have been scaled equivalently.

Second, replace the P_i input with two inputs, A_i and B_i. The weight of the P_i input was two after being doubled in step 1. Thus, together, A_i and B_i contribute a weight of two also. By setting the weight of each to "1," their weight will only be two if both are "1," achieving the AND operation. Now, the inputs A_i and B_i will either produce a total of zero, one, or two to the threshold. When the sum is two, it behaves like P_i="1" in the original XOR threshold gate. When the sum is less than two, the inputs behave like P_i="0."

The half adder with the AND operation incorporated has the same complexity as a full adder. It uses seven memristors, two MOSFETs, and two GOTO pairs and removes the need for a separate AND gate per input. It also removes a step from the delay because the delay of the AND is encapsulated into the threshold-gate delay.

The full adders in the array multiplier can also incorporate the AND operation into the adder. The only difference is that there is a fourth input for the carry-in signal C_i. The integrated full adder is shown in Figure 7.17.

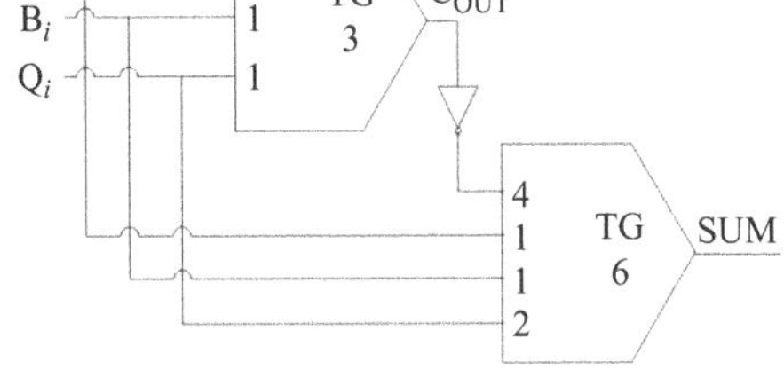

Figure 7.16 Diagram for an AND gate coupled into a threshold-based half adder in the array multiplier

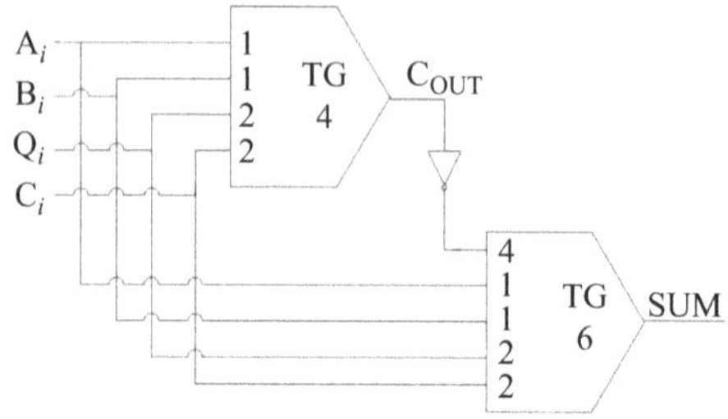

Figure 7.17 Threshold-gate implementation of full adders optimized for an array multiplier

Table 7.13 Modified threshold-gate XOR operation for a full adder in an array multiplier

Input A_i, B_i, Q_i, C_i	**Expected sum output**	**Threshold check**	**Threshold-gate sum output**
0000	0	$4 < 6$	0
0001	1	$6 = 6$	1
0010	1	$6 = 6$	1
0011	0	$4 < 6$	0
0100	0	$5 < 6$	0
0101	1	$7 > 6$	1
0110	1	$7 > 6$	1
0111	0	$5 < 6$	0
1000	0	$5 < 6$	0
1001	1	$7 > 6$	1
1010	1	$7 > 6$	1
1011	0	$5 < 6$	0
1100	1	$6 = 6$	1
1101	0	$4 < 6$	0
1110	0	$4 < 6$	0
1111	1	$6 = 6$	1

In this case, the number of memristors required for the updated full adder is equivalent to the number of memristors in an AND gate followed by a normal threshold full adder. The modified threshold gate does have one fewer GOTO pair since the AND is no longer performed externally. The first gate performs a majority function on (A_i AND B_i), Q_i, and C_i to produce C_{out}. Since the threshold of the gate is four, the threshold is satisfied if both Q_i and C_i are "1," if A_i, B_i, and Q_i are "1" or if A_i, B_i, and C_i are "1." This correctly performs the AND operation in addition to the majority function.

The second gate performs the XOR operation. Now check the desired outcomes against the threshold-gate outputs. All possible input combinations and their expected and actual XOR outputs are shown in Table 7.13.

The output of the threshold gate is identical to the desired behavior, and the XOR gate is correctly implemented. This XOR gate is used for the full adders in all of the rows except the first row of the array multiplier.

Some of the adders in the first row of the multiplier cannot use this modified adder. The first $N - 1$ adders use the result of two AND operations as inputs P_i and Q_i. It is not possible to incorporate both AND operations into a single threshold gate. The threshold gate cannot differentiate which inputs correspond to which AND. For example, consider trying to integrate the second AND into the threshold gate with inputs D_i AND F_i. There are no possible weights to correctly evaluate C_{out}. There are no input weights that will satisfy the equation. One seemingly viable operation is to give D_i AND F_i the same weights as A_i and B_i. This threshold gate can be seen in Figure 7.18.

In this scenario, the gate would not be able to differentiate between the case in which only A_i and B_i were 1 (should result in $P_i = 0$ and $Q_i = 1$ output for the ANDs) and the case in which only A_i and D_i were 1 (should result in $P_i = 0$ and $Q_i = 0$ outputs for the ANDs). Both scenarios would behave equivalently to the case in which one of the ANDs resolved to 1. Table 7.14 shows the desired vs. the actual outputs of the C_{out} threshold gate if two inputs D_i and F_i replaced Q_i to perform a second embedded AND in the full adder. For brevity purposes, only some of the input combinations are given to highlight the issues with the proposed gate.

In all cases shown in Table 7.14, the desired carry-out signal is "0," but the threshold gate mistakenly produces a "1." Similar issues exist in the resolution of the sum output. Thus, for each of the initial $N - 1$ adders, one of the inputs uses an external threshold gate to perform its AND operation.

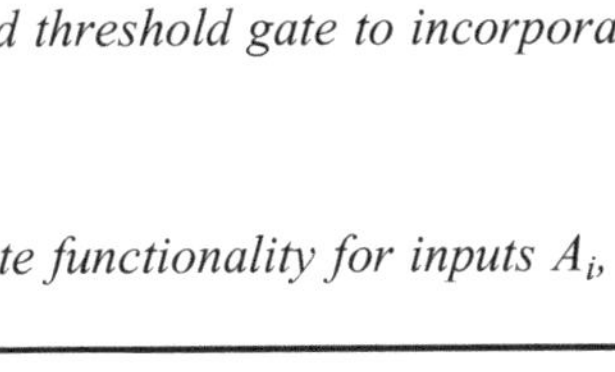

Figure 7.18 Proposed threshold gate to incorporate two AND operations

Table 7.14 Threshold-gate functionality for inputs A_i, B_i, D_i, F_i, C_i

Input A_i, B_i, D_i, F_i, C_i	**Expected C_{out} output**	**Threshold check**	**Threshold-gate C_{out} output**
01011	0	4=4	1
01101	0	4=4	1
10011	0	4=4	1
10101	0	4=4	1

The total complexity for the array multiplier is now as follows:

Full adders:

$$= N^2 - 2N \text{ (9 memristors + 2 MOSFETs + 2 GOTO pairs)}$$
$$= 9N^2 - 18N \text{ memristors} + 2N^2 - 4N \text{ MOSFETs} + 2N^2 - 4N \text{ GOTO pairs}$$

Half adders:

$$= N \text{ (7 memristors + 2 MOSFETs + 2 GOTO pairs)}$$
$$= 7N \text{ memristors} + 2N \text{ MOSFETs} + 2N \text{ GOTO pairs}$$

AND gates:

$$= N\text{(2 memristors + 1 GOTO pair)}$$
$$= 2N \text{ memristors} + N \text{ GOTO pairs}$$

Total complexity:

$$= 9N^2 - 9N \text{ memristors} + 2N^2 - 2N \text{ MOSFETs} + 2N^2 - N \text{ GOTO pairs}$$

The total complexity is reduced to $9N^2 - 9N$ memristors, $2N^2 - 2N$ MOSFETs, and $2N^2 - N$ GOTO pairs. The total delay remains $5N - 5$ gate delays since the critical path delay cannot remove the need for the AND gate as described above. A full schematic for a 4-bit implementation of the threshold-gate array multiplier is shown in Figure 7.19.

Execution traverses from the left to the right. The first 4-bit adder uses two full adders and two half adders and the remaining two 4-bit adders use three full adders and one half adder. Three of the adders in the first row use one AND gate to generate an input and incorporate the other AND gate into the adder itself. The remaining adders incorporate one AND gate into the adder and receive inputs from the other adders as described.

7.2.3.2 CSTG implementation

The CSTG implementation for an N-bit array multiplier is similar to the GOTO implementation presented and can use many of the same optimizations. The main novelty and benefit found for the GOTO threshold-gate implementation was its ability to incorporate the initial N^2 AND gates into the adders themselves and reduce the design complexity. This is an inherent benefit of threshold gates and still applicable when the design is based on CSTG. This allows the two designs to be nearly identical at the gate level.

One key advantage of the CSTG threshold gates over GOTO threshold gates is the fact that they produce the threshold gate output and its inverse. Recall that this occurs due to the series of MOSFETs at the threshold output of a CSTG gate. As explained for the ripple-carry adder, this removes the need for the inverters on the carry-out signals. Since there are $N^2 - N$ adders, this is a total savings of $2N^2 - 2N$ MOSFETs. The updated schematic with the inverters removed is shown in Figure 7.20.

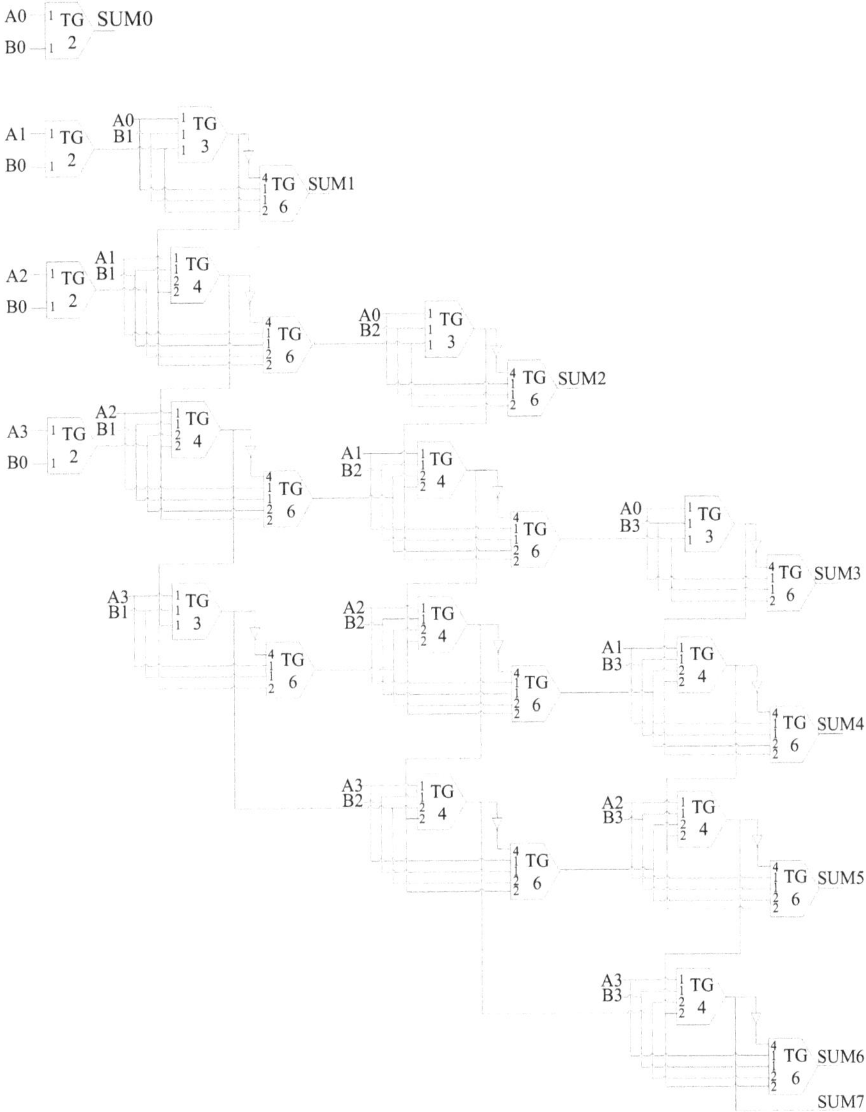

Figure 7.19 GOTO threshold gate implementation of a 4-bit array multiplier

Execution is identical, but the number of gate delays is reduced since the inverters have been removed from the critical path. Now, the critical path is the delay through an AND gate, N steps through the first ripple-carry adder, two steps through $N-3$ full adders, and then $N+1$ steps through the final ripple-carry adder. The total delay is $4N-4$ gate delays rather than $5N-5$ in the GOTO implementation.

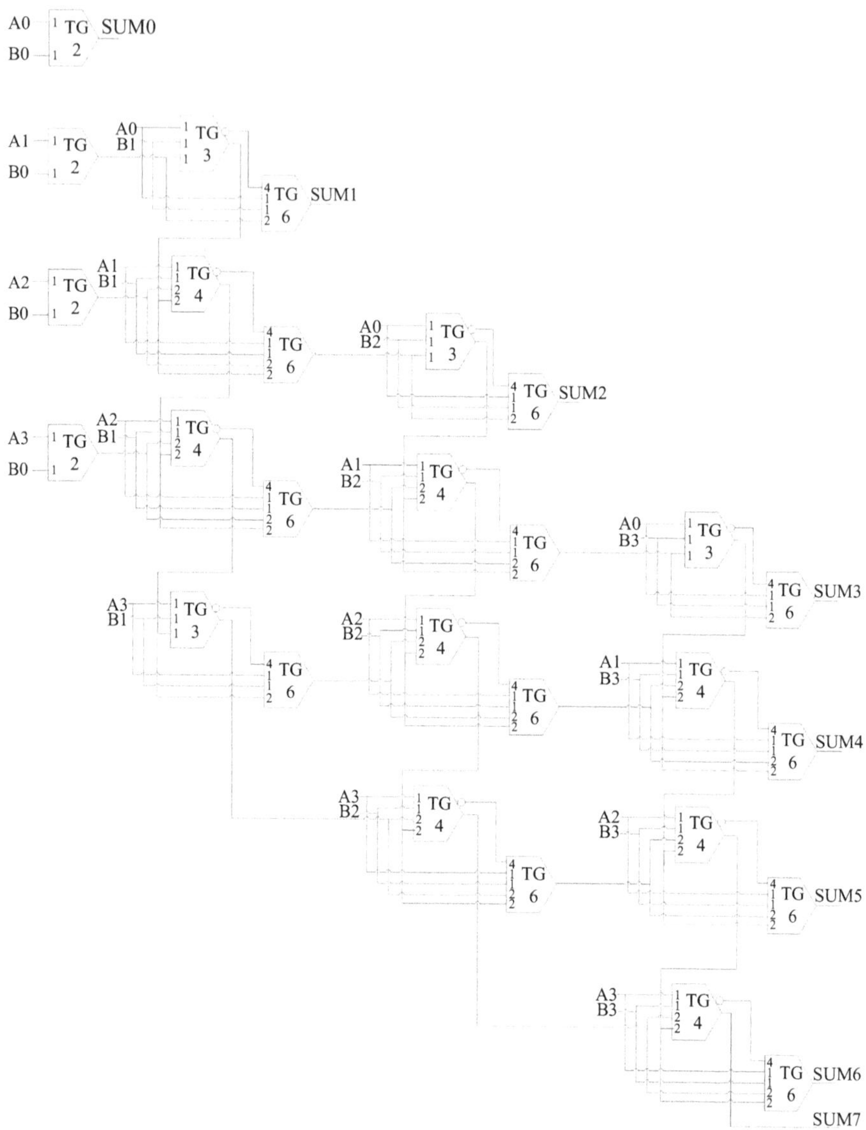

Figure 7.20 CSTG threshold gate implementation of a 4-bit array multiplier

The improvement in delay comes at the cost of far greater complexity for the gates. The total complexity is

Full adders:

$$= N^2 - 2N \text{ (9 resistors + 9 memristors + 26 MOSFETs)}$$
$$= 9N^2 - 18N \text{ resistors} + 9N^2 - 18N \text{ memristors} + 26N^2 - 52N \text{ MOSFETs}$$

Half adders:

$$= N\ (7 \text{ resistors} + 7 \text{ memristors} + 22 \text{ MOSFETs})$$
$$= 7N \text{ resistors} + 7N \text{ memristors} + 22N \text{ MOSFETs}$$

AND gates:

$$= N\ (2 \text{ resistors} + 2 \text{ memristors} + 8 \text{ MOSFETs})$$
$$= 2N \text{ resistors} + 2N \text{ memristors} + 8N \text{ MOSFETs}$$

Total complexity:

$$= \quad 9N^2 - 9N \text{ resistors} + 9N^2 - 9N \text{ memristors} + 26N^2 - 22N \text{ MOSFETs}$$

The number of memristors is the same as the GOTO implementation, but the amount of CMOS has increased by over 10×. Even with the removal of the inverters, the design sees a large increase in MOSFETs. It is left to the designer to trade off the delay benefits with the complexity consequences.

7.2.4 MAD gate array multiplier

This section presents a MAD-based implementation of an array multiplier. Initially, the design is constructed from the components discussed in previous sections. Then, optimizations for the MAD context are applied to achieve improved delay and component area in the multiplier design.

The optimized MAD ripple-carry adder presented in the previous chapter required $4N$ memristors, $3N+1$ drivers, and $13N$ switches for an N-bit array multiplier. $N-1$ rows of the ripple-carry adder are used as a starting point for the array. Each of the N^2 AND gates in the initial product matrix requires three memristors and two switches. Three drivers are also needed to operate all of the AND gates in parallel. For this initial design, the total complexity is $7N^2-4N$ memristors, $3N^2-2N+2$ drivers, and $15N^2-13N$ switches. The delay of the critical path is through an AND gate, a copy operation, $N-1$ full adders in the first row, a copy operation, and full adder traversal for $N-3$ rows vertically, and finally N full adders in the final row.

$$\begin{aligned}
&1 \text{ (AND)} \\
&+1 \text{ (copy)} \\
&+N - 1 \text{ (first ripple-carry adder)} \\
&+1 \text{ (copy)} \\
&+2(N - 3) \text{ (full adders with a copy operation)} \\
&+N \text{ (final ripple-carry adder)} \\
&= 4N - 5 \text{ steps}
\end{aligned}$$

This delay is already very competitive and faster than the previously presented approaches. However, both the delay and area can be reduced even further for the array multiplier.

First, as was done for the other designs, N of the full adders are replaced with half adders to match the array multiplier architecture. The full adders and half adders take the same form except the half adder can remove the switches corresponding to the carry-in signal. Specifically, the C_{out} memristor only requires one switch corresponding to the AND operation on the A and B inputs. The sum memristor only requires two switches for the XOR operation on the A and B inputs. The half adder is essentially an instance of a MAD AND gate and a MAD XOR gate, both using the same inputs A and B. The MAD half adder is shown in Figure 7.21.

Since there are N half adders, this replacement process removes $6N$ switches from the design.

Second, the copy operations can be removed from the design. MAD copy operations are cheaper than IMPLY copy operations, requiring a single step instead of two. However, the design can still benefit from their removal. There are a total of $N-1$ copy operations along the critical path of the design, contributing $N-1$ steps to the overall delay. All these steps can be removed from the critical path.

The full adders are optimized to store their sum into the input memristor in the adder, which uses this value as an input. The process is the same as described for the shift-and-add multiplier. Thus, each sum will be passed into either the A or B input in the "next" full adder in the chain by applying V_{set} to the memristor in the next full adder rather than the current full adder. Now, almost every intermediate adder can remove its local sum memristor. These full adders now only require three memristors.

The full adders in the final row of the array do not perform this optimization. They require their sum memristor since they are the final full adders in the design

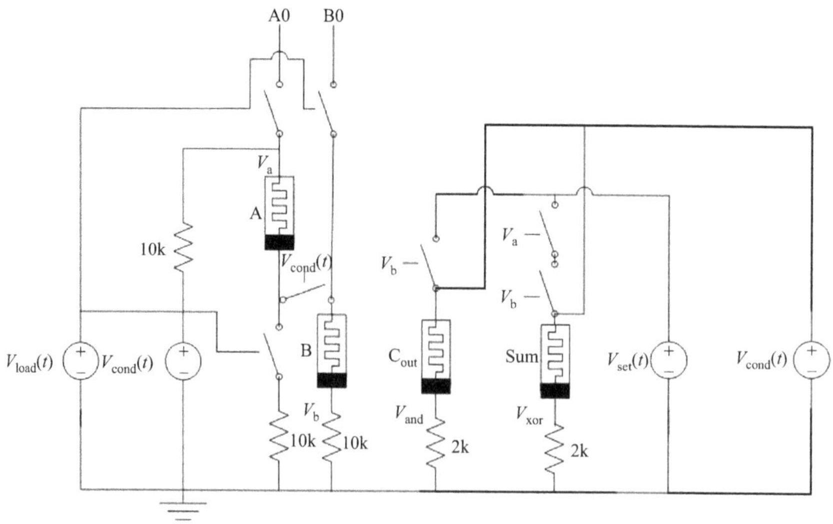

Figure 7.21 MAD half adder

and must hold the final product result. Note, if this design is incorporated into a larger block, it may be possible to remove the sum memristors in the final full adders and store the result directly into another block's memristors in the complete system. In all, $(N-2)(N-1)-1$ full adders can perform the optimization, removing N^2-3N+1 memristors from the full adders.

The optimization is only performed on one of the half adders for the same reason. Only the final half adder in the first row of the design does not compute a final product bit. This adder can store its sum result into the next full adder in the array. This removes one more memristor from the design.

The carry-out memristor in each adder cannot be removed in the same manner. The carry-out signal needs to be stored in a memristor so that both the p and n terminals can be sensed for voltages. The voltages of the nodes on the carry-out signals are used to drive switches in the subsequent adders. There is not a resident memristor in the subsequent adder to copy the carryout into. Thus, it must be stored in the local adder.

The one exception to this rule is the *N*th adder in each ripple-carry adder and the full adder that feeds the half adder in the first row. The carryout of these adders serves as one of the A or B inputs to the next adder; thus, it can be stored into the next full adder in its respective input memristor. For these adders, the C_{out} memristor can be removed from the full adder itself. $N-1$ adders can remove their carry-out memristors. This removes memristors from the design but does not remove any switches or drivers. Rather, it just moves these switches and drivers onto the next adder's logic.

Each of these steps removes unnecessary memristors due to shared inputs and outputs. The first step removed N^2-3N+2 memristors, and the second removed $N-1$. The total savings is N^2-2N+1 memristors.

Last, a cross-approach optimization is performed. So far in the designs, the focus is on a single approach to memristor-based logic for a given design. However, this is not necessary. Instead, the benefits of each design can be leveraged in a single design.

Here, the N^2 AND gates are implemented with hybrid-CMOS gates. The hybrid-CMOS approach is the most efficient at single-gate implementations that do not need to be incorporated into a logic-in-memory context. They have a gate delay of one and fewer components. In fact, the hybrid-CMOS AND gate is the least expensive out of all the approaches that have been discussed.

The hybrid-CMOS technique suffers in high fan-out or high concatenation situations. In the array multiplier, the N^2 AND gates execute once and only once and have a fan-out of one. Each gate's result is stored in a memristor in a single full adder. Thus, the concatenation of the signals lifetime is a single gate. Thus, all of the AND operations can be performed with hybrid-CMOS gates and then used directly as the inputs to the B inputs during the initialization sequence of the inputs into the array. For the adders that use AND gates for both of their inputs, the second AND gate will feed the A input during the initialization sequence. The resultant schematic for the proposed MAD-based array multiplier is shown in Figure 7.22.

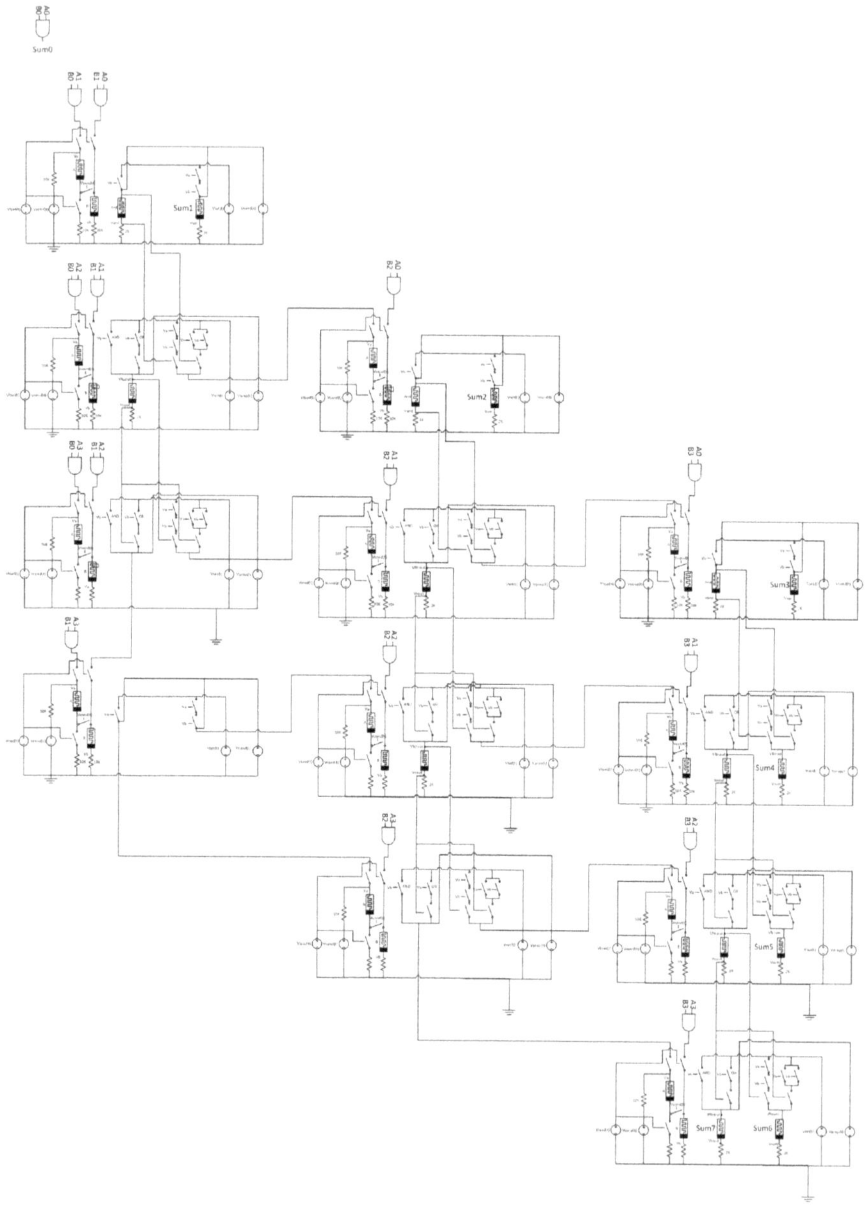

Figure 7.22 4-bit MAD array multiplier schematic

This design decision removes $3N^2$ memristors, $2N^2$ switches, and three drivers from the design. It replaces these with $2N^2$ memristors for the hybrid-CMOS gates. The net reduction in hardware is N^2 memristors, $2N^2$ switches, and three drivers.

The final design requires $5N^2 - 2N - 1$ memristors, $3N^2 - 2N - 1$ drivers, and $13N^2 - 13N$ switches. Together, all of the optimizations reduce the delay to

Table 7.15 Delay and complexity comparisons of the proposed array multipliers

	CMOS	IMPLY	Hybrid-CMOS	Threshold gates-GOTO	MAD
Area	$40N^2 - 46N$ MOSFETs	$7N^2 - 8N + 9$ memristors	$16N^2 - 22N$ memristors	$9N^2 - 9N$ memristors	$5N^2 - 2N - 1$ memristors
		$8N^2 - 9N + 9$ switches	$12N^2 - 14N$ MOSFETs	$2N^2 - 2N$ MOSFETs	$13N^2 - 13N$ switches
		$8N^2 - 9N + 9$ drivers		$2N^2 - N$ GOTO pairs	$3N^2 - 2N - 1$ drivers
Delay	$10N - 11$	$25N - 42$	$13N - 15$	$5N - 5$	$3N - 3$

essentially one step per full adder on the critical path. The total delay is one step for the initial AND, $N - 1$ for the initial ripple-carry adder, one step each for $N - 3$ full adder traversals, and N steps for the final ripple-carry adder. The total number of steps is $3N - 3$ steps for an N-bit multiplication.

Another benefit of this design is its ability to be pipelined. Similar to the ripple-carry adder, the array multiplier can be pipelined at the granularity of a full adder. Since the delay of a full adder has not changed for the array multiplier, it is still possible to pipeline multiplications such that a new multiplication begins every four steps. Once an adder has computed its C_{out} and sum bits, whether it be into a local memristor or another adder's memristor, it can reset its memristors and accept its next two inputs.

Alternatively, the hardware can be pipelined within a single multiplication. This pipelining option was discussed for the IMPLY approach. Rather than pipeline multiplications through the adders, the design will have a single row of adders that the computation is pipelined through. To accommodate this, the drivers of the design must change. Now, when each full adder computes its sum and carryout, its sum will be stored into the full adder before it. This full adder will logically represent the "next" adder in the traditional array multiplier.

To accommodate this, each full adder now takes two steps. The first step computes and stores the sum. The second step resets the internal memristors to accept the next full adder's value. Now, the delay of an N-bit full adder is $2N$ steps. Since the design must logically implement $N - 1$ ripple-carry adders, the total delay for an N-bit multiplication is $2N^2 - 2N$ steps. Although this is significantly higher, the complexity is reduced to a single row of the adder with the N^2 hybrid-CMOS gates.

7.2.5 Array multiplier analysis and comparison

Table 7.15 shows the delay and complexity of each of the proposed array multipliers and the traditional CMOS design. For the threshold gate approach, the GOTO implementation is used for the comparisons.

All of the implementations significantly reduce the number of components as compared to the CMOS design, some to less than 1/3. The IMPLY and MAD implementations require the fewest memristors and about equivalent counts for the driver logic components. When the overhead of the switches and drivers is considered, the threshold-gate implementation requires the fewest components.

The hybrid-CMOS implementation requires the most components, including a significant number of MOSFETs. This high amount of CMOS is due to the frequent use of intermediate signal buffering for the carryout and sum bits. However, the number of MOSFETs is less than 1/3 the number of MOSFETs in the traditional CMOS design, and the total number of components is less than 75 percent of the CMOS design. Thus, even with the signal buffering, the hybrid-CMOS implementation has a competitive component count.

The threshold-gate and MAD implementations are significantly faster than the CMOS and other approaches. The two designs are over two and three times faster than the traditional CMOS design, three to four times faster than the hybrid-CMOS design, and five to eight times faster than the IMPLY design. The MAD design is the fastest and requires only $3N-3$ steps. This is partially due to the use of the hybrid-CMOS gates as the AND gates. The threshold-gate implementation requires $5N-5$ gate delays, which is also a very low delay.

The hybrid-CMOS and IMPLY approaches require more steps than the CMOS approach. The IMPLY approach requires the greatest number of steps with $24N-35$, over twice as many steps as traditional CMOS.

If the delays are pipelined, the IMPLY and MAD designs have improved throughputs. Table 7.16 shows the effective latency for a multiplication for the pipelined implementations of each design, where applicable.

The latencies shown are for the pipelined designs, which pipeline across multiplications and use the full array multiplier architecture.

The MAD design is capable of beginning a new multiplication every four steps and the IMPLY design can every 22 steps. The only consequence is more complicated driver logic, but no additional delay or hardware. The threshold, hybrid-CMOS, and CMOS approaches cannot be pipelined so the effective latency is identical to the unloaded, raw latency.

7.3 Dadda multipliers

This chapter presents memristor-based Dadda multipliers for the IMPLY, hybrid-CMOS, threshold, and MAD contexts.

Depending on the size of N, there are a different number of stages in the multiplication. Recall that the Dadda multiplier reduces the height of the multiplication matrix from N to two through a series of adder stages. Once the height is 2,

Table 7.16 Effective latency for an N-bit multiplication in pipelined designs

	CMOS	IMPLY	Hybrid-CMOS	Threshold	MAD
Delay	$10N-11$	22	$13N-15$	$5N-5$	4

the final addition executes. The size of this final addition is $2N-2$ bits. Let H_x represent the height of the multiplication in stage X. For a multiplication with S stages, $H_1 = N$ and $H_s = 2$. To determine the height of each intermediate stage, the following recursive equation is used:

$$H_{(x-1)} = \text{floor}(1.5 * H_x)$$

Until $H \geq N$. This will determine how many stages are needed. For an 8-bit multiplication, the stages have the heights shown in Table 7.17.

This chapter will present 8-bit designs for Dadda multipliers. However, the same principles and optimizations can be extended to any size multiplier.

7.3.1 IMPLY Dadda multiplier

Dadda multipliers are constructed from a number of instantiations of full adders and half adders. Given the components and optimizations that have been discussed in previous sections, it is straightforward to design a baseline IMPLY Dadda multiplier design. Then, design-specific optimizations are performed to improve its performance metrics.

The first stage of the Dadda multiplier performs N^2 AND operations to produce the input multiplication matrix to the adders in subsequent stages. The AND operations are performed in three steps with four memristors. The next stages use a combination of full adders and half adders for their execution. The optimized full adders presented in the first chapter have a delay of 21 steps from the inputs to the sum output. The adder requires eight memristors and their drivers. The half adders consist of an XOR and AND operation for a total of seven steps and seven memristors. These components make up the entirety of all of the stages in the Dadda multiplier except the final stage. The final stage's addition uses the optimized ripple-carry adder that requires $2N + 19$ steps and $7N + 1$ memristors for an N-bit addition.

Each stage of the multiplier uses half adders and full adders to achieve the necessary height in each stage. The total number of adders in an N-bit Dadda multiplier is

Full adders: $N^2 - 4N + 3$
Half adders: $N - 1$

Table 7.17 Height of the multiplication matrix at each stage of an 8-bit Dadda multiplier

Stage	**Height**
1	8
2	6
3	4
4	3
5	2

For an 8-bit multiplication, the total complexity is

Full adders:

$$= 35\,(8 \text{ memristors} + 8 \text{ drivers} + 8 \text{ switches})$$
$$= 280 \text{ memristors} + 280 \text{ drivers} + 280 \text{ switches}$$

Half adders:

$$= 7\,(7 \text{ memristors} + 7 \text{ drivers} + 7 \text{ switches})$$
$$= 49 \text{ memristors} + 49 \text{ drivers} + 49 \text{ switches}$$

AND gates:

$$= 64\,(4 \text{ memristors} + 4 \text{ drivers} + 4 \text{ switches})$$
$$= 256 \text{ memristors} + 256 \text{ drivers} + 256 \text{ switches}$$

Ripple-carry adder:

$$= 7N + 1 \text{ memristors} + 7N + 1 \text{ drivers} + 7N + 1 \text{ switches}$$
$$= 99 \text{ memristors} + 99 \text{ drivers} + 99 \text{ switches}$$

Copy operations:

$$= 42\,(2 \text{ memristors} + 2 \text{ drivers} + 2 \text{ switches}) + 64 \text{ memristors} + 64 \text{ drivers} + 64 \text{ switches}$$
$$= 148 \text{ memristors} + 148 \text{ drivers} + 148 \text{ switches}$$

Total complexity:

$$= 832 \text{ memristors} + 832 \text{ drivers} + 832 \text{ switches}$$

The initial AND operation requires three steps plus two steps to copy the outputs to the next stage. Each of the first four stages has a delay of a full adder, 21 steps, plus two steps to copy the outputs to the next stage's inputs. Notice that there is no benefit to having a half adder execute quickly. The critical path is still determined by the delay of the full adders. The final 14-bit ripple-carry adder requires 47 steps for a total delay of 144 steps.

The majority of the delay for an N-bit Dadda multiplier is spent in the intermediate stages of reduction. This becomes more evident as the size of N increases. The remaining substantial element of the delay is the final N-bit addition. However, the designer can select any adder implementation for this final adder, from the simple, long delay ripple-carry adder, to the fast, complex carry-lookahead adder. Thus, the final stage of the adder can be optimized to decrease its delay at the expense of increased complexity. It is left as an exercise to calculate the complexity and delay of the Dadda multiplier if other adder architectures are implemented in the final stage.

The intermediate reduction stages are more difficult to optimize. This is due to the fact that the delay is simply the delay of a full adder from input to output. The full adder has already been heavily optimized in its introduction for the ripple-carry adder. However, some optimization tricks can still be performed now that the full adders are not used in the context of a ripple-carry adder.

First, the adders leverage the fact that the inputs can be overwritten unlike standard IMPLY-based addition. In the Dadda multiplier, the inputs to the adders are the results of AND operations or previous intermediate adders. Each output has exactly one destination adder. Thus, it can be used and overwritten during their use as an input in the next adder. Previously, the full adder was designed with the assumption that the inputs could not be overwritten.

Second, the adder no longer needs to incorporate steps to propagate the carry signals since the adders are not used in the context of a ripple-carry adder. There is no need to propagate the carry-out signal before the full adder completes. For similar reasons, the sum and carry-out signals are prioritized equally in terms of resolution. The updated execution steps for a half adder with inputs P and Q are shown in Table 7.18.

The half adder requires seven steps and seven memristors. It takes a nearly identical form to the half adder shown for the array multiplier. Since the half adder has a significantly shorter delay, it is not on the critical path and does not need to be optimized. Attention is focused on the full adders. The full adder is more noticeably changed. Table 7.19 shows the updated execution steps for a full adder with inputs P, Q, and R.

R+ represents a copy of the R value. The explanation of how to obtain this value appears later in this section. The full adder requires 15 steps and 11 memristors. The 11 memristors are the 3 input memristors, R+, and 7 internal memristors M0 to M6.

Notice that the full adder uses two different expressions to obtain the two XOR operations. Since the full adder can overwrite its values, both expressions that have been discussed in the text for IMPLY XOR gates are equally viable depending on the available operands.

In a standard full adder, the critical path lies along the carry propagation and thus the delay for the carry-out bit is minimized. However, in the design of a Dadda multiplier, both the carry-out and sum bits of the full adders are along the critical path because both must be computed before the next stage can begin. Thus, the execution steps are slightly different from the standard full adder to accomplish a mutually minimal delay for the two outputs.

Similar to the other designs that use IMPLY operations, the copy operations between the stages are removed. Instead of storing the carryout and sum directly

Table 7.18 Execution steps for an IMPLY Dadda half adder

Step	Half adder	Goal
1	Q IMP 0 → M0	NOT Q
2	P IMP 0 → M1	$\overline{\text{P}}$
3	P IMP $\overline{\text{Q}}$	P NAND Q
4	(P IMP $\overline{\text{Q}}$) IMP 0 → M2,C_{out}	P AND Q = C_{out}
5	$\overline{\text{P}}$ IMP Q	P OR Q
6	Q IMP M2	P XNOR Q
7	M2 IMP 0 → Sum	P XOR Q = Sum

Table 7.19 Execution steps for an IMPLY Dadda full adder

Step	Full adder	Goal
1	Q IMP 0 → M0	
2	P IMP 0 → M1	
3	$\overline{P}$ IMP $\overline{Q}$	
4	Q IMP $\overline{P}$	P NAND Q
5	P IMP Q	
6	($\overline{P}$ IMP $\overline{Q}$) IMP 0 → M2,M3	
7	(P IMP Q) IMP (($\overline{P}$ IMP $\overline{Q}$) IMP 0)	P XOR Q
8	R IMP 0 → M4	
9	M2 IMP $\overline{R}$ → $\overline{R}$	(P XOR Q) NAND R
10	$\overline{R}$ IMP 0 → M5	(P XOR Q) AND R
11	$\overline{P}$ IMP M5→ M5	C_{out}
12	M3 IMP R → R	
13	R^+ IMP M3	
14	M3 IMP 0 → M6	
15	R IMP M6 → M6	Sum

into the local full adder, they are stored into the subsequent full adders in the next stage that use them as inputs. Specifically, steps 10, 11, 14, and 15 can be performed on memristors in the next full adders, which use the sum and carryout as inputs.

Steps 10 and 11 produce the carry-out memristor in the local full adder's M5, and steps 14 and 15 produce the sum memristor in the local full adder's M6. Instead, let M5 be a memristor in the subsequent full adder, which uses the carryout from this adder. Instead, these values are directly stored into the next stage of the Dadda multiplier. M5 is changed to be a memristor in the subsequent stage in the full adder, which uses it as an input. Similarly, M6 is changed to be a memristor in the full adder, which uses the sum from the current adder.

This does not remove any steps, but removes two memristors, M5 and M6, from each full adder. Instead, M5 and M6 are either the P, Q, or R memristor in their next full adder.

The same reasoning can be used to remove the C_{out} and sum memristors in the half adder. This reduces the complexity of the half adder to seven steps and five memristors.

Now with the complete design, how each full adder obtains the value of R+ can be discussed. The R+ memristor is produced without any additional copy operations or overhead. Consider the steps, which produce R. This value comes as an input from another full adder or AND operation in the multiplier.

For a full adder, either steps 10 and 11 produced R for the next full adder or steps 14 and 15 did. For example, let steps 10 and 11 in full adder *X* resolve the R value into a memristor in full adder *Y*. There can be two destination memristors, M5 and R+, in these two steps. Both exist in the next full adder. In this way, the copy of the R value, R+, is produced in parallel with R and incurs no extra delay. Essentially, steps 10 and 11 are changed as shown in Table 7.20.

Table 7.20 Modified adder steps to obtain R+ in full adder

10	$\overline{R}$ IMP 0 → M5,M7	(P XOR Q) AND R
11	$\overline{P}$ IMP M5 → M5,M7	C_{out}

Table 7.21 Standard IMPLY A AND B execution steps

Step	Functionality	Description
1	B IMP 0 → M0	NOT B
2	A IMP (B IMP 0)	A NAND B
3	(A IMP (B IMP 0)) IMP 0 → M1	A AND B

Full adder *X* produces R and R+ for its destination full adder *Y*. When full adder *Y* begins execution, it has its R and R+ memristors already. If steps 14 and 15 produce R, the same modification is performed. If an AND operation produces R, then the final step will perform onto two memristors in the same fashion. Regardless, this does not introduce any memristors to the design because the M7 memristors in full adder *X* are equivalent to the R+ memristor in full adder *Y*, which has already been counted in the complexity.

Some of the adders in the design use the output of an AND gate as one or more of the inputs. The design for the array multiplier showed that it is possible to incorporate an AND operation into an IMPLY adder. For the adders, which require an AND input, this optimization is used. A standard IMPLY AND operation for inputs A and B is shown again in Table 7.21 as a refresher.

The result of the third step, M1, will exist in the full adder, which uses the output of this AND gate. This is done in a similar fashion as described above for full adder outputs.

Second, notice that step two produces the inverse of the result in step three. In both the full adder and half adder, the first step in execution produces the inverse of the input. Thus, step two of the AND gate is essentially step one in the full adder, which uses its result. The effective latency of the AND gate is just two steps. To avoid all of the stages incurring the delay for the AND operation, all these AND operations can perform in parallel in the first stage regardless of the stage they are used.

Only the input memristors A and B are introduced since M0 and M1 are both memristors in the next full adder. M0 and M1 in the AND gate are essentially memristors Q and M0, P and M1, or R and M4 in the destination full adder. Thus, the N^2 AND gates introduce two steps and $2N^2$ memristors to the design.

The total delay of the IMPLY Dadda multiplier is two steps for the AND operation, 15 steps for each stage, and $2N + 19$ steps for the final N-bit addition. Thus, the total delay for an N-bit Dadda multiplier is $15S + 2N + 21$ steps.

Although this design does not perform pipelining, pipelining is possible with IMPLY. In order to pipeline this multiplier, it is necessary to pipeline across the reduction stages and across the individual bits of the final adder. In this configuration, the adders, which use an AND gate as an input, have the longest delay at 17 steps. Thus, every 18 steps a new multiplication can begin.

The area has been reduced as a secondary priority to the delay since IMPLY-based implementations tend to have greater delay than other approaches. Numerous area optimizations were performed by reusing memristors within components and across components to reduce the total complexity to

Full adders:

$$= 35\ (9 \text{ memristors} + 9 \text{ drivers} + 9 \text{ switches})$$
$$= 315 \text{ memristors} + 315 \text{ drivers} + 315 \text{ switches}$$

Half adders:

$$= 7\ (5 \text{ memristors} + 5 \text{ drivers} + 5 \text{ switches})$$
$$= 35 \text{ memristors} + 35 \text{ drivers} + 35 \text{ switches}$$

AND gates incorporated into adders:

$$= 2N^2 \text{ memristors} + 2N^2 \text{ drivers} + 2N^2 \text{ switches}$$
$$= 128 \text{ memristors} + 128 \text{ drivers} + 128 \text{ switches}$$

Ripple-carry adder:

$$= 7N + 1 \text{ memristors} + 7N + 1 \text{ drivers} + 7N + 1 \text{ switches}$$
$$= 99 \text{ memristors} + 99 \text{ drivers} + 99 \text{ switches}$$

Total complexity:

$$= 577 \text{ memristors} + 577 \text{ drivers} + 577 \text{ switches}$$

Note this is less than 70 percent of the original complexity.

7.3.2 Hybrid-CMOS Dadda multiplier

This section presents an optimized hybrid-CMOS-based implementation for a Dadda multiplier. The Dadda multiplier is especially well suited for the hybrid-CMOS approach due to its low fan-out. The outputs of the AND gates and the adders all have exactly one destination adder. Inside each adder, the value is used a maximum of two times, removing the need for signal restoration due to fan-out.

Also, carry propagation buffering is less necessary in the Dadda multiplier since the intermediate stages use disjoint full adders that are not in a chain. Signal restoration is still necessary for long series of propagations across stages. For example, the final N-bit ripple-carry adder requires buffering in the same manner described for ripple-carry adders. However, for the remainder of the multiplier, it was found that it is sufficient to buffer the output signals after every two adder stages rather than after every stage. This removes nearly half of the buffers that would otherwise be present.

The schematic for the 8-bit hybrid-CMOS Dadda multiplier design is shown in Figure 7.23.

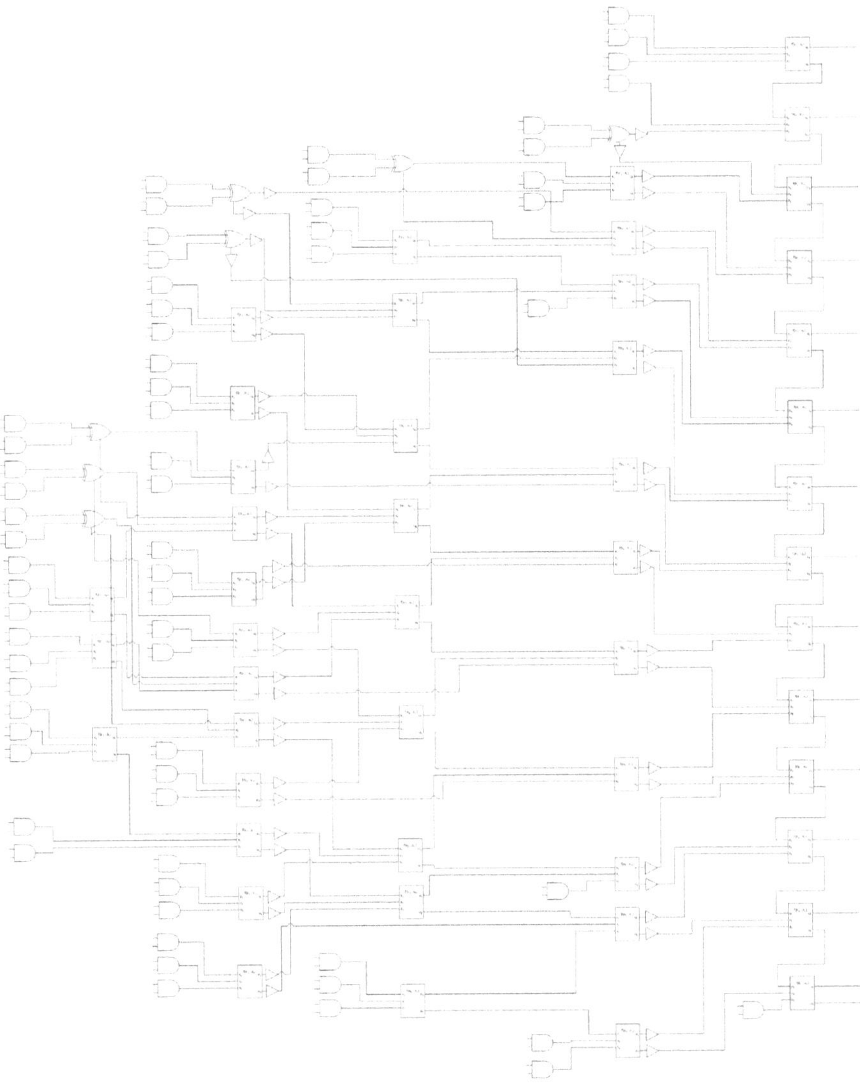

Figure 7.23 8-bit hybrid-CMOS Dadda multiplier schematic

In the second stage, there are 12 full adders and 2 half adders for a total of 28 outputs. Each of these outputs is buffered for a total of 112 MOSFETs. Similarly, stage 4 has 11 full adders and 1 half adder for a total of 96 MOSFETs. The remainder of the component area of the design comes from the N^2 AND gates and the adders. The N^2 AND gates require $2N^2$ memristors. The 35 full adders require 14 memristors and 4 MOSFETs. Each half adder is implemented with an optimized XOR gate as described for the hybrid-CMOS array multiplier design.

Each half adder requires six memristors and two MOSFETs. The total component area is

Full adders:

$$= 35(14 \text{ memristors} + 4 \text{ MOSFETs})$$
$$= 490 \text{ memristors} + 140 \text{ MOSFETs}$$

Half adders:

$$= 7(6 \text{ memristors} + 2 \text{ MOSFETs})$$
$$= 42 \text{ memristors} + 14 \text{ MOSFETs}$$

Restoration buffers:

$$= [2 * (12 + 11) + 2 * (2 + 1)] * 4$$
$$= 208 \text{ MOSFETs}$$

14-bit ripple-carry adder

$$= 14(14 \text{ memristors} + 12 \text{ MOSFETs})$$
$$= 196 \text{ memristors} + 168 \text{ MOSFETs}$$

Total complexity:

$$= 728 \text{ memristors} + 530 \text{ MOSFETs}$$

The signal restoration buffers for the individual adders and the final ripple-carry adder contribute about half of the CMOS components in the design. This highlights the significance of the consequence of using signal restoration buffers in a memristor-based design.

In terms of delay, the initial AND requires one step, each intermediate stage requires six, two stages require a step for signal restoration, and the final N-bit addition requires $3N + 4$ steps. Note that N here is the size of the final addition, not the size of the multiplication. The total delay for a multiplication with S stages is $6S + 3N + 7$. For an 8-bit Dadda multiplier with a 14-bit final adder, this is 73 steps. Even with the CMOS signal buffers, the overall delay is much shorter than the IMPLY approach. The final carry propagation adder accounts for over half of the total delay. If a faster carry propagating adder was used in the final stage, this delay could be further decreased.

7.3.3 Threshold gate Dadda multiplier

The threshold gate approach to Dadda multipliers extends the idea of column reduction to leverage the properties of threshold gates. However, it takes an unconventional form as compared to other implementations. It still adheres to the principles originally outlined by Dadda [62].

One of the main characteristics of threshold gates is their ability to reduce the computation height of their respective logic block. The Dadda multiplier has the same goal: reduce the computation height using half adders and full adders in a maximally efficient way. Thus, threshold gates are especially well suited for a Dadda multiplier. Together, they can achieve even greater depth reduction.

In the traditional Dadda multiplier design, the number of remaining additions is reduced in each stage by using full adders that reduce the height from three to two. This chapter presents an extension on this idea that still focuses on column reduction but uses gates with more inputs than three. In this way, greater reduction can occur in each stage. Full adders are restricted to have specifically three inputs and two outputs. Raw threshold gates are not restricted to any input or output quantity.

This approach redesigns the traditional circuitry for Dadda multipliers, removing the final *N*-bit addition and removing the concept of full adder stages. Instead, the design is split up into stages of individual counters. A conceptual figure for the Dadda multiplier is shown in Figure 7.24.

The 64 AND gates are represented by the dot matrix. Each dot represents an (A_i AND B_k) output. In this figure, counters are used for the main building blocks rather than simple full adders and half adders. Execution begins on the right and traverses left through the adders in the same fashion that normal additions do.

The reduction is performed on $\log_2$(columns) of the addition at a time. For an 8-bit multiplication, the dot matrix can be visualized as eight rows of 15-bit values because the total number of columns in the matrix is 15. Any td missing a dot in the row is equivalent to a "0" and does not contribute to the addition. For a multiplication with 15 rows, the reduction is performed on 3 columns at a time.

Consider the first three columns. To perform an addition on these columns, let column 0 have weight 1, column 1 have weight 2, and column 2 have weight 4. This is no different than the weights assigned in a standard binary addition of 3-bit values. With this weighting scheme, one of the inputs to the first counter has weight

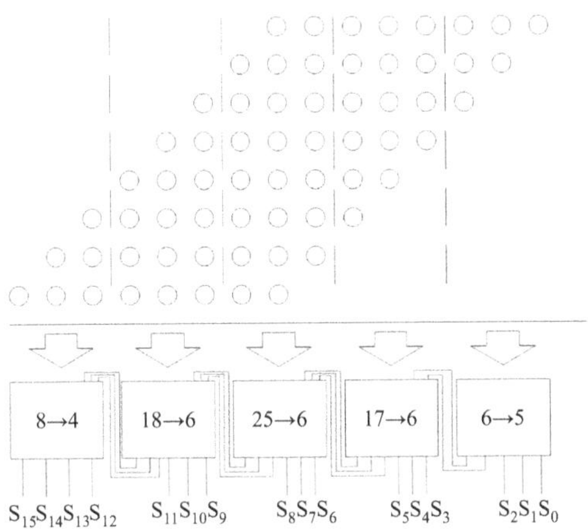

Figure 7.24 8-bit threshold gate Dadda multiplier diagram

1, two have weight 2, and three have weight 4, respectively. If all of the valid inputs are b1, the maximum sum is

$$\begin{array}{ccccc} & & 1 & 1 & 1\\ & & 1 & 1 & 0\\ + & & 1 & 0 & 0\\ \hline 1 & 0 & 0 & 0 & 1\end{array}$$

This is equivalent to ‘h7 + ‘h6 + ‘h4 = ‘h11 or ‘d17. Five bits are needed to hold this unsigned value as an output of the counter. There is no possible valid input combination that will produce an output larger than this given the format of the dot matrix. Thus, a 6 → 5 counter is sufficient for the first three columns. The first number in the name of the counter represents the number of single bit inputs and the second number represents the number of bits for the output.

Since this counter belongs to the first three columns of the multiplication, it can only compute three bits of the final sum. The two most significant bits of the output bits overflow into columns four and five of the matrix. Thus, these bits must be incorporated into the next counter. These two bits are used as additional inputs for the next counter in columns four and five, respectively. Now the next three columns have these two inputs and the preexisting 15 inputs from the matrix. The maximum sum of these binary values is

$$\begin{array}{cccccc} \multicolumn{4}{l}{\text{Carry-ins:}} & 1 & 0\\ & & 1 & 1 & 1 & \\ & & 1 & 1 & 1 & \\ & & 1 & 1 & 1 & \\ & & 1 & 1 & 1 & \\ \hline 1 & 1 & & & & \\ + & & & 1 & & \\ \hline =1 & 0 & 1 & 0 & 0 & 0\end{array}$$

This is equivalent to ‘h2 + ‘h7 + ‘h7 + ‘h7 + ‘h7 + ‘h6 + ‘h4 = h28. This output requires six bits. Thus, the second counter is a 17→6 counter. This pattern continues until the final columns are reached, and the final product is resolved. The remaining three counters are 25 → 6, 18 → 6, and 9 → 4. The counter for the middle columns is generally the largest since its columns are the most complete. The counters for the most significant columns and the least significant columns are the smallest by similar logic.

The counters operate in series since they propagate carry signals between each other. However, the intermediate adder stages and final ripple-carry addition are removed. Thus, the overall delay is decreased.

7.3.3.1 Counter structure

Threshold gates are used for the underlying structure of the counters found in this section. There are two approaches to threshold gate counter implementations—Kautz and Minnick methods. Prior work showed the utility of each of these approaches to threshold-gate-based counters and multipliers [55]. A summary of the complexity for each of these approaches for standard counter sizes is shown in Table 7.22.

A diagram showing the construction of a 7 → 3 counter in each of these methods is given in Figure 7.25.

A description and analysis of each approach is given in the following two subsections. The two approaches are also applied to the 8-bit Dadda multiplier.

7.3.3.2 Minnick method for counters

In the Minnick method, the number of gates for an $X \rightarrow Y$ counter is

$$\text{Gates} = 2^{y-1} + y - 2$$

Also, in order to obtain the correct weights, the first tier of inputs are all inverted for a total of $2^{y-1} - 1$ NOT gates. It is also important to consider the complexity of the threshold gates in order to determine the true component count of the implementation. In the Minnick method, all of the counter inputs feed into all of the threshold gates in the counter. Additionally, some of the gates propagate their results to other gates internally. The total number of inputs for the counter is

$$\text{Inputs} = X\left(2^{y-1} + y - 2\right) + 2^{y} - y - 1$$

Table 7.22 Component area of threshold-gate counters

	3 → 2 Counter area	**7 → 3 Counter area**	**15 → 4 Counter area**
Kautz	7 memristors 2 GOTOs 2 MOSFETs	24 memristors 3 GOTOs 4 MOSFETs	66 memristors 4 GOTOs 6 MOSFETs
Minnick	7 memristors 2 GOTOs 2 MOSFETs	39 memristors 5 GOTOs 6 MOSFETs	161 memristors 10 GOTOs 14 MOSFETs

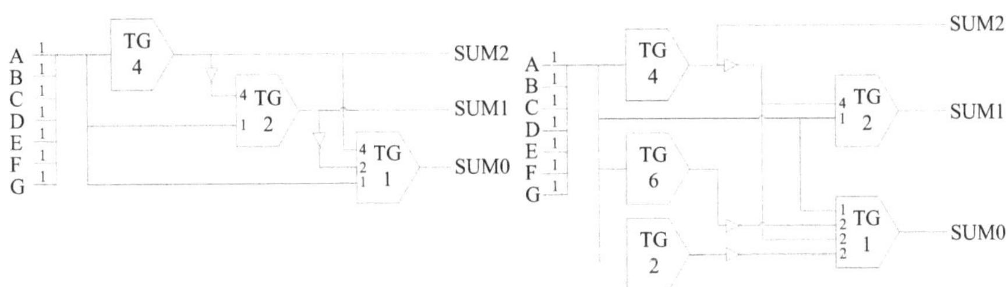

Figure 7.25 Threshold gate design of a 7 → 3 counter used (a) Kautz and (b) Minnick methods

For GOTO threshold gates, there is one GOTO pair per gate and one memristor per input. Thus, the total component area of an $X \rightarrow Y$ threshold gate is

$$\left[X\left(2^{y-1}+y-2\right)+2^{y}-y-1\right] \text{ memristors} + 2^{y-1}+y-2 \text{ GOTO pairs} + 2^{y}-2 \text{ MOSFETs}$$

For an 8-bit Dadda multiplier, the complexity for the counters becomes

$6 \rightarrow 5$ counter:

$$\begin{aligned} &= [6(19)+26] \text{ memristors} + 19 \text{ GOTO pairs} + 30 \text{ MOSFETs} \\ &= 140 \text{ memristors} + 19 \text{ GOTO pairs} + 30 \text{ MOSFETs} \end{aligned}$$

$17 \rightarrow 6$ counter:

$$\begin{aligned} &= [17(36)+57] \text{ memristors} + 36 \text{ GOTO pairs} + 62 \text{ MOSFETs} \\ &= 669 \text{ memristors} + 36 \text{ GOTO pairs} + 62 \text{ MOSFETs} \end{aligned}$$

$25 \rightarrow 6$ counter:

$$\begin{aligned} &= [25(36)+57] \text{ memristors} + 36 \text{ GOTO pairs} + 62 \text{ MOSFETs} \\ &= 957 \text{ memristors} + 36 \text{ GOTO pairs} + 62 \text{ MOSFETs} \end{aligned}$$

$18 \rightarrow 6$ counter:

$$\begin{aligned} &= [18(36)+57] \text{ memristors} + 36 \text{ GOTO pairs} + 62 \text{ MOSFETs} \\ &= 705 \text{ memristors} + 36 \text{ GOTO pairs} + 62 \text{ MOSFETs} \end{aligned}$$

$9 \rightarrow 4$ counter:

$$\begin{aligned} &= [9(10)+11] \text{ memristors} + 10 \text{ GOTO pairs} + 14 \text{ MOSFETs} \\ &= 101 \text{ memristors} + 10 \text{ GOTO pairs} + 14 \text{ MOSFETs} \end{aligned}$$

Total counter complexity:

$$= 2{,}572 \text{ memristors} + 137 \text{ GOTO pairs} + 230 \text{ MOSFETs}$$

When the original AND gate matrix is taken into account, the complexity becomes

Counter complexity:

$$= 2{,}572 \text{ memristors} + 137 \text{ GOTO pairs} + 230 \text{ MOSFETs}$$

AND gates:

$$\begin{aligned} &= 2N^2 \text{ memristors} + N^2 \text{ GOTO pairs} \\ &= 128 \text{ memristors} + 64 \text{ GOTO pairs} \end{aligned}$$

Total complexity:

$$= 2{,}700 \text{ memristors} + 201 \text{ GOTO pairs} + 230 \text{ MOSFETs}$$

Notice that this component area is very high compared to other discussed implementations. This is largely due to the fact that Minnick threshold counters use the inputs to the counter as inputs to all of the first-level gates. Thus, for high-input counters like those in a Dadda multiplier, the number of memristors grows quickly.

Although the Minnick method to threshold gates can have high complexity, it generally reduces the delay of the design. The delay of any Minnick counter is three steps, regardless of the number of inputs or outputs. In fact, the most significant bit of the threshold gate is computed in a single step. This is especially beneficial when the most significant bit is on the critical path as it is in the adder and multiplier contexts. The delay for a y-bit output Z in a Minnick counter can be expressed as

$$\begin{aligned}\text{Delay} &= 3 \text{ for } Z_0, Z_1, \ldots, Z_{y-2} \\ &= 1 \text{ for } Z_{y-1}\end{aligned}$$

For an 8-bit Dadda multiplier, the delay consists of a single-gate delay for the AND gates in the dot matrix and three delays per counter for a total delay of 16 steps.

7.3.3.3 Kautz method for counters

In general, the Kautz method requires fewer components, but a longer delay. There is exactly one threshold gate per output bit of the counter. For an $X \rightarrow Y$ counter,

$$\text{Gates} = Y$$

All but the least significant gate must be inverted for a total of $Y-1$ NOT gates. All of the inputs feed each of the threshold gates. Additionally, the outputs of the threshold gates serve as inputs to all of the less significant bits in the counter. The total number of inputs is

$$\text{Inputs} = XY + 1/2(Y-1)Y$$

For a GOTO threshold gate implementation, the component complexity is:

$$[XY + 1/2(Y-1)Y] \text{ memristors} + Y \text{ GOTO pairs} + 2Y - 2 \text{ MOSFETs}$$

For an 8-bit Dadda multiplier, the complexity for the counters becomes

$6 \rightarrow 5$ counter:

$= 40$ memristors + 5 GOTO pairs + 8 MOSFETs

$17 \rightarrow 6$ counter:

$= 117$ memristors + 6 GOTO pairs + 10 MOSFETs

$25 \rightarrow 6$ counter:

$= 165$ memristors + 6 GOTO pairs + 10 MOSFETs

$18 \rightarrow 6$ counter:

$= 123$ memristors + 6 GOTO pairs + 10 MOSFETs

$9 \rightarrow 4$ counter:

$= 42$ memristors + 4 GOTO pairs + 6 MOSFETs

Total counter complexity:

$$= 487 \text{ memristors} + 27 \text{ GOTO pairs} + 44 \text{ MOSFETs}$$

When the original AND gate matrix is taken into account, the complexity becomes

Counter complexity:

$$= 487 \text{ memristors} + 27 \text{ GOTO pairs} + 44 \text{ MOSFETs}$$

AND gates:

$$= 2N^2 \text{ memristors} + N^2 \text{ GOTO pairs}$$
$$= 128 \text{ memristors} + 64 \text{ GOTO pairs}$$

Total complexity:

$$= 615 \text{ memristors} + 91 \text{ GOTO pairs} + 44 \text{ MOSFETs}$$

Note that this complexity is significantly less than the Minnick method. Figure 7.26 shows an 8-bit threshold gate Dadda multiplier implemented with the Kautz method.

In the Kautz implementation, the most significant bit of the counter output has a single delay and the remaining output bits have a delay one greater than the next most significant bit of the output. Both of these delays can be seen in the $7 \rightarrow 3$ counters in Figure 7.25. Naively, an 8-bit Dadda multiplier, would require $1 + 5 + 6 + 6 + 6 + 4 = 28$ cycles. However, the counters do not need to be entirely serialized. The most significant bits resolve sooner than the less significant bits and can be passed on as soon as they are complete to the next adder. Thus, the effective latency of each counter becomes the number of steps to produce the bits forwarded to the next counter. The delay becomes $1 + 2 + 3 + 3 + 3 + 4 = 16$ steps. This is equivalent to the delay for the Minnick method.

7.3.3.4 Counter comparison

Table 7.23 shows the comparison for the proposed two approaches and the traditional approach.

Depending on the application and optimization criteria, either the Kautz or Minnick approach should be selected. The Kautz approach requires about 1/2 of the GOTO pairs and about 1/4 the number of memristors for the 8-bit design. Also, it requires about 1/5 the number of MOSFETs, the more expensive component in the design. The trade-off changes slightly for multipliers of other sizes.

Depending on the exact size of the multiplier and its counters, either the Kautz or Minnick method will have a shorter delay. The Kautz method is able to improve its speed since it computes its most significant bits first. However, the Minnick method benefits from resolving all, but its most significant bit in a constant three steps.

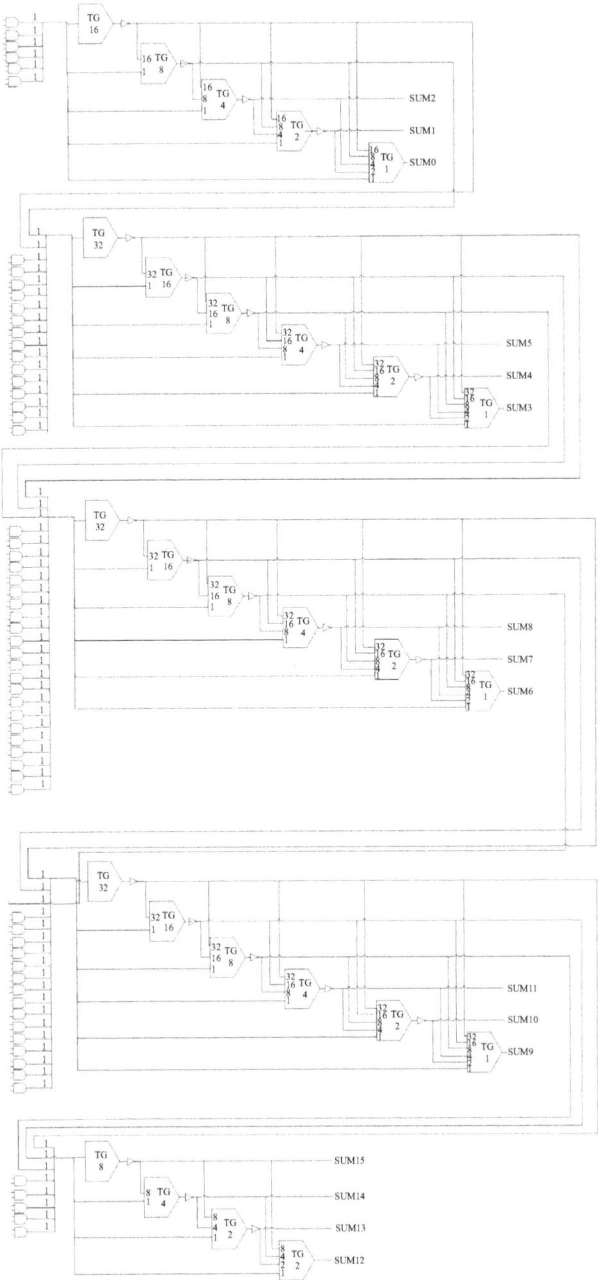

Figure 7.26 8-bit threshold gate Dadda multiplier with Kautz counters

Table 7.23 Complexity and delay comparisons for two 8-bit threshold gate Dadda multiplier designs

	Kautz	Minnick
Delay	16	16
Complexity	615 memristors 91 GOTO pairs 44 MOSFETs	2,700 memristors 201 GOTO pairs 230 MOSFETs

7.3.4 MAD gate Dadda multiplier

The final Dadda multiplier in this chapter is a MAD-based multiplier. At a high-level, the design consists of N^2 MAD AND gates, 35 full adders, 7 half adders, and finally an N-bit ripple-carry adder.

First, the N^2 MAD AND gates are instantiated. Each gate requires two input memristors and an output memristor. Thus, there are a total of $3N^2$ memristors in first stage for the dot matrix.

Recall that the IMPLY Dadda multiplier optimized each individual full adder and half adder to store its results into the destination adder rather than locally whenever possible. This optimization can be applied to many of the full and half adders in the MAD Dadda multiplier design too. The main exception is for the adders, which feed their output as a carry-in signal to the next adder.

As explained for the array multiplier, the carry-in signal to a MAD full adder does not have a resident memristor in the adder. Instead, it is expected to be stored elsewhere in a memristor and its terminals sensed to drive switches in the adder. Thus, for all adder outputs, which serve as carry inputs to the subsequent stages, the result will be stored into the local adder. It is desirable to minimize the number of places this occurs in the design.

In an 8-bit Dadda multiplier, there are 35 full adders and 7 half adders. Some of the full adders have at least one input that comes from an AND gate in the dot matrix. Other adders get all of their inputs from outputs of other adders.

First, consider the full adders, which have at least one input from an AND gate. Call one of these full adders FA0. The input from the AND gate is sitting in the result memristor of the AND gate. This input is selected to serve as the carry-in signal to FA0. The memristor already exists in the AND gate to perform the sensing of the p and n terminals to drive the carry-in switches in FA0. The other two inputs will be directly stored into FA0 from their previous full adders.

Now consider the full adders, which get all of their inputs from adder outputs. Call one of these full adders FA1. FA1 needs to have one of its inputs serve as the carry-in value. As explained, this value must be stored in the previous adder. The other two inputs can be directly stored into the input memristors in FA1.

In this way, the maximum number of memristors can be removed from the output adders. In an 8-bit Dadda multiplier, 17 adders receive all of their inputs from adder outputs. Thus, 17 memristors must be placed in local adders to store the carry-in signals as described. The remaining full adders only need the two input memristors. Since there are 42 total adders in the multiplier, the total memristor requirements for the intermediate stages becomes

$$\text{Memristors} = (42\text{ adders}) * 2 + 17 = 101$$

Although memristors are removed from the designs, the corresponding switch logic cannot be removed or simplified. The logic still needs to exist; it is just now acting on a different result memristor as described for the array multiplier. The number of switches remains 7 per half adder and 13 per full adder for a total of

$$\text{Switches} = 7 * 7 + 13 * 35 = 504$$

Notice that every adder executes in lockstep in each intermediate stage of the design. The same drivers can be used across all adder instantiations in the stage to drive the V_{cond}, V_{reset}, and V_{set} signals. Only four drivers are needed per stage. The total number of drivers for the intermediate stages in an 8-bit Dadda multiplier is 16.

In the final ripple-carry adder, all of the full adders need to store the final product. Thus, no sum memristors can be removed from this block. This ripple-carry adder will have identical structure to the optimized ripple-carry adder presented previously. It requires 56 memristors, 43 drivers, and 182 switches.

Consider the N^2 AND gates in the initial stage of the multiplier. For similar logic as described above, some of the result memristors in the AND gates can be removed. There are 35 total full adders in the design and that 17 of them get all their inputs from adders. This means 18 full adders get at least one of their inputs from an AND gate. Thus, 18 of the AND gates must hold their results in a local memristor to be sensed as a carry-in signal. The other 46 AND gates are able to remove their result memristor and store the result directly into an adder in a subsequent stage.

The AND gates share one driver to load the input values in parallel, one driver to read the input voltages, one driver to set the output memristors, and one driver to read the outputs for a total of four drivers. The number of switches is four per input circuitry in the AND gates and one per output memristor. The total number of switches is $4N^2+18$. The total complexity of the design is

Adder stages:

$$= 101\text{ memristors} + 504\text{ switches} + 16\text{ drivers}$$

Ripple-carry adder:

$$= 56\text{ memristors} + 182\text{ switches} + 43\text{ drivers}$$

AND gates:

$$= 146\text{ memristors} + 274\text{ switches} + 4\text{ drivers}$$

Total complexity:

$$= 303\text{ memristors} + 960\text{ switches} + 63\text{ drivers}$$

The delay through each full adder and half adder is a single step. The first stage performs the AND gates and stores the results into the corresponding AND gate result memristors or adders. At $t = T$, the current stage performs the V_{set} operation and the input memristors A and B in all of the stage's adders are populated. The input memristors, which use an AND gate as an input, are loaded a priori. At $t = T + 1$, V_{cond} is applied to the inputs, and the carry-in signals are sensed from the previous stage(s) to compute the output sum and carry-out values. These values are stored into the input memristors in the next stage. Each stage completes in a single cycle.

The final 14-bit ripple-carry adder takes another 14 steps, and the AND gates take 2 steps for a total of 20 steps. Note that 2/3 of the delay is attributed to the final adder. Thus, as the size of the multiplication increases, a faster adder should be considered at the cost of complexity.

7.3.4.1 Challenges to pipelining

In theory, the Dadda multiplier can also benefit from pipelining. Since execution is based on drivers, after a stage of adders completes its portion of a multiplication, it can begin a subsequent multiplication. The design can be pipelined across intermediate stages and across the individual bits of the final ripple-carry adder. Thus, each pipeline stage has the equivalent critical path delay of a full adder.

However, the current design has some issues that must be addressed. Recall that some of the adders require the carry-in bit from an adder in a previous stage. However, it is not necessarily the previous stage. It is possible that an adder in the final stage requires a carry-in signal from an adder in the first stage. These values cannot be reset for the next multiplication until they have been used. Thus, either additional "placeholder" memristors must be added for these values to be copied to for later stages or the operations must be rearranged such that inputs always feed the next stage in which case pipelining can be performed.

If these modifications are performed on the design, it is possible to pipeline the Dadda multiplier at the granularity of an adder stage. The longest stage is the first stage that performs an AND operation. In the first step, the inputs are loaded. In the second step, the outputs are computed. In the third step, the memristors are reset. A new multiplication can begin every three steps.

7.3.5 *Dadda multiplier analysis and comparison*

Table 7.24 shows the complexity and delay requirements for each of the presented Dadda multiplier designs. For the threshold approach, the Kautz method is shown since it has a significantly lower component area and maintains an equivalent delay for the 8-bit implementation.

The IMPLY, threshold, and MAD-based approaches require the fewest components. The threshold gate approach requires no drivers or switches and has the fewest total number of components. It requires very minimal CMOS due to its ability to reuse inverted values. In all, it uses about 1/3 the number of components that the CMOS design uses.

Table 7.24 Complexity and delay for the 8-bit Dadda multiplier designs

108	**CMOS**	**IMPLY**	**Hybrid-CMOS**	**Threshold-GOTO**	**MAD**
Complexity	2,204 MOSFETs	577 memristors	728 memristors 530 MOSFETs	615 memristors	303 memristors 960 switches
		577 drivers 577 switches		91 GOTO pairs 44 MOSFETs	63 drivers
Delay	57	109	73	16	20

The IMPLY and MAD approaches are fairly equivalent in terms of complexity. The MAD approach requires fewer memristors and reduces the number of drivers by almost 90 percent. However, it uses nearly 400 more switches than the IMPLY approach.

The hybrid-CMOS approach benefits from not needing driver circuitry. This removes the switches and drivers present in the IMPLY and MAD designs. However, it requires the most memristors with 728. Additionally, the significant use of signal restoration buffers increases the amount of CMOS in the design to 530 MOSFETs, making it the largest of the memristor-based designs. However, this is still about one-fourth of the number of MOSFETs in the CMOS design and fewer overall components.

The MAD-based approach significantly reduces the delay as compared to the IMPLY approach from 109 steps to just 20. This is also almost 1/3 of the CMOS delay. The threshold-gate approach requires the fewest steps at just 16, over a 3× improvement as compared to CMOS. The hybrid-CMOS approach aligns closely with the traditional CMOS delay, with the additional delays for the signal restoration buffers after some of the intermediate stages.

As described earlier, because of its driver-centric nature, the IMPLY approach can pipeline numerous multiplications. Once each stage of the multiplication finishes its execution and stores its result into the subsequent adder in the next stage, it can execute its work on the next multiplication. Similarly, the final N-bit addition can be pipelined across bits. The resultant throughput is now just a factor of the slowest stage in the pipeline, 18 steps per multiplication for the IMPLY approach.

The MAD approach has a couple limitations for pipelining due to the optimizations that have been performed on the multiplier. However, at a fundamental level, pipelining is also possible for the MAD design, beginning a new multiplication every three steps.

Chapter 8
Memristor-based divider designs

This chapter presents completed work in the design and implementation of memristor-based dividers. IMPLY, hybrid-CMOS, threshold-gate, and MAD approaches to binary-restoring, SRT, and Goldschmidt dividers are described and analyzed in terms of complexity and delay. Completed schematics and simulations are also given.

8.1 Binary-restoring dividers

Memristor-based binary performing restoring dividers are presented for the IMPLY, hybrid-CMOS, threshold, and MAD contexts. Recall that binary-restoring dividers execute by performing iterative subtractions of the divisor from the dividend. The divisor consists of a shift register and adder. For the discussed designs, let the width of the dividend and divisor be N.

8.1.1 IMPLY binary-restoring divider

Binary-restoring dividers offer a low complexity in comparison to other divider architectures. This simplicity comes at the cost of higher delay since the hardware is used iteratively to perform the division. Generally, IMPLY-based designs require lower area than other approaches, making the same trade-off to favor complexity at the cost of delay. This commonality makes IMPLY a prime candidate for the design approach to a binary-restoring divider since they share a common goal.

It is natural to construct the divider in the order that the division occurs. First, the divisor is copied into the divider to calculate its 2's complement. The 2's complement of the divisor is needed to perform the subtraction. The subtraction is performed each iteration of the divider by adding the 2's complement of the divisor to the dividend.

To obtain the 2's complement, the N-bit value is inverted with a NOT operation, and 1'b1 is added to it. In the context of IMPLY, the divisor can be directly NOT'ed into the input operands of the adder in one step. Assume the input operands of the adder, A and B, are reset to 0, and the value of the divisor d sits in memristors d_{N-1} to d_0. By applying an IMPLY operation to memristors d_{N-1} to d_0 and B_{N-1} to B_0, the inverse of d ends up in B. V_{cond} is applied to the d_{N-1} to d_0 memristors, and V_{set} is applied to the B_{N-1} to B_0 memristors. Next, the least significant memristor of

A is set to "1" by driving a V_{set} voltage to the memristor. This represents the 1'b1 added to the B memristors to get the 2's complement. Now the addition is performed. For this design, a ripple-carry adder is used for the addition.

8.1.1.1 Logic simplifications

Note that the carry-out of the adder can be ignored for all of the iterations. This simplifies the logic in the final bit of the adder to only compute the sum equation, removing one memristor and four steps from the critical path. The total delay for each subtraction iteration is $2N + 16$ steps. This is crucial to the critical path of the divider as will be described. The updated steps are shown in Table 8.1.

The steps associated with resolving the carry-out signal and propagating the signal to the next adder are removed.

More optimizations can be performed given the fact that the 2's complement of the divisor is used in all subsequent additions in the design. The result of the 2's complement is stored directly into the input memristors rather than the sum memristors when it is computed. When the 2's complement computation is performed, the V_{set} voltage is applied to the input memristors instead. In the final sum operation (step 21 in a standard full adder and step 18 in Table 8.1), the B_i memristor is used rather than the sum memristor M1 in each bit i.

Once the 2's complement is stored in the B input memristors, it stays resident for the remainder of the division since it is used in every addition.

The remaining subtraction iterations can also perform this optimization. The dividend is reduced by subtracting the divisor in each iteration. Thus, the running

Table 8.1 Steps for an IMPLY full adder with no carry-out

Step	Functionality	Goal
1	A IMP 0	A→M2
2	B IMP 0	B→M3
3	B	M3→M1
4	(A IMP 0) IMP (B IMP 0)	M2→M3
5	A IMP B	A→M1
6	((A IMP 0) IMP (B IMP 0)) IMP 0	M3→M0
7	A XOR B	M1→M0
8		FALSE(M1,M2,M3)
9	(A XOR B) IMP 0	M0→M2
10	C_{in} IMP ((A XOR B) IMP 0)	C_{in}→M2
11		
12		
13	C_{in} IMP 0	C_{in} → M1
14	C_{in} OR (A XOR B)	M1→M0
15	(C_{in} OR (A XOR B)) IMP 0	M0→M3
16	(C_{in} NAND (AXORB)) IMP ((C_{in} OR (A XOR B)) IMP 0)	M2→M3
17		False M1
18	(C_{in} NAND (AXORB)) IMP ((C_{in} + (A XOR B)) IMP 0) IMP 0 = SUM	M3→M1

sum is used as the A input in each iteration. The sum can be stored into the A memristors rather than the sum memristors for each of the division iterations. This removes copy steps that would otherwise be required to move the sum to the inputs at the end of each iteration.

Now, in the step before the final sum IMPLY operation, the A input memristors are reset. Originally, the sum memristors were reset. No steps or memristors are removed for this optimization.

In the traditional ripple-carry adder, the first three steps perform NOT and COPY operations on the input operands as recreated in Table 8.2.

However, in the context of this divider, the B operand remains constant as the 2's complement of the divisor. Thus, steps two and three can be performed immediately after the full adder completes its subtraction, and the remainder of the full adders execute.

All of the full adders sit idly for many steps of a ripple-carry addition. Thus, during any two of these idle steps, a full adder can reset its memristors and perform steps 2 and 3. In this way, the latency of these steps is hidden by the remainder of the division iteration after the subtraction. This removes another two steps from all subsequent additions.

Note that it is not possible to remove these steps entirely. The values of memristors M1 and M3 are reset and overwritten after steps two and three. Thus, they must be repopulated prior to each subtraction iteration.

8.1.1.2 Shift logic

The shift functionality can be incorporated into the adder to remove the need for any additional hardware and reduce the delay by overlapping operations. This can be done in the same manner as the shift-and-add multiplier. The structure of these two designs is very similar.

Each iteration, the running result needs to be shifted left before it is used as the A operand in the next iteration. The shift operation can be performed by storing the sum of each full adder f into the next full adder $f+1$ rather than itself. Thus, the reset operation before the final sum IMPLY step will reset this updated memristor. This is identical to the shift functionality described in previous designs. Bit 0 of the A operand is simply reset to 0, and no sum is stored into it.

The shift operation for the divider is slightly more complicated than the shift operation for the shift-and-add multiplier. The shift-and-add multiplier performs a right shift. In general, this is a safer operation for an IMPLY design because the result is being stored into a block that was "earlier" in execution and has

Table 8.2 Initial steps for a pipelined IMPLY full adder

Step	Functionality	Goal
1	A IMP 0 → M2	NOT A
2	B IMP 0 → M3	NOT B
3	(B IMP 0) IMP 0 → M1	Copy of B

completed. For example, in the shift-and-add multiplier, full adder i stores its sum into full adder $i - 1$.

For the left shift in the binary-restoring divider, the result is stored into a block later in execution, which may be still executing. For example, full adder i stores its sum into full adder $i + 1$. However, full adder $i + 1$ is still executing its addition steps when full adder i is ready to store its sum.

Before the sum can be stored into the full adder $i + 1$, full adder $i + 1$ must reset its input memristor A. This can be performed in the same step that memristors M1, M2, and M3 are reset so that no additional delay is incurred. It is guaranteed that the A input memristor is no longer in use at that step based on the ripple-carry adder execution steps. Thus, the sum from full adder f can be stored into the A memristor in $i + 1$ when i completes, even if full adder $i + 1$ is still executing its original addition.

8.1.1.3 Multiplexer logic

Additionally, the result of the addition is a conditional store unlike the shift-and-add multiplier. In the shift-and-add multiplier, the driver logic was unchanged but updated to drive a different memristor. However, in the binary-restoring divider, a simple logic change does not work for the final V_{set} operation. If the result of the subtraction is negative (MSB = 1), the original value is stored, and if it is positive (MSB = 0), the new result is stored. Thus, additional logic must be performed before the final sums are shifted and stored back into the adder.

Because of this complication, full adder $i + 1$ cannot receive the shifted value from i until it computes the most significant bit of the subtraction. If the most significant bit is a "1," the current iteration's value of A needs to be preserved and shifted. In this case, the value in full adder $i + 1$'s A memristor needs to be moved to the A memristor in full adder $i + 2$. If the most significant bit is a "0," the new sum needs to shifted and stored into the A memristors.

This decision cannot be determined until the final adder in the ripple-carry adder computes its value. Thus, all full adders $F_0, \ldots, F_{n-1}$ must wait to perform the final sum operation until this logic is complete. During these idle cycles, the full adders can reset all but their B memristors. It is safe to reset the A memristors since they definitely receive a new value for the next iteration. However, the design must ensure that the A memristors have been read by the multiplexer logic before they are reset.

The multiplexer logic for this divider is responsible for selecting the input operand A for each iteration. The multiplexer logic uses the most significant bit of the subtraction result as a select line. The inputs to the multiplexer are the shifted value of the current A memristors and the shifted value of the current sum memristors. The output is stored back into the A memristors to begin the next iteration. The steps for this multiplexer are shown in Table 8.3.

Let S be the MSB of the sum, which lies in the Nth full adder. Let A be the bit of the dividend in each full adder, held in the A input memristor. Let Sum be the computed sum output from the adder that inputs into the multiplexer.

Table 8.3 IMPLY mux and shift execution steps

Step	Mux functionality	Mux goal
1	(NOT S) IMP 0 → M0	S lies in all adders
2	M0 IMP (NOT A)	S NAND A
3	(NOT Sum) IMP 0 → M1	Sum
4	Sum IMP M0	Sum NAND NOT S
5	M0 IMP 0→A	Sum AND NOT S
6	(S NAND A) IMP (Sum AND NOT S)	(S AND A) OR (NOT S and Sum)

The six steps execute in parallel across all full adders. The inverse of the Nth sum is used as indicated by NOT S in step 1. This is because the inverse is ready a step earlier than the sum as shown in Table 8.3. Similarly, the inverse of the sum bit in the N full adders is used. Now, the last two steps to compute the sum in the N full adders are no longer necessary. Instead, the multiplexer can begin. The updated steps for the full adders with the multiplexer incorporated are shown in Table 8.4.

The multiplexer only requires three memristors so the full adder memristors can be reused without any additional components. The total delay for a full adder is 29 steps. Note that the number of steps could be reduced by adding more memristors. However, the steps do not contribute to the critical path. Recall that all of the full adders except the last sit idle for at least two steps while the remainder of the full adders execute. Thus, the added steps are hidden. The complexity of each of the full adders is seven memristors.

In step 27, the result of the IMPLY operation is stored into the A memristor of the next full adder $i+1$. This way, the next addition is ready to begin as usual when the multiplexer completes. The A memristor was previously cleared with the other internal memristors in step 10. The multiplexer logic is strategically designed to never use the value of the dividend itself, only its inverse. Thus, the A memristor can be cleared and ready in step 27. The A memristors are also logically separated from the remainder of the adder during steps 27 and 28. This allows each full adder i to both read memristors M1 and M0 and have the A memristor set by full adder $i-1$.

Note that S and Sum are the same value in the final full adder. The final full adder is also the only adder, which cannot reset its internal memristors during any idle steps before the MSB is computed. However, the final full adder can still be optimized to remove the three steps associated with the carry-out signal. Thus, the final full adder completes in 26 steps.

The total complexity of the N-bit modified ripple-carry adder is $7N$ memristors. At first glance, the total delay is $2N+24$ steps. However, recall that steps two and three can be performed during the idle steps after the previous subtraction. Thus, only $2N+22$ steps per iteration are on the critical path.

Last, the design must place one bit of the quotient into the quotient register on each iteration. Again, let S be the MSB of the sum in the Nth full adder. If S is a "1," the quotient bit is a 0, else it is a 1. Thus, the quotient bit is essentially NOT S.

Table 8.4 Steps for an optimized IMPLY parallel add operation with multiplexer

Step	Functionality	Goal
1	A IMP 0	A→M2
2	B IMP 0	B→M3
3	B	M3→M1
4	(A IMP 0) IMP (B IMP 0)	M2→M3
5	A IMP B	A→M1
6	((A IMP 0) IMP (B IMP 0)) IMP 0	M3→M0
7	B IMP (A IMP 0)	B→M2
8	B IMP (A IMP 0) IMP 0	M2→M4
9	A XOR B	M1→M0
10		FALSE(M1,M2,M3,A)
11	(A XOR B) IMP 0	M0→M2
12	C_{in} IMP ((A XOR B) IMP 0)	C_{in}→M2
13	C_{in} AND (A XOR B) + AB→ C_{OUT}	M2→M4
14	*Next bit is reading C_{in}*	
15	C_{in} IMP 0	C_{in} → M1
16	C_{in} OR (A XOR B)	M1→M0
17	*Next bit is reading C_{in}*	
18	(C_{in} OR (A XOR B)) IMP 0	M0→M3
19	(C_{in} NAND (AXORB)) IMP ((C_{in} OR (A XOR B)) IMP 0)	M2→M3
20	Reset all except M3	
21	A IMP 0 → M0	NOT A
22	Reset A	
23	(NOT S) IMP 0 → M1	S lies in all adders
24	M1 IMP M0	S NAND A
25	M3 IMP 0 → M2	Sum
26	M2 IMP M1	Sum NAND NOT S
27	M1 IMP 0→A	Sum AND NOT S
28	M0 IMP A	(S AND A) OR (NOT S and Sum)
29	Reset all except A and B	

The first step of the multiplexer already performs NOT S in parallel across the full adders. Now this step will also be performed on the quotient bit memristor. For each iteration i, the quotient bit will be stored into the $(N+1-i)$th bit of the quotient register to effectively perform the shifts.

8.1.1.4 Complexity and delay

Due to the ripple-carry adder simplifications, the total delay for a single subtraction is $2N+22$ steps. The total delay for the divider is $(N+1)(2N+22)$ steps for the subtraction iterations. The initial 2's complement operation takes $2N+16$ steps and a single step for the NOT operation. Thus, the total delay is $(N+2)(2N+22)-6$ steps. Note that it is not possible to pipeline divisions since none of the bits are able to complete until the final bit in the adder completes.

The total complexity of the design is very minimal. In fact, the total complexity is just the complexity of the adder. Recall that the main components of a binary-restoring divider are an adder for computing the 2's complement, an adder for computing the iterative subtractions, an N-bit multiplexer, and registers for holding the running values in each iteration. The optimizations in this section have removed the registers and consolidated the multiplexer and 2's complement logic into the adder. Thus, the total complexity is $7N + 1$ memristors $+ 7N + 1$ switches $+ 7N + 1$ drivers. The circuitry looks identical to the traditional ripple-carry adder.

8.1.2 Hybrid-CMOS binary-restoring divider

The low fan-out of the signals in the binary-restoring divider make hybrid-CMOS gates a practical option for a design because minimal signal restoration buffering is needed. The proposed design also minimizes the need for these signal restoration buffers by placing them at optimal granularity in the design.

The design consists of two N-bit adders, an N-bit multiplexer, and a $2N$-bit shift register. The quotient shift register is initialized to the dividend value. The first adder computes the 2's complement of the divisor. This is entirely combinational in the hybrid-CMOS design so the result is not latched anywhere in the design. Thus, a second adder is needed for the remaining iterations so that the 2's complement is preserved.

The second adder is used for each subtraction across the iterations. One of its inputs is the 2's complement adder's output, and one of the inputs is the most significant N bits of the quotient shift register. The output of the second adder stores itself back into the most significant bits of the shift register, and the quotient bit for the iteration is stored into bit 0 of the shift register.

When this store operation is performed, the most significant bit of the sum is used to determine if the A inputs of the adder or the newly computed sum are stored into the shift register. As described previously, if the most significant bit is a "0," the result is positive, and the result of the subtraction is stored into the register. If the most significant bit is a "1," the result is negative, and the register is not changed. The multiplexer performs this logic. Last, the left shift is performed in the register.

8.1.2.1 Complexity and cost

The adders each require $14N$ memristors and $4N$ MOSFETS without any signal restoration buffers. The multiplexer requires $6N$ memristors and two MOSFETs for the single inverter on the select line. Both these complexities have been described for previous hybrid-CMOS designs.

The hybrid-CMOS shift-and-add multiplier showed that a shift register requires $(22N - 6)$ memristors $+ (26N - 4)$ MOSFETs. The register does not require any signal restoration buffers due to the frequent use of MOSFETs for NOT gates. For a $2N$-bit register, this is $(44N - 6)$ memristors $+ (52N - 4)$ MOSFETs.

Both the adders need signal restoration buffers on their carry-out signals as they propagate through the bits of the adder. Also, the sum outputs of the second adder must be buffered due to the series of gates in the multiplexer and shift logic that the signals must propagate through before reaching the shift register. The $2N$ carry-out buffers require $8N$ MOSFETs, and the N sum output buffers require $4N$ MOSFETs. Due to the CMOS inverters in the adders, buffers between the first two adders are not necessary. Similarly, the concatenation of gates in the multiplexer logic is not long enough to warrant signal buffers. The CMOS signal restoration complexity is $12N$ MOSFETs. The total complexity for the hybrid-CMOS binary-restoring divider is

2 N-bit ripple-carry adders:

$$\begin{aligned} &= 2(14N \text{ memristors} + 4N \text{ MOSFETS}) \\ &= 28\,N \text{ memristors} + 8N \text{ MOSFETs} \end{aligned}$$

$2N$-bit shift register:

$$= (44N - 6) \text{ memristors} + (52N - 4) \text{ MOSFETs}$$

N-bit multiplexer:

$$= 6N \text{ memristors} + 2 \text{ MOSFETs}$$

Signal-restoration buffers:

$$= 12N \text{ MOSFETs}$$

Total complexity:

$$= (78N - 6) \text{ memristors} + (72N - 2) \text{ MOSFETs}$$

The gate delay of the design is equivalent to the delay of a hybrid-CMOS adder, followed by $N+1$ iterations of an adder, multiplexer, and shift operation addition. One execution of a hybrid-CMOS ripple-carry adder takes $3N+4$ steps. The shift operation on the $2N$-bit register requires six steps—two to load and four to shift. The multiplexer takes three cycles. The delay per iteration is $3N+13$ steps. The initial 2's complement addition takes $3N+4$ steps. The total delay is

Delay:

$$\begin{aligned} &3N + 4 \\ &+(N+1)(3N+13) \\ &= (N+2)(3N+13) - 9 \text{ steps} \end{aligned}$$

8.1.3 Threshold-gate binary-restoring divider

The design of a binary-restoring divider using threshold gates combines the presented adder, latch, and multiplexer designs from previous sections. The GOTO pair implementations for the threshold gates are used since they use fewer components than the CSTG approach. This aligns with the low complexity of the binary-restoring divider.

The N-bit ripple-carry adder requires $N+2$ gate delays and $7N$ memristors, $2N$ GOTO pairs, and $2N$ MOSFETs. The threshold-gate shift-and-add multiplier design also introduced a shift register cell that was capable of loading and shifting. It was shown that a 2N-bit shift register uses 32N memristors, 10N GOTO pairs, and 4 MOSFETs. However, the shift register is modified to incorporate the multiplexer logic too. Figure 8.1 shows a diagram for the updated flip-flops in the shift register.

The least significant N bits in the shift register do not need the multiplexer logic. They are used for storing the quotient bits. Thus, they will take the form of the standard flip-flop shown in Figure 8.1(a). These registers execute as presented for the shift-and-add multiplier, each accepting its quotient bit in the ith iteration.

The most significant N bits use the updated flip-flop with the multiplexer incorporated as shown in Figure 8.1(b). The load capability has been enhanced to execute the multiplexer and load its output. The select line of the multiplexer is the MSB line in Figure 8.1(b). The two inputs to the multiplexer are denoted as ORIG and SUM, representing the current value of the A memristors and the result of the adder.

The multiplexer logic is in the top three threshold gates in Figure 8.1(b). The multiplexer executes when the CLK and the LD signal are high. If the most significant bit of the sum is a "1," the original value held in A will be stored in the shift register. This is done with the first threshold gate. If the most significant bit of the sum is a "0," the output of the third threshold gate will be the updated sum bit. The second threshold gate remains unchanged and represents the scenario in which the LD signal is low, and a shift is occurring on the D input from the previous bit.

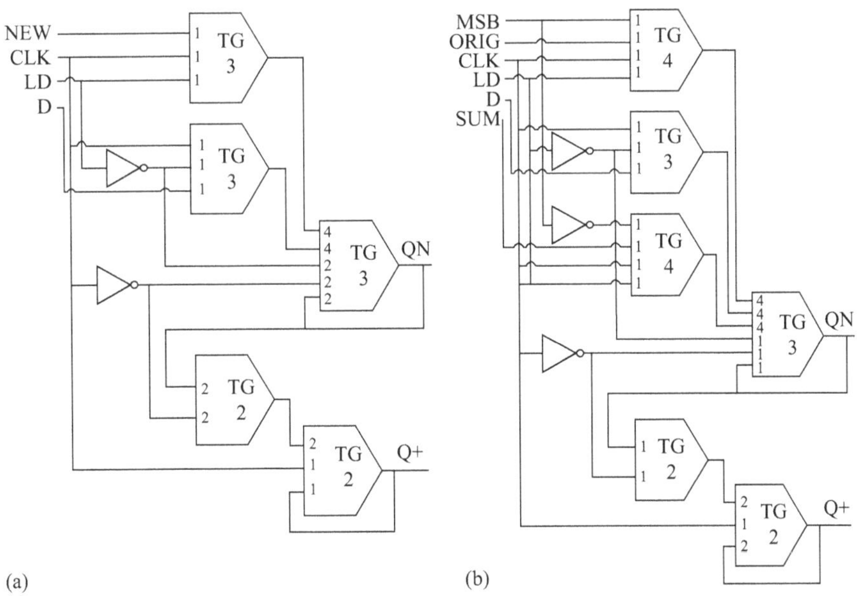

Figure 8.1 Threshold-gate-based shift register cells (a) without and (b) with multiplexer logic

The first three threshold gates are mutually exclusive. Exactly one of them can have an output of "1." Each feeds as an input to the fourth threshold gate. Additionally, the fourth threshold gate has three inputs that are identical to the shift register in Figure 8.1(a). In all, there are four scenarios that can set the output QN to one, and all of the scenarios are mutually exclusive. This ensures correctness in the fourth threshold gate. The remainder of the gate is identical to the original shift register in Figure 8.1(a).

The updated shift register cell has a higher complexity than the original. The additional two threshold gates introduce a total of six threshold-gate inputs, one threshold-gate outputs, and one NOT gate. The NOT gate inverts the MSB input and can be shared across all N shift register cells. Each input adds a memristor, each output adds a GOTO pair, and a NOT gate adds two MOSFETs. In all, the N-bit shift register/multiplexer requires

2N-bit shift register:

$$\begin{aligned}&32N \text{ memristors} + 10 \text{ NGOTO pairs} + 4 \text{ MOSFETs}\\&+ N(6 \text{ memristors} + 1 \text{ GOTO pair}) + 2 \text{ MOSFETs}\\&= 38N \text{ memristors} + 11N \text{ GOTO pairs} + 6 \text{ MOSFETs}\end{aligned}$$

The remainder of the design consists of the two N-bit adders. The first computes the 2's complement of the divisor, and the second iteratively performs the subtractions for the division. Alternatively, the first adder could be removed, and the 2's complement addition could be performed in the second adder. However, this requires considerable logic on the inputs and the outputs of the adder. An additional register would also be needed to store the 2's complement of the divisor so it is not lost. This logic and register would require more complexity than a second adder, so a second adder is used instead.

The total complexity including the adders is

Two N-bit adders:

$$\begin{aligned}&= 2(7N \text{ memristors} + 2N \text{ GOTO pairs} + 2N \text{ MOSFETs})\\&= 14N \text{ memristors} + 4N \text{ GOTO pairs} + 4N \text{ MOSFETs}\end{aligned}$$

Total complexity:

$$= 52N \text{ memristors} + 15N \text{ GOTO pairs} + 4N + 6 \text{ MOSFETs}$$

In terms of delay, the critical path of the design is the delay for the 2's complement and the iterative division. The delay for the initial 2's complement addition is the expected $N+2$ delays. Each iteration of the division consists of an addition, multiplexer logic, and a shift operation. The delay for each of these additions is again $N+2$ delays. The delay of the updated shift register with the multiplexer maintains the original delay of the shift register: six gate delays for a shift and two gate delays for a load. Each iteration takes $N+10$ delays. Thus the total delay is

$$\begin{aligned}\text{Delay} &= (N+2) + (N+1)(N+10)\\&= (N+2)(N+10) - 8 \text{ steps}\end{aligned}$$

8.1.4 MAD-gate binary-restoring divider

The MAD approach to a binary-restoring divider allows for heavy optimizations. Many of the optimizations performed on the shift-and-add multiplier can be applied to the binary-restoring divider too. As a result, the complete design only requires a single N-bit ripple-carry adder and an N-bit register.

The N-bit register holds the current quotient as it is resolved one bit at a time across the iterations. Note that this N-bit register replaces the original design's $2N$-bit shift register. Not only does the register have half the width, it does not need the shift logic. The support and adaptations for this in the design are the same as described for the shift-and-add multiplier. Fundamentally, the shift logic can be removed by changing the destination drivers for store ops. The N bits of the shift register for the running sum can be removed by storing the running sum directly back into the adder itself.

The binary-restoring divider also has the 2's complement and multiplexer. Neither of these logic blocks were necessary for the shift-and-add multiplier. This section will focus on optimizations that can be performed on these logic blocks. Specifically, the multiplexer, shift, and 2's complement logic can all be performed within a single MAD adder. The driver circuitry must change slightly from the standard ripple-carry adder to accommodate these optimizations.

First, focus on the initial 2's complement operation. The adder is loaded with the inverse of the divisor in the B memristors and 0 in the A memristors. Then, a carry-in of 1 is driven to bit 0 and the 2's complement completes in N steps. The value is then copied back to the B memristors using a reset and single-step MAD copy operation. The value in the B memristors will then remain through all of the iterations of the addition. In all, the 2's complement requires $N + 3$ steps for the initialization, addition, and copy. No changes to the circuitry are required to enable this since it is a standard addition.

Second, the adder is adapted to logically shift the result of the subtraction left as described for the shift-and-add multiplier. The sum from each bit i's full adder will be (conditionally) stored into the input operand A in bit $i + 1$ instead. This poses the same challenge as the IMPLY design because the destination memristor is in a later pipeline stage that may not be reset or idle yet. Also, as described for the IMPLY design, the adder must determine the final sum bit before resolving the sum into the input memristors in any of the bits.

In each iteration, the final full adder bit in the ripple-carry adder executes and resolves its most significant bit in N steps. This sum in turn can drive the resolution of the remaining sum bits as they are shifted and stored back into the A memristors. Before this can occur, the input memristors A must be cleared so they can accept the new value.

Let each full adder reset all of its internal memristors except the A and sum memristors during its idle stages. All of the full adders except the Nth have at least one idle step to perform this reset. In step $N + 1$, all of the full adders copy their A memristors to the C_{out} memristor in the full adder. This is purely to move the value out of the A memristors which serve as the destination during the shift operation. In step $N+2$, the A memristors are reset, and in step $N+3$, the shift operation occurs.

If the most significant bit has a sum of 0, the drivers drive the result of the sum in the $(i-1)$th adder into the A memristor in the ith adder. If the most significant bit has a sum of 1, the drivers drive the input A memristor in the $(i-1)$th adder into the A memristor in the ith adder. Either way, a left shift operation executes.

The driver logic must be adapted to support this multiplexer logic. A V_{cond} read voltage is applied to both the C_{out} memristor (holds A value now), and the sum memristor in the $(i-1)$th adder at the same time a V_{set} voltage is applied to the A memristor in the ith adder. The V_{set} voltage will be gated by two parallel paths, each with two switches.

The first path's gates will be driven by the n-terminal of the Nth full adder's sum memristor and the p-terminal of the $(i-1)$th full adder's sum memristor. This essentially performs $\overline{\text{MSB}_n}$ AND Sum_{i-1}.

The second path's gates will be driven by the p-terminal of the Nth full adder's sum memristor and the p-terminal of the $(i-1)$th full adder's A value in the C_{out} memristor. This essentially performs MSB_N AND A_{i-1}. The schematic for this logic for two consecutive full adders is shown in Figure 8.2.

In this way, the multiplexer and shift logic are incorporated into the adders. This adds four switches to each full adder in the design for a total of $17N$ switches. After step $N+3$, the updated dividend value exists in all of the full adders.

The quotient bit is stored in the same manner as described for the IMPLY design. An MSB of "1" is stored as a "0" in the quotient memristor, and an MSB

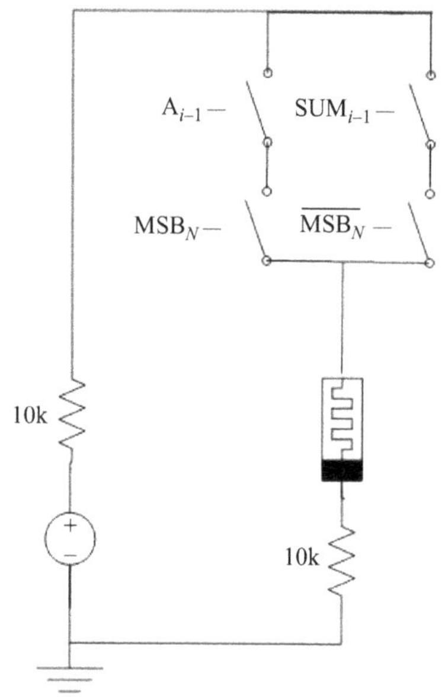

Figure 8.2 Logic for performing the shift operation in a MAD binary-restoring divider

of "0" is stored as a "1." This is achieved with a NOT operation on the MSB in parallel to the shift operation. This updates the quotient bit according to the result's sign. Now, the design simply resets all working memristors in preparation for the next iteration.

The delay of the divider is straightforward to compute. The initial 2's complement uses $N+3$ steps for the addition. After all of the optimizations performed on the shift and multiplexer logic, each iteration of the division also only incurs the delay of the addition, $N+3$ steps. The total delay for the binary-restoring divider is $(N+2)(N+3)$. Note that it is not possible to pipeline divisions since the circuitry is already pipelined for the individual iterations. The full schematic for an 8-bit restoring divider is shown in Figure 8.3.

8.1.5 Binary-restoring divider analysis and comparison

Table 8.5 summarizes the complexity and delay of each of the proposed binary-restoring dividers as well as the CMOS design.

In terms of delay, the IMPLY and CMOS approaches have roughly equivalent delays, and the threshold and MAD approaches have roughly equivalent gate delays. The hybrid-CMOS design has the highest delay, on the order of $3N$ per iteration. The IMPLY design has about double the delay of the threshold and MAD-based designs, but it is still competitive with the CMOS delay.

The proposed MAD-based design has the shortest delay by a small margin, requiring just three additional steps per iteration over the MAD ripple-carry adder to perform the shift and multiplexer logic. Each iteration takes less than half as many steps as CMOS and IMPLY and less than 1/3 as many steps as hybrid-CMOS. The threshold-gate design only requires seven more steps per iteration for the shift and multiplexer functionality, offering a competitive design delay too.

The IMPLY- and MAD-based approaches require the fewest components. Both designs have roughly equivalent component count. The IMPLY design requires fewer switches, and the MAD design requires fewer memristors and drivers. The MAD-based approach requires less than half the number of drivers as the IMPLY design and almost half the number of memristors. For both designs, the component count is less than 20 percent of the number of components in the CMOS design.

The threshold-gate implementation has significantly more components than the IMPLY and MAD designs but is still less complex than the CMOS design. The majority of its complexity count comes from memristors, which are relatively low in area. The high component count is partially due to the shift logic requiring a substantial number of components compared to other approaches.

The hybrid-CMOS implementation has the highest component count. It is the only implementation presented which requires more components than the CMOS design. This is largely due to the inefficiencies of the shift register and other combinational logic. For the binary-restoring divider, it may not be desired to use the hybrid-CMOS approach.

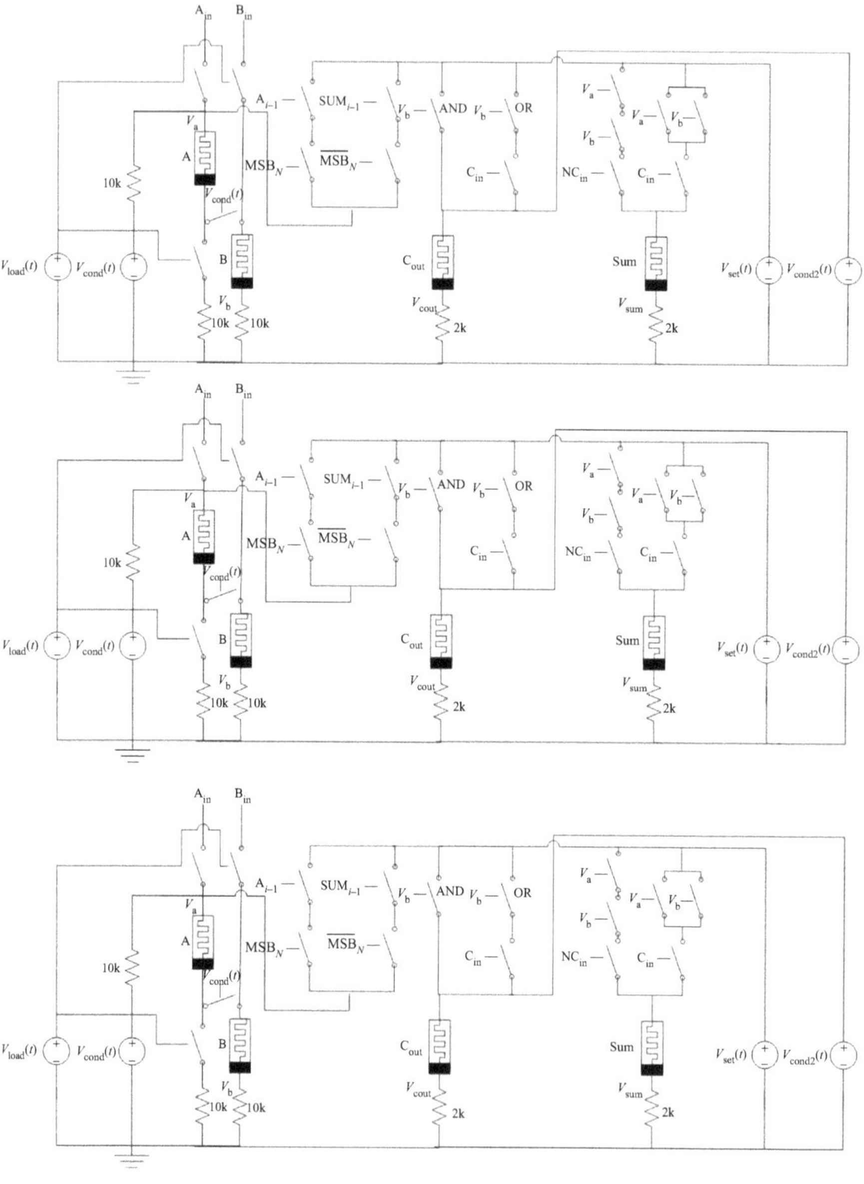

Figure 8.3 MAD binary-restoring divider schematic

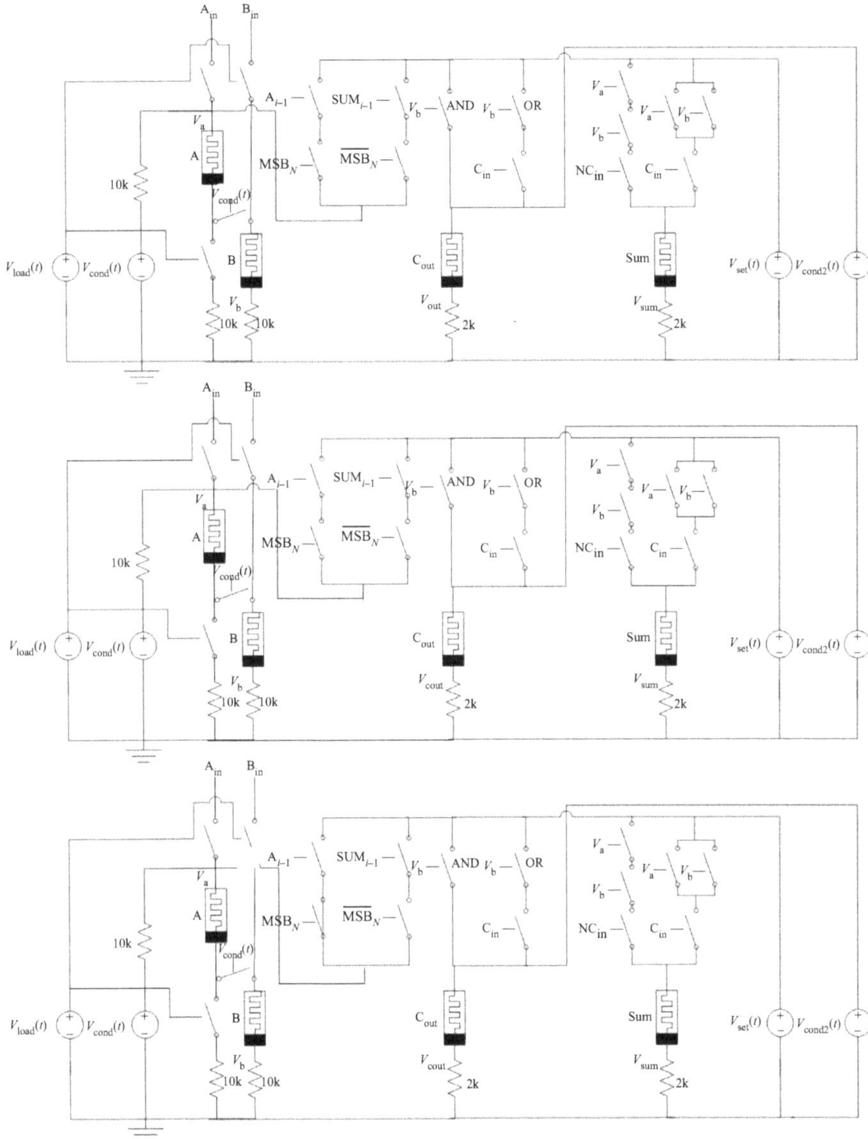

Figure 8.3 (Continued)

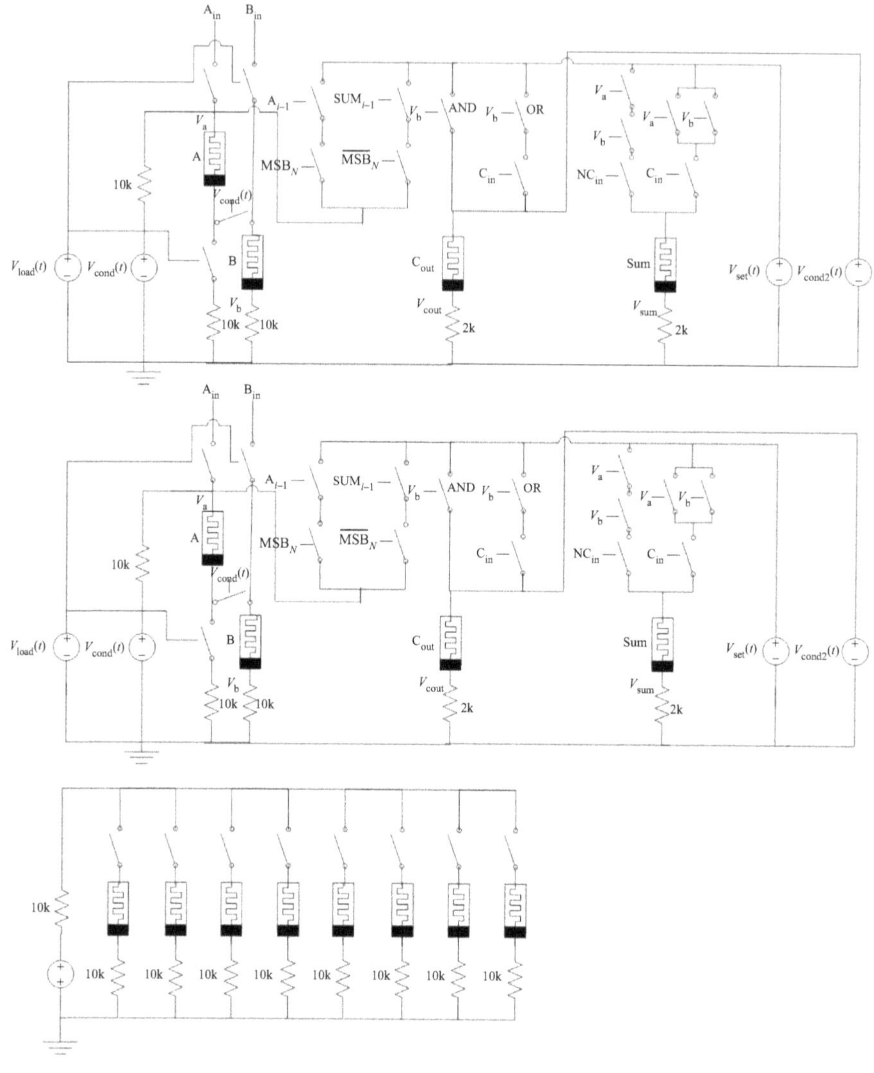

Figure 8.3 (Continued)

Table 8.5 Complexity and delay comparisons for the proposed binary-restoring divider designs

	CMOS	**IMPLY**	**Hybrid-CMOS**	**Threshold**	**MAD**
Delay	$(N+3)(2N+9)-5$	$(N+2)(2N+22)-6$	$(N+2)(3N+13)-9$	$(N+2)(N+10)-8$	$(N+2)(N+3)$
Area	$122N+22$ MOSFETs	$7N+1$ memristors $7N+1$ switches $7N+1$ drivers	$78N-6$ memristors $72N-2$ MOSFETs	$52N$ memristors $15N$ GOTO pairs $4N+6$ MOSFETs	$5N$ memristors $4N+1$ drivers $18N$ switches

8.2 SRT dividers

This section presents four approaches to SRT dividers constructed from memristors. First, an IMPLY approach is implemented and optimized, followed by a hybrid-CMOS, threshold-gate, and MAD-gate implementation. Each of the designs is compared and analyzed against the traditional CMOS design.

The SRT divider is similar to the binary-restoring divider presented in the previous chapter. One key difference is that the SRT divider stores the quotient in binary signed digit form rather than binary. This poses challenges to the hardware to represent three values with a single bit: 1, -1, and 0. Instead, the SRT divider is enhanced to perform on-the-fly conversion as described in the introduction of SRT division.

8.2.1 IMPLY SRT divider

Begin with the IMPLY binary-restoring divider as the initial design and modify it to perform SRT division instead. In this way, all of the optimizations in relation to memristor reuse, shift functionality and 2's complement addition are already present. Only the multiplexer logic and the quotient conversion logic need to be modified. Let the SRT divider perform X iterations before converging.

First, the shift operation on the running dividend P in an SRT divider is not a conditional shift. The dividend register necessarily shifts left every iteration to perform its multiplication by two. The shift operation does not need to wait for the last full adder to complete before it can start. In the SRT divider, the shift operation performs in a ripple manner through the adder like the addition does. This simplifies the SRT divider in comparison to the binary-restoring divider.

In the SRT divider, each full adder i can complete and store its result as soon as it executes. The only requirement is that the destination full adder, $i+1$, has cleared its A input memristor to accept the store operation from full adder i. To accommodate this, each full adder $i+1$ will reset its A memristor during the same reset stage that the other internal memristors are reset (step 10 in a standard full adder).

This reset step occurs before the previous full adder i has computed its sum. Thus, full adder i does not need to waste any idle steps once it is ready to store the sum in full adder $i+1$. In this way, the reset operation does not contribute to the critical path of the adder. Now, full adder i can perform its final step by applying the V_{set} voltage to the A memristor in full adder $i+1$. No multiplexer logic is needed on the A memristors now.

8.2.1.1 Multiplexer logic

Second, the SRT division must retain the divisor and its 2's complement, not just the 2's complement as was done in the binary-restoring divider. Depending on the running remainder in P, an iteration will either add 0, the divisor D, or the 2's complement of the divisor, D′ to P. Recall that the equations are as follows:

$$P^{+} = \begin{cases} P + D' \text{ if } P > \frac{1}{2} \\ P + D \text{ if } P < -\frac{1}{2} \\ \text{else } P + 0 \end{cases}$$

To determine this, the 2-to-1 multiplexer in the binary-restoring divider must be changed to a 4-to-1 multiplexer. Also, it will now be used on the B input memristors rather than the A input memristors since the A input memristors are used to hold the running sum P in this divider.

Let the four inputs to the multiplexer be D, E, F, and G and the select lines be S_{n-1} and S_{n-2} to represent the most significant two bits of the shifted sum S for the iteration. The multiplexer implements the following equation:

$$\begin{aligned} \text{Out} = {} & \left(\overline{S_{n-1}} \text{ AND } \overline{S_{n-2}} \text{ AND}\right) \text{ OR } \left(S_{n-1} \text{ AND } \overline{S_{n-2}} \text{ AND E}\right) \text{ OR} \\ & \left(\overline{S_{n-1}} \text{ AND } S_{n-2} \text{ AND F}\right) \text{ OR } \left(S_{n-1} \text{ AND } S_{n-2} \text{ AND G}\right) \end{aligned}$$

Now the inputs D and G are selected to be 0, input E to be the divisor and F to be the 2's complement of the divisor. However, even more extensive reduction can be done to simplify the 4-to-1 multiplexer in the context of an SRT divider. Since inputs D and G are 0, the multiplexer equation simplifies to

$$\text{Out} = \left(S_{n-1} \text{ AND } \overline{S_{n-2}} \text{ AND E}\right) \text{ OR } \left(\overline{S_{n-1}} \text{ AND } S_{n-2} \text{ AND F}\right)$$

The execution steps for this simplified 4-to-1 multiplexer are given in Table 8.6.

This multiplexer only requires nine steps and four additional memristors. The E, F, and select line memristors are also required, but these already exist in the design. This is now very competitive with the original 2-to-1 multiplexer that has been replaced from the binary-restoring divider.

The memristors can be strategically selected to also avoid having to copy the result of the multiplexer back into the adder. Consider the result memristor M3 for

Table 8.6 Optimized IMPLY implementation of a 4-to-1 mux for SRT

Step	Operation	Functionality
1	S_{n-2} IMP 0 → M0	$\overline{S_{n-2}}$
2	S_{n-1} IMP 0 → M1	$\overline{S_{n-1}}$
3	(S_{n-1} IMP 0) IMP 0 → M2	S_{n-1}
4	(NOT S_{n-2}) IMP (NOT S_{n-1})	S_{n-1} NAND $\overline{S_{n-2}}$
5	E IMP ((NOT S_{n-2}) IMP (NOT S_{n-1})) → M1	$\left(S_{n-1} \text{ AND } \overline{S_{n-2}}\right)$ NAND E
6	(S_{n-2}) IMP (S_{n-1})	$\overline{S_{n-1}}$ NAND S_{n-2}
7	F IMP ((S_{n-2}) IMP (S_{n-1})) → M2	$\left(\overline{S_{n-1}} \text{ AND } S_{n-2}\right)$ NAND F
8	M2 IMP 0 → M3	$\left(\overline{S_{n-1}} \text{ AND } S_{n-2}\right)$ AND F
9	M1 IMP M3	$\left(\left(S_{n-1} \text{ AND } \overline{S_{n-2}}\right) \text{ AND E}\right)$ OR $\left(\left(\overline{S_{n-1}} \text{ AND } S_{n-2}\right) \text{ AND F}\right)$

each bit in the adder. If M3 is selected as the B memristor in the full adder, then the result of the multiplexer will end up in the B memristor at the end of execution. This is the memristor that the adder expects the divisor input to be in each iteration. Thus, no copy operation is needed from the multiplexer to the adder each iteration. It also removes the need for the M3 memristor in each bit of the multiplexer. Now the multiplexer requires three memristors—M0, M1, and M2.

Also, to enable this 4-to-1 multiplexer to act in parallel across the N bits of the adder, the first IMPLY steps of the design have been selected to create inverses and copies of the selection bits S_{n-1} and S_{n-2} inside each full adder. This is necessary to preserve the original values for each of the adders. It also allows each full adder to operate separately from the other full adders.

Note that the select lines to the multiplexer, S_{n-1} and S_{n-2}, are the most significant bits of the shifted sum result from the previous iteration. This value sits in the A memristors as described at the beginning of this section. Thus, S_{n-1} and S_{n-2} are equivalent to A_{n-1} and A_{n-2}.

Is it possible to reduce the number of memristors for the multiplexer by reusing some of the memristors in the design? Each full adder has seven memristors. The A memristors are dedicated to holding the result of the shift operations and represent the running dividend as it decreases. The B memristors are dedicated to holding the result of the multiplexer operation and represent the value being added or subtracted from the dividend in that iteration.

Four of the memristors, M0 to M3, in each full adder are only used for intermediate results needed internally to the adder. As each full adder resolves its sum in the ripple-carry adder, it can clear these four memristors. Since the 4-to-1 multiplexer only needs three memristors for its functionality, it can reuse these memristors. The multiplexer can now perform entirely within the adder.

When an adder completes in step X, it will reset memristors M0 to M3 in step $X+1$. Then, when the adder fully completes, each adder can perform the multiplexer steps in parallel. Each full adder will use the A memristors in full adders $N-1$ and $N-2$ as select lines, the divisor and 2's complement bits as inputs E and F, and the full adder memristors M0 to M2 as the multiplexer memristors M0 to M2. Again the result of the multiplexer will store into the B memristor in the full adder to enable the next iteration. Finally, all of the memristors can be reset except for the input memristors. The iteration is complete.

So far, the total delay of each iteration is the delay of the ripple-carry adder and the delay of the 4-to-1 multiplexer since the shift operation does not contribute any delay on the critical path. The addition with shift takes $2N+19$ steps, resetting M0 to M3 takes one step, the multiplexer takes nine steps, and the final reset of the memristors in preparation for the next iteration takes one step. The total delay is $2N+30$ steps per iteration. This does not include the delay for the quotient conversion.

The SRT divider implementation incorporates a shift operation, a 4-to-1 multiplexer, and an adder all inside of the same $7N+1$ memristors that originally were dedicated to the ripple-carry adder. This is an example of the key benefits of memristor logic design. The only consequence is the additional switches and driver logic.

The total complexity is the same as the original ripple-carry adder with three additional N-bit registers. These registers hold the quotient, the divisor, and its 2's complement. The divisor and the 2's complement registers are essentially the "E" and "F" bits for the multiplexer. These were not necessary in the binary-restoring divider since the divisor was not used and the 2's complement could be stored directly as an operand in the full adder. Thus, the total complexity of the current design is $10N+1$ memristors, $10N+1$ drivers, and $10N+1$ switches.

8.2.1.2 Quotient conversion logic

Last, logic is required for converting the quotient from binary signed digit to binary and correcting both the remainder and quotient as required. As described in the introduction to the SRT divider, it is expensive to perform the conversion of the quotient at the end of the division. The latency of the conversion contributes directly to the critical path. For IMPLY implementations, the steps are serialized, and the expense of the conversion if magnified.

Instead, the SRT divider will perform on-the-fly conversion during each iteration j of the division. The on-the-fly conversion has to calculate the values of $Q[j]$ and $QM[j]$ as described in the introduction to dividers. This conversion also performs the correction on the remainder and quotient. Recall the original equations for on-the-fly conversion in the introduction to SRT dividers:

$$Q[j+1] = \begin{cases} (Q[j], q_{j+1}) \text{ if } q_{j+1} \geq 0 \\ (QM[j], (2-|q_{j+1}|)) \text{ if } q_{j+1} < 0 \end{cases}$$
$$QM[j+1] = \begin{cases} (Q[j], q_{j+1}-1) \text{ if } q_{j+1} > 0 \\ (QM[j], (1-|q_{j+1}|)) \text{ if } q_{j+1} \leq 0 \end{cases}$$

j is the iteration number, q_{j+1} is the quotient bit for the iteration, QM represents the final quotient, and Q is an intermediate value used solely for conversion. During each iteration j, j bits of the final quotient should be recalculated. The most significant $j-1$ bits are assigned to the current $(j-1)$-bit value of $Q[j]$ or $QM[j]$ value. Only the least significant bit of Q and QM must be calculated during each iteration.

Note that $q_{j+1} < 0$ corresponds to the case where $\{S_{n-1}, S_{n-2}\} = 10$, $q_{j+1} > 0$ corresponds to the case where $\{S_{n-1}, S_{n-2}\} = 01$, and $q_{j+1} == 0$ corresponds to the case where $\{S_{n-1}, S_{n-2}\} = 00$ or 11. Thus, in terms of a radix-2 SRT divider, the equations for $Q[j+1]$ and $QM[j+1]$ become

$$Q[j+1] = \begin{cases} (Q[j], q_{j+1}) \text{ if } \{S_{n-1}, S_{n-2}\} \in \{00,01,11\} \\ (QM[j], (2-|q_{j+1}|)) \text{ if } \{S_{n-1}, S_{n-2}\} == \{10\} \end{cases}$$
$$QM[j+1] = \begin{cases} (Q[j], q_{j+1}-1) \text{ if } \{S_{n-1}, S_{n-2}\} == \{01\} \\ (QM[j], (1-|q_{j+1}|)) \text{ if } \{S_{n-1}, S_{n-2}\} \in \{00,10,11\} \end{cases}$$

By using these equations and the relationship between S_{n-1}, S_{n-2} and q_{j+1}, the equations for $Q[j+1]_0$ and $QM[j+1]_0$ are simplified to

$$Q[j+1]_0 = \begin{cases} q_{j+1} \text{ if } \{S_{n-1}, S_{n-2}\} \in \{00{,}01{,}11\} \\ 1 \text{ if } \{S_{n-1}, S_{n-2}\} == \{10\} \end{cases}$$
$$QM[j+1]_0 = \begin{cases} 0 \text{ if } \{S_{n-1}, S_{n-2}\} == \{01\} \\ 1 - |q_{j+1}| \text{ if} \{S_{n-1}, S_{n-2}\} \in \{00{,}10{,}11\} \end{cases}$$

These conditional equations can be further simplified to Boolean expressions. Recall that q_{j+1} is 1 when $\{S_{n-1}, S_{n-2}\} = $ "01." Thus, $Q[j+1]_0$ is "1" if $\{S_{n-1}, S_{n-2}\} \in \{01, 10\}$ and 0 otherwise. Similar logic can be used for $QM[j+1]_0$. In terms of Boolean gates, the equations for $Q[j+1]_0$ and $QM[j+1]_0$ become

$$Q[j+1]_0 = \left(S_{n-1}\,\overline{S_{n-2}}\right) \text{ OR } \left(\overline{S_{n-1}}\,S_{n-2}\right)$$
$$QM[j+1]_0 = \left(S_{n-1}S_{n-2}\right) \text{ OR } \left(\overline{S_{n-1}}\,\overline{S_{n-2}}\right)$$

The original equations have been translated and simplified to a much more compact expression. With these simplified equations, $Q[j]_0$ and $QM[j+1]_0$ are computed with IMPLY operations as shown in Table 8.7.

Notice that $QM[j+1]_0$ is simply the inverse of $Q[j+1]_0$. The value of $QM[j+1]_0$ does not need to be recomputed from the select lines: Only one step is needed to resolve $QM[j+1]_0$ once $Q[j+1]_0$ is available. The quotient conversion equations for the remaining $j-1$ bits can also be solved with IMPLY operations in parallel. The equations for $Q[j]$ and $QM[j]$ simplify to

$$Q[j+1] = \begin{cases} (Q[j]) \text{ if } \{S_{n-1}, S_{n-2}\} \in \{00{,}01{,}11\} \\ (QM[j]) \text{ if } \{S_{n-1}, S_{n-2}\} == \{10\} \end{cases}$$
$$QM[j+1] = \begin{cases} (Q[j]) \text{ if } \{S_{n-1}, S_{n-2}\} == \{01\} \\ (QM[j]) \text{ if } \{S_{n-1}, S_{n-2}\} \in \{00{,}10{,}11\} \end{cases}$$

For the purposes of the IMPLY implementation, the equations for $Q[j]$ and $QM[j]$ are rewritten as the following:

$$Q[j+1] = \begin{cases} (Q[j]) \text{ if } \{S_{n-1}, S_{n-2}\}\,! = \{10\} \\ (QM[j]) \text{ if } \{S_{n-1}, S_{n-2}\} == \{10\} \end{cases}$$
$$QM[j+1] = \begin{cases} (Q[j]) \text{ if } \{S_{n-1}, S_{n-2}\} == \{01\} \\ (QM[j]) \text{ if } \{S_{n-1}, S_{n-2}\}\,! = \{01\} \end{cases}$$

Table 8.7 Execution steps for calculating $Q[j]_0$ and $QM[j+1]_0$ for iteration j

Step	**Operation for $Q[j]$**	**Functionality**
1	S_{n-2} IMP 0	NOT S_{n-2}
2	(S_{n-2} IMP 0) IMP 0	Copy S_{n-2}
3	S_{n-1} IMP S_{n-2} → M0	S_{n-1} NAND $\overline{S_{n-2}}$
4	S_{n-2} IMP S_{n-1} → M1	$\overline{S_{n-1}}$ NAND S_{n-2}
5	M0 IMP 0 → M2	S_{n-1} AND $\overline{S_{n-2}}$
6	M1 IMP M2	$Q[j+1]_0$
7	M2 IMP 0 → M3	$QM[j+1]_0$

Now, only two Boolean combinations of S_{n-1}, S_{n-2}, "01" and "10" need to be computed, rather than all four. The IMPLY steps for resolving $Q[j]$ and $QM[j]$ each iteration are shown in Table 8.8.

The resolution of the most significant $j-1$ bits takes nine steps. Q and QM can be resolved in parallel using four memristors total. Notice that the operations are identical except the inverses of the select lines used for Q are used for QM, respectively.

The steps for performing on-the-fly conversion can occur in parallel with the steps for the 4-to-1 multiplexer. Recall that the multiplexer requires nine steps. This is identical to the steps needed to resolve the $j-1$ most significant bits of Q and QM and greater than the number of steps required for $Q[j]_0$ and $QM[j+1]_0$. Thus, the overhead of the quotient conversion is hidden by the latency of the addition. Also, the next iteration of the addition could begin even if the quotient were not yet converted since the addition iterations are not dependent on the running quotient changes.

The quotient conversion needs to compute two N-bit values for Q and QM for a total of $2N$ values. The values for the most significant $j-1$ bits require four memristors, and the values for the least significant bit require four also. There are also two N-bit registers to hold the resolved Q and QM each iteration. In all, each bit requires six memristors for a total of $6N$ memristors for the quotient conversion. The total complexity of the design is now $16N+1$ memristors, $16N+1$ drivers, and $16N+1$ switches and the delay remains $2N+30$ steps per iteration.

The only remaining logic is for the correction of the remainder. If the remainder is negative at the end of the division, the remainder and quotient must be corrected. However, this correction can be performed in the existing circuitry by altering the multiplexer driver behavior.

On the final correction, rather than using both S_{n-2} and S_{n-1}, only S_{n-1} is used to select either the divisor or the value 0. If S_{n-1} of the remainder is 1, the divider needs to add the divisor back to the remainder. In this case, the divisor is selected as operand B. If S_{n-1} of the remainder is 0, the remainder is positive and 0 can be added as operand B. A final addition now corrects the remainder as necessary. This takes the same delay as a standard operation, or $2N+19$ steps. Because the design just alters the driver behavior in the final iteration, no additional memristors are required. The overall delay of the final design is now $(X+2)\ (2N+30)-22$.

8.2.2 Hybrid-CMOS SRT divider

The hybrid-CMOS approach to an SRT divider is nearly identical to the traditional CMOS approach at the gate level. It is also very similar to the binary-restoring divider but has enhancements for a more complex multiplexer and quotient conversion logic.

The circuit consists of three 8-bit adders, a 4-to-1 8-bit multiplexer, three 2-to-1 8-bit multiplexers, an 8-bit shift register, and two 8-bit registers. The first adder computes the 2's complement of the divisor. The 4-to-1 multiplexer selects whether the divisor or 2's complement of the divisor should be used as the input operand to

Table 8.8 Execution steps for calculating Q[j] and QM[j] for iteration j

	Operation for *Q*[*j*]	Functionality	Operation for *QM*[*j*]	Functionality
1	S_{n-2} IMP 0	NOT S_{n-2}	S_{n-1} IMP 0	NOT S_{n-1}
2	(S_{n-2} IMP 0) IMP 0	Copy S_{n-2}	(S_{n-1} IMP 0) IMP 0	Copy S_{n-1}
3	S_{n-1} IMP S_{n-2} → M0	S_{n-1} NAND $\overline{S_{n-2}}$	S_{n-2} IMP S_{n-1} → M2	S_{n-2} NAND $\overline{S_{n-1}}$
4	M0 IMP 0 → M1	S_{n-1} AND $\overline{S_{n-2}}$	M2 IMP 0 → M3	S_{n-2} AND $\overline{S_{n-1}}$
5	*QM*[*j*] IMP M0	*QM*[*j*] NAND (S_{n-1} AND $\overline{S_{n-2}}$)	*Q*[*j*] IMP M2	*Q*[*j*] NAND (S_{n-2} AND $\overline{S_{n-1}}$)
6	*Q*[*j*] IMP M1	*Q*[*j*] NAND (S_{n-1} NAND $\overline{S_{n-2}}$)	*QM*[*j*] IMP M3	*QM*[*j*] NAND (S_{n-2} NAND $\overline{S_{n-1}}$)
7	Reset *Q*[*j*]	Reset *Q*[*j*]	Reset *QM*[*j*]	Reset *QM*[*j*]
8	M1 IMP 0 → *Q*[*j*]	*Q*[*j*] AND (S_{n-1} NAND $\overline{S_{n-2}}$)	M3 IMP 0 → *QM*[*j*]	*QM*[*j*] AND (S_{n-2} NAND $\overline{S_{n-1}}$)
9	M0 IMP *Q*[*j*]	*QM*[*j*] AND (S_{n-1} AND $\overline{S_{n-2}}$) OR *Q*[*j*] AND (S_{n-1} NAND $\overline{S_{n-2}}$)	M2 IMP *QM*[*j*]	*Q*[*j*] AND (S_{n-2} AND $\overline{S_{n-1}}$) OR *QM*[*j*] AND (S_{n-2} NAND $\overline{S_{n-1}}$)

the next addition. The subsequent adder performs the iterative additions to execute the division. Upon completion of each addition, the result of the addition is stored into the shift register and shifted left. Finally, the upper two bits are used as select lines to feed the multiplexer and adder. The 2-to-1 multiplexers and two registers are used for the quotient conversion logic, and the final adder is used for the remainder correction.

As described for the binary-restoring divider, the delay through the shift register and adder is $3N + 10$ steps, $3N + 4$ for the addition, and six for the load and shift. The delay through a 4-to-1 multiplexer is normally five steps. However, since the first select line bit S_{n-2} is resolved three steps before the addition completes, the first tier of gates in the multiplexer logic can complete while the last bit of the adder resolves S_{n-1}. Thus, the effective latency on the critical path is the delay of a 2-to-1 multiplexer, or three steps. The total delay of each iteration is $3N + 13$ steps.

The on-the-fly conversion is performed using the two 8-bit registers and the two 2-to-1 8-bit multiplexers. One of the registers holds $Q[j]$, and one holds $QM[j]$. At the end of each iteration, the top two bits of the shift register are used to select the most significant $j-1$ bits of either the current $Q[j]$ or $QM[j]$ as the new most significant $j-1$ bits of the updated $Q[j]$ or $QM[j]$. They are also used to produce the updated least significant bit of each register. These conversions are performed using the following Boolean equations:

$$\begin{aligned}
Q[j+1]_0 &= \left(\mathrm{S}_{n-1}\,\overline{\mathrm{S}_{n-2}}\right) \text{ OR } \left(\overline{\mathrm{S}_{n-1}}\,\mathrm{S}_{n-2}\right)\\
QM[j+1]_0 &= \left(\mathrm{S}_{n-1}\,\mathrm{S}_{n-2}\right) \text{ OR } \left(\overline{\mathrm{S}_{n-1}}\,\overline{\mathrm{S}_{n-2}}\right)\\
Q[j]_{[j-1:1]} &= \left(QM[j] \text{ AND } \left(\mathrm{S}_{n-1}\,\overline{\mathrm{S}_{n-2}}\right)\right) \text{ OR } \left(Q[j] \text{ AND } \overline{\left(\mathrm{S}_{n-1}\,\overline{\mathrm{S}_{n-2}}\right)}\right)\\
QM[j]_{[j-1:1]} &= \left(Q[j] \text{ AND } \left(\overline{\mathrm{S}_{n-1}}\,\mathrm{S}_{n-2}\right)\right) \text{ OR } \left(QM[j] \text{ AND } \overline{\left(\overline{\mathrm{S}_{n-1}}\,\mathrm{S}_{n-2}\right)}\right)
\end{aligned}$$

These are the same as discussed for the IMPLY operation. Both $Q[j+1]$ and $QM[j+1]$ are resolved with 2-to-1 multiplexers. Consider $Q[j+1]$. The select line for the most significant $j-1$ bits of the multiplexer for $Q[j+1]$ will be the value of $\left(\mathrm{S}_{n-1},\,\overline{\mathrm{S}_{n-2}}\right)$. To obtain this value, a single NOT gate and a 2-input AND gate are used. However, the inverses of both select lines are already available since they are used for the 4-to-1 multiplexer. Thus, the only additional complexity is a single gate. The result of the AND gate is the select line to the 2-to-1 multiplexer for Q.

If the select line $\left(\mathrm{S}_{n-1},\,\overline{\mathrm{S}_{n-2}}\right)$ evaluates to "1," the value of $QM[j]$ will be stored into $Q[j+1]$. If the select line $\left(\mathrm{S}_{n-1},\,\overline{\mathrm{S}_{n-2}}\right)$ evaluates to "0," the value of $Q[j]$ will be stored into $Q[j+1]$. Similarly, the select line for the most significant $j-1$ bits of the multiplexer for $QM[j]$ will be the value of $\left(\overline{\mathrm{S}_{n-1}}\,\mathrm{S}_{n-2}\right)$. This also uses a single AND gate. If the select line $\left(\overline{\mathrm{S}_{n-1}}\,\mathrm{S}_{n-2}\right)$ evaluates to "1," the value of $Q[j]$ will be stored into $QM[j+1]$. If the select line $\left(\overline{\mathrm{S}_{n-1}}\,\mathrm{S}_{n-2}\right)$ evaluates to "0," the value of $QM[j]$ will be stored into $QM[j+1]$.

The logic for the least significant bits is different. For these, the inverted values of S_{n-1} and S_{n-2} and the values of $\left(\mathrm{S}_{n-1}\,\overline{\mathrm{S}_{n-2}}\right)$ and $\left(\overline{\mathrm{S}_{n-1}}\,\mathrm{S}_{n-2}\right)$ produced for the other bits of Q and QM can be used directly. The only additional complexity is two

AND gates and two OR gates. Figure 8.4 shows the logic for the two 8-bit multiplexers for resolving $Q[j+1]$ and $QM[j+1]$ each iteration.

The total delay for the quotient conversion is four steps. The most significant bits of Q and QM require four steps, and the least significant bits require three steps. However, recall that this delay does not lie on the critical path of the design. This conversion can occur in parallel with the multiplexer feeding the adder.

Last, the remainder must be conditionally corrected at the end of the division iterations. This can be computed with another 2-to-1 multiplexer and an 8-bit adder. The multiplexer will use the most significant bit of the remainder as the select line. If the select line is a "1," the result was negative and the multiplexer will select the divisor as the output operand. If the select line is a "0," the result is positive and the multiplexer will select "0" as the output operand. Finally, the last addition executes. This final correction step takes three steps for the multiplexer and $3N+4$ steps for the addition. Thus, the total delay of the design is $(X+2)(3N+13)-15$.

The total complexity becomes

Ripple-carry adder:

$$= 3(14N \text{ memristors} + 12N \text{ MOSFETs})$$
$$= 42N \text{ memristors} + 36N \text{ MOSFETs}$$

4-to-1 multiplexer:

$$= 12N \text{ memristors and } 4 \text{ MOSFETs}$$

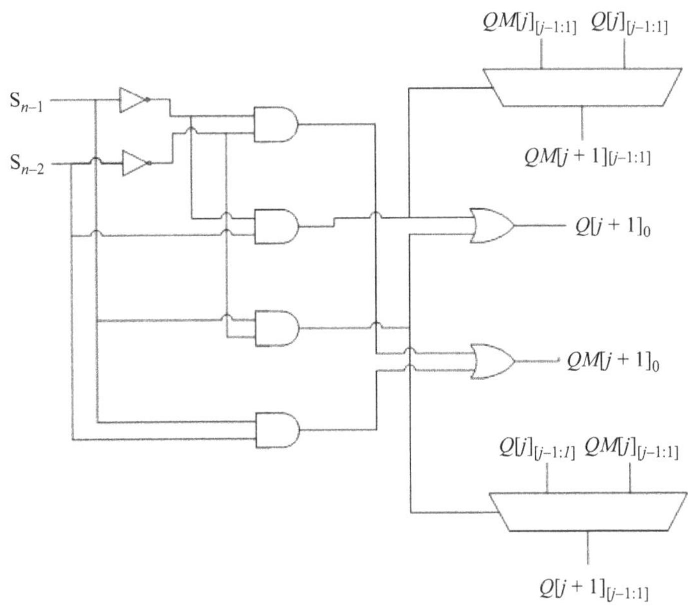

Figure 8.4 Multiplexer and Boolean logic for on-the-fly quotient conversion

Shift register:

$$= (22N - 6) \text{ memristors} + (26N - 4) \text{ MOSFETs}$$

Quotient conversion:

$$\begin{aligned} &= 2(16N \text{ memristors} + 20N \text{ MOSFETs}) \quad (2\ N\text{-bit registers}) \\ &\quad + 2(6N \text{ memristors} + 2 \text{ MOSFETs}) \quad (2\ N\text{-bit multiplexers}) \\ &\quad + 12 \text{ memristors} \quad (\text{Select line Boolean logic}) \\ &= (44N + 12) \text{ memristors} + (40N + 4) \text{ MOSFETs} \end{aligned}$$

Total complexity:

$$= (120N + 6) \text{ memristors} + (102N + 4) \text{ MOSFETs}$$

8.2.3 Threshold-gate SRT divider

The proposed threshold-gate-based design for an SRT divider uses the discussed optimized ripple-carry adder along with similar circuitry as the threshold-gate binary-restoring divider. Slight modifications to the adder and multiplexer logic are performed to enable radix-2 SRT division. GOTO pairs are used for the implementation in this section.

The traditional ripple-carry adder is used in the SRT for the 2's complement computation since this is a normal addition. A modified ripple-carry adder is used for the iterative additions. This modified adder incorporates the 4-to-1 multiplexer for deciding. The subsequent logic in the SRT design is optimized to incorporate the multiplexer with minimal additional delay and area. Figure 8.5 shows the design of a full adder with the proposed combined multiplexer-adder functionality.

The two threshold gates on the left of the schematic and their inputs represent the 4-to-1 multiplexer. The remainder of the schematic is the full adder. Rather than the traditional full adder with inputs A, B, and carry-in, this adder has the inputs A, B0, B1, and C_{in}. The A and C_{in} inputs are the same as the traditional ripple-carry adder. B0 and B1 are the two outputs of the multiplexer logic.

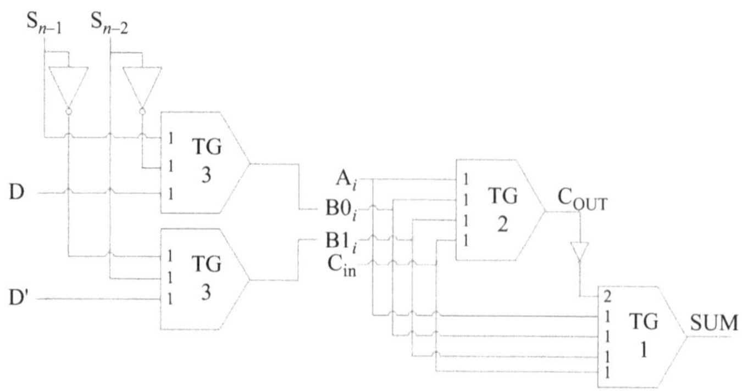

Figure 8.5 Threshold-gate multiplexer-adder functionality for SRT division

B0 is the output of the first threshold gate in the multiplexer. This gate is essentially a 3-input AND gate because it will only meet its threshold to output a "1" if all of its inputs are a "1." Essentially, this threshold gate implements

$$\mathrm{B0} = \mathrm{S}_{n-1} \text{ AND } \overline{\mathrm{S}_{n-2}} \text{ AND D}$$

where S_{n-1} and S_{n-2} are the select lines to the multiplexer, and D is the bit of the divisor corresponding to the full adder. Recall the select lines come from the two most significant bits in the shift register, which will be discussed later in the section. Effectively, B0 will be D if $\{\mathrm{S}_{n-1},\mathrm{S}_{n-2}\} = \{10\}$, else it will be 0. B0 encapsulates the logic for selecting the divisor as the B input for the iteration.

Similarly, the second threshold gate in the multiplexer is a 3-input AND gate. Its Boolean expression is

$$\mathrm{B1} = \overline{\mathrm{S}_{n-1}} \text{ AND } \mathrm{S}_{n-2} \text{ AND } \mathrm{D}'$$

where D′ is the bit of the 2's complement of the divisor corresponding to this full adder. Effectively, B0 will be D′ if $\{\mathrm{S}_{n-1},\mathrm{S}_{n-2}\} = \{01\}$ else it will be 0. B1 encapsulates the logic for selecting the 2's complement of the divisor as the B input for the iteration.

Note that the events B0 == "1" and B1 == "1" are mutually exclusive. It is not possible for the select lines to satisfy both gates concurrently. At most, one of the threshold gates will output a nonzero value. If the select lines do not satisfy either of the gates, both will output a "0." This corresponds to the two scenarios in the multiplexer logic that select "0" as the B input for the iteration. Now, B0 and B1 can feed the full adder logic.

The $\mathrm{C_{out}}$ and Sum threshold gates behave essentially the same as they do in the standard full adder. The only difference is that the B input is now split across B0 and B1. However, at most, one of these inputs will contribute to the threshold of these gates. Thus, the correct functionality is preserved. Essentially, at least one of the inputs B0 and B1 will be idle in the circuit for a given iteration. At most, one of the inputs B0 and B1 will be active in which case it will serve as the "B" input for that iteration as determined by the multiplexer logic.

This optimized full adder, which includes the multiplexer, requires $15N$ memristors, $4N$ GOTO pairs, and $2N+4$ MOSFETs. The multiplexer adds two steps to the critical path of each iteration for an updated delay of $N+4$ steps.

This divider also uses the shift register cell with load capability presented for the binary-restoring divider. It was shown that a $2N$-bit shift register uses $32N$ memristors, $10N$ GOTO pairs, and 4 MOSFETs. The delay for a shift operation is two steps to load and six steps to shift. The total delay for each iteration is now $N+12$ steps.

8.2.3.1 Quotient conversion logic

Last, the design must incorporate on-the-fly quotient conversion. First, $2N$ additional flip-flops for the $Q[j]$ and $QM[j]$ values are necessary. For these registers, the latches shown for the shift-and-add multiplier are reused. Each N-bit register

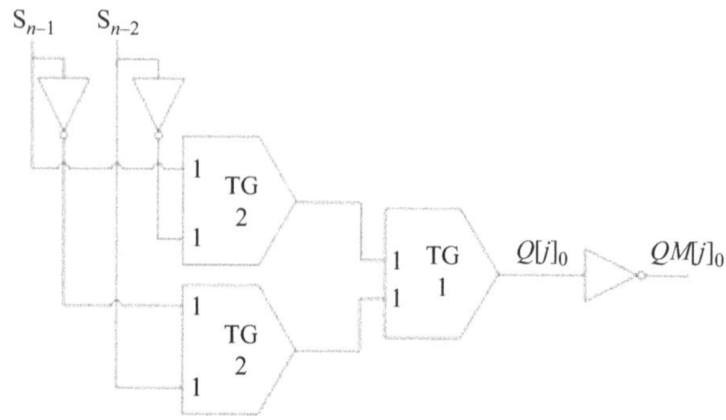

Figure 8.6 Threshold-gate circuits for $Q[j]_0$ and $QM[j]_0$ registers

requires 7*N* memristors and 2*N* GOTO pairs. Second, the logic to compute *QM* and *Q* must be added to the design. From the description of the IMPLY SRT divider, the equations for $QM[j]_0$ and $Q[j]_0$ can be simplified to

$$\begin{aligned} Q[j+1]_0 &= (S_{n-1}\,\overline{S_{n-2}}) \text{ OR } (\overline{S_{n-1}}\,S_{n-2}) \\ QM[j+1]_0 &= (S_{n-1}S_{n-2}) \text{ OR } (\overline{S_{n-1}}\,\overline{S_{n-2}}) \end{aligned}$$

The threshold-gate circuit to compute these Boolean expressions is shown in Figure 8.6.

The threshold gates essentially behave as standard 2-input AND and OR gates. The first two threshold gates are AND gates corresponding to $(S_{n-1}\,\overline{S_{n-2}})$ and $(\overline{S_{n-1}}S_{n-2})$, and the third threshold gate is an OR gate corresponding to the OR operation between these two values.

Note that for bit 0, rather than evaluating both *Q*[*j*] and *QM*[*j*] in parallel circuitry, the result of *Q*[*j*] is simply inverted to obtain *QM*[*j*]. This is done because the evaluation of these registers is not on the critical path, so an additional gate delay can be sacrificed in favor of lesser area. The delay for this circuitry is four steps, and it requires six memristors, three GOTO pairs, and two MOSFETs. The inverters on the select lines are already present for the multiplexer logic and do not add any complexity to the quotient conversion logic. The total complexity for this circuitry is six memristors, three GOTO pairs, and two MOSFETs. The logic for the remaining $j-1$ bits of the quotient uses the conversion logic shown in Figure 8.7.

Again, since the quotient conversion logic is not on the critical path, a longer delay design is chosen over a lower area design. Note that the inverters on the select lines are reused from the 4-to-1 multiplexer. Also, the first two threshold gates, which evaluate $(S_{n-1}\,\overline{S_{n-2}})$ and $(\overline{S_{n-1}}S_{n-2})$, can be reused from the logic for the least significant bit circuity and shared across all of the $N-1$ instances of the logic. Similarly, the inverses of these values can be shared across all of the $N-1$ instances of the logic. Thus, the total contributed area for the circuity in Figure 8.7 is $(N-1)$ (12 memristors + 6 GOTO pairs) and four MOSFETs. The total delay is five steps. Since the on-the-fly conversion is not on the critical path, the total delay of each iteration remains $N+12$ steps.

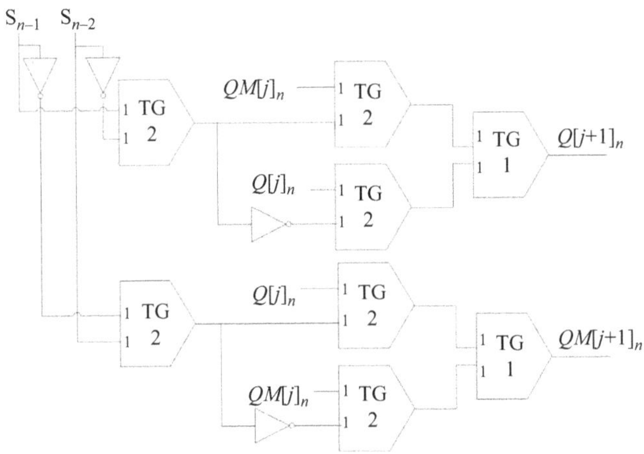

Figure 8.7 Threshold-gate circuits for $Q[j]_n$ and $QM[j]_n$ registers

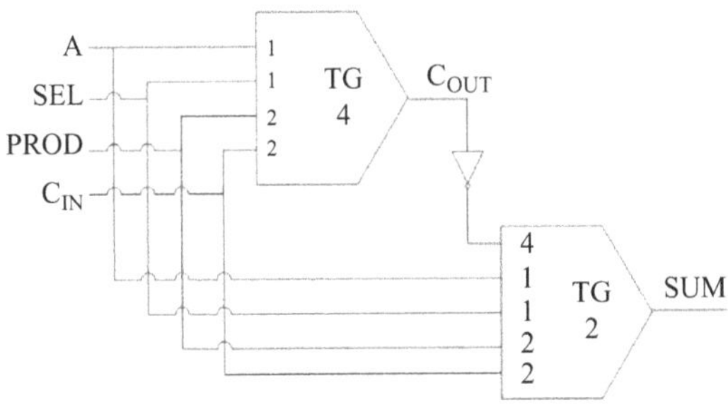

Figure 8.8 Threshold-gate 1-bit adder design optimized for use in a shift-and-add multiplier

Last, the remainder correction logic must be considered. The design will use the same logic as the hybrid-CMOS design. If the most significant bit of the result of the last iteration is a 1, the divisor will be added to the result, else nothing will change. This calculation uses a 2-to-1 N-bit multiplexer and an N-bit adder. These units can be combined like they were for the shift-and-add multiplier. The circuitry for this multiplexer and adder is shown again in Figure 8.8.

For the binary-restoring divider, the A input corresponds to the divisor bit, the SEL input corresponds to the most significant bit of the result, PROD corresponds to the remainder, and CIN is the standard carry-in. This effectively performs the final remainder correction. It requires the same delay as a standard ripple-carry adder, $N + 2$ gate delays.

The total delay of the threshold-gate SRT divider consists of the initial 2's complement operation, the iterative additions, and the final remainder correction. The total delay is $(X + 2)(N + 12)-20$ steps.

The total complexity is calculated as follows:

2's complement adder:

$$= 7N \text{ memristors} + 2N \text{ GOTO pairs} + 2N \text{ MOSFETs}$$

Multiplexer + adder:

$$= 15N \text{ memristors} + 4N \text{ GOTO pairs} + 2N + 4 \text{ MOSFETs}$$

2N-bit shift register:

$$= 32N \text{ memristors} + 10N \text{ GOTO pairs} + 4 \text{ MOSFETs}$$

Quotient conversion:

2N-bit registers:

$$= 14N \text{ memristors} + 4N \text{ GOTO pairs}$$

Logic:

$$\begin{aligned} &= (N-1)(12 \text{ memristors} + 6 \text{ GOTO pairs}) + 4 \text{ MOSFETs} \\ &\quad + 6 \text{ memristors} + 3 \text{ GOTO pairs} + 2 \text{ MOSFETS} \\ &= (12N-6) \text{ memristors} + (6N-3) \text{ GOTO pairs} + 6 \text{ MOSFETs} \end{aligned}$$

Remainder correction adder:

$$= 9N \text{ memristors} + 2N \text{ GOTO pairs} + 2N \text{ MOSFETs}$$

Total complexity:

$$= (89N-6) \text{ memristors} + (28N-3) \text{ GOTO pairs} + (6N+14) \text{ MOSFETs}$$

8.2.4 MAD-gate SRT divider

This section presents a MAD-based implementation of an SRT divider. First, an optimized ripple-carry adder will be enhanced to perform a left-shift of the result at the end of each iteration. Second, the design is modified to hold both the 2's complement of the divisor and the divisor in N-bit registers. Third, a 4-to-1 multiplexer must be added to select either the divisor, the 2's complement of the divisor, or the value 0 as the B operand in each iteration. Finally, the logic for on-the-fly quotient conversion must be added.

First, the shift functionality is added to the adder. As described for the binary-restoring divider, it is possible to perform a left-shift operation without any additional memristors or delay to the adder design. This is done by performing the final set operation for the sum resolution by applying the V_{set} voltage to a destination memristor in the next full adder rather than the local full adder.

The additional logic to make this a conditional store in the binary-restoring divider is no longer needed in the SRT divider. Thus, each bit full adder f only needs to wait for full adder $f+1$ to reset its A memristor operand before performing the final store. However, this is a special case for the MAD adder. The A memristor in full adder $f+1$ is required to compute its own carry-out and sum. Thus, each full adder f cannot store its sum into full adder $f+1$ until $f+1$ resets its A memristor, but $f+1$ cannot reset its A memristor until it stores its own sum into full adder $f+2$. This creates a forward dependency across the adders. Thus, it is not possible to perform this left shift seamlessly.

Instead, each full adder f in the MAD ripple-carry adder will store the sum value directly into its own sum memristor. Then, once the next full adder completes and resets its A memristor, f will perform a copy operation of its sum memristor into the A memristor. This copy step can occur during the idle steps while the remainder of the adder completes. One consequence is that the sum memristors cannot be removed as is traditionally done when performing iterative additions. The updated steps for the first three full adders in the ripple-carry adder are shown in Table 8.9.

Assume that the original inputs are loaded into A and B a priori. In the first step, full adder 0 resolves its sum and carry-out signal. In step 2, full adder 0 propagates its carry-out to full adder 1 and full adder 1 resolves its sum and carry-out signal. In step 3, full adder 1 propagates its carry-out signal to full adder 2, and full adder 2 resolves its sum and carry-out signal. These steps are identical to the traditional ripple-carry adder. Now, each full adder performs a third step after they have propagated their carry signal that resets the A memristor. Then, the sums are copied one by one from each full adder to the next full adder's A memristor.

The updated adder with support for the left-shift operation takes $N+3$ steps. The shift operation essentially adds three steps to the critical path of the adder. The complexity remains unchanged.

Next, the registers are added for the divisor and its 2's complement. Each register requires N bits for a total complexity of $2N$ memristors, $2N$ switches, and $2N$ drivers. These registers must simply support read and write operations, so their circuitry is minimal and straightforward. The same ripple-carry adder can be used

Table 8.9 Steps for full adders in a MAD shift adder for an SRT divider

Step	Full adder 0	Full adder 1	Full adder 2
1	Resolve C_{out} and sum		
2	Propagate C_{out}	Resolve C_{out} and sum	
3	Reset A	Propagate C_{out}	Resolve C_{out} and sum
4	Resolve shifted sum into A	Reset A	Propagate C_{out}
5	Copy from sum	Copy into A from full adder 0	Reset A
6		Copy from sum	Copy into A from full adder 1

to complete the 2's complement of the divisor. When the store operation for the sum occurs in each full adder, it will apply V_{set} to the corresponding memristor in the 2's complement register rather than a memristor in the local adder. This functionality has been described for many previous designs.

Third, the multiplexer logic is added to select between the divisor, 2's complement of the divisor and 0 for the B operand in each iteration. The logic to load the B operand memristor is enhanced as shown in Figure 8.9.

S_{n-2} and S_{n-1} represent the select lines from the result of the previous iteration, D represents the divisor, and D′ represents the 2's complement of the divisor. Recall that the values of S_{n-2} and S_{n-1} are equivalent to the values in the A memristors of the most significant two full adders once the shift has been performed. The multiplexer can sense these two A memristors to obtain the values of the select lines for the multiplexer.

The 4-to-1 multiplexer really only has two inputs, D and D′. The other two inputs are 0 and correspond to a nop (NULL operation) in the MAD adder. V_{set} will be applied to the B memristor if $\left(S_{n-2}\ \text{AND}\ \overline{S_{n-1}}\ \text{AND}\ D'\right)$ OR $\left(\overline{S_{n-2}}\ \text{AND}\ S_{n-1}\ \text{AND}\ D\right)$. This is equivalent to the Boolean expression for the multiplexer.

The multiplexer logic adds $6N$ switches to the adder design and requires a single step. The read voltage V_{cond} is applied to A_{n-2} and A_{n-1}, while the V_{set} voltage is applied to the B memristors in all N full adders. At the same time, the individual memristors in the N-bit registers for D and D′ are also driven with the

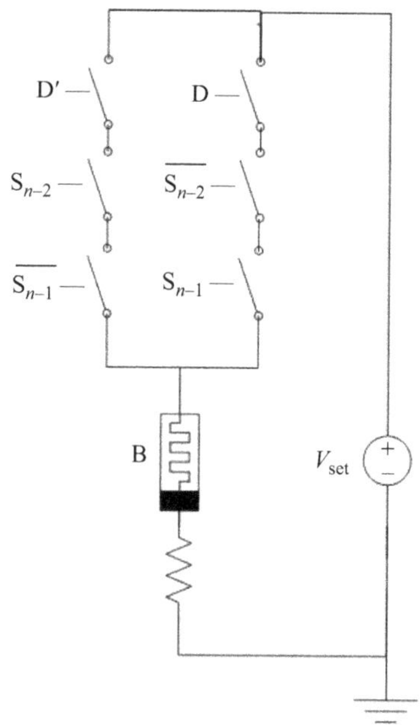

Figure 8.9 Multiplexer logic for the 4-to-1 multiplexer in the SRT divider

read voltage V_{cond}. The total complexity of the entire design so far is 6N memristors, $5N+1$ drivers, and 21N switches, and the delay is $N+4$ steps per iteration.

Next, the on-the-fly quotient conversion is performed. The MAD-based approach to the on-the-fly conversion is able to greatly reduce the overhead in terms of both area and delay. To perform the on-the-fly conversion, the design has N memristors for the $Q[j]$ value and N memristors for the $QM[j]$ value. These memristors are necessary since they hold the values of Q and QM across iterations. They cannot be optimized out.

The Q and QM values can be computed in parallel with the 4-to-1 multiplexer functionality. This is convenient because both the multiplexer and the Q and QM values depend on the two most significant bits of the value S. These bits can be read once by both logic blocks in parallel.

Recall the equations for Q and QM in iteration j.

$$Q[j+1]_0 = \left(\mathrm{S}_{n-1}\,\overline{\mathrm{S}_{n-2}}\right) \text{ OR } \left(\overline{\mathrm{S}_{n-1}}\,\mathrm{S}_{n-2}\right)$$

$$QM[j+1]_0 = \left(\mathrm{S}_{n-1}\mathrm{S}_{n-2}\right) \text{ OR } \left(\overline{\mathrm{S}_{n-1}}\;\overline{\mathrm{S}_{n-2}}\right)$$

$$Q[j]_{[j-1:1]} = \left(QM[j] \text{ AND } \left(\mathrm{S}_{n-1}\,\overline{\mathrm{S}_{n-2}}\right)\right) \text{ OR } \left(Q[j] \text{ AND } \overline{\left(\mathrm{S}_{n-1}\,\overline{\mathrm{S}_{n-2}}\right)}\right)$$

$$QM[j]_{[j-1:1]} = \left(Q[j] \text{ AND } \left(\overline{\mathrm{S}_{n-1}}\,\mathrm{S}_{n-2}\right)\right) \text{ OR } \left(QM[j] \text{ AND } \overline{\left(\overline{\mathrm{S}_{n-1}}\,\mathrm{S}_{n-2}\right)}\right)$$

Consider $Q[j]_{[j-1:1]}$. If the most significant bits of the sum are 2’b10, then $Q[j]_{[j-1:1]}$ is set to $QM[j]_{[j-1:1]}$. To perform this, the drivers of the $Q[j]_{[j-1:1]}$ memristors are driving a V_{reset} signal followed by a V_{set} signal, and the voltages of the terminals of the $QM[j]_{[j-1:1]}$ memristors are read by driving a V_{cond} signal. Otherwise, the value of $Q[j]$ keeps the $Q[j]$ value.

Now consider $QM[j]_{[j-1:1]}$. If the most significant bits of the sum are 2’b01, then the drivers of the $QM[j]_{[j-1:1]}$ memristors are driven a V_{reset} signal followed by a V_{set} signal, and the voltages of the terminals of the $Q[j]_{[j-1:1]}$ memristors are read by driving a V_{cond} signal. An example of this logic for $QM[j]$ and $Q[j]$ are shown in Figure 8.10.

All of the bits of QM can share drivers and all of the bits of Q can share drivers since all bits perform in parallel. Thus, only six drivers are introduced to the design. Note that at most one of these events can occur so there is no risk of clearing out a memristor that would need to be read by its counterpart. This not only removes memristors intended for holding duplicates of these values prior to clearing the values but also allows for parallelizing the resolution of $QM[j]$ and $Q[j]$. Each memristor for the most significant $j-1$ bits requires seven switches for a total of $14(N-1)$ switches.

The resolution logic for $Q[j]_0$ and $QM[j]_0$ is simpler, sensing the same bits of the results as drivers, but using XOR and XNOR gates, respectively. Thus, each memristor for the least significant bit of both QM and Q requires six switches. They can also share the drivers from the most significant bits of Q and QM. The total complexity of the on-the-fly quotient conversion is 2N memristors, $14N-2$ switches, and 6 drivers.

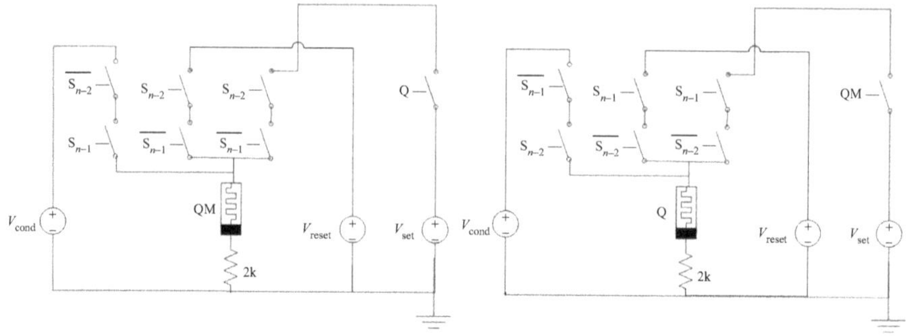

Figure 8.10 On-the-fly conversion for a bit of QM[j] and Q[j]

The on-the-fly conversion can complete in two steps—one for the reset operation and the other for the set operation. These steps are overlapped with the multiplexer and addition functionality, so the latency is hidden and does not lie on the critical path. Even the latency of the final iteration is hidden by the remainder correction.

The final remainder check and correction can be completed using the same single adder and 4-to-1 multiplexer. The delay is N steps. Before the addition, the B operand must be reset and loaded with either D or 0 depending on the most significant bit of the result S_{n-1}. If S_{n-1} is "0," B remains "0," and if S_{n-1} is "1," B is set to D. Essentially, $B = S_{n-1}$ AND D. This can be achieved with the circuitry for the 4-to-1 multiplexer in Figure 8.9 by driving S_{n-2} with "0." This is achieved by only applying the V_{cond} read signal the most significant bit S_{n-1} and not S_{n-2}. The remaining behavior is identical to the sense circuitry for the multiplexer as described. The complete schematic is shown in Figure 8.11.

The right column is the N-bit adder with the multiplexer logic. The left column shows the two N-bit registers for the divisor and 2's complement of the divisor. The bottoms of the left column are the N bits for the Q and N bits for the QM registers. Although not shown, drivers are shared across many of the instantiations as described. For example, all of the bits of the QM register can actually share their drivers since they act in lockstep.

The total complexity of the design comprises the registers, adder (with multiplexer logic), and on-the-fly quotient conversion. The registers, adder, and multiplexer require $6N$ memristors, $21N$ switches, and $5N+1$ drivers, respectively. The quotient conversion requires $2N$ memristors, $14N-2$ switches, and 6 drivers. Thus, the total complexity is $8N$ memristors, $5N+7$ drivers, and $35N-2$ switches.

The delay of each iteration is $N+4$ steps. Because of the optimizations performed, the on-the-fly quotient conversion can perform in parallel with the multiplexer logic and the first step of the adder without interference. The delay of the 2's complement addition and the remainder correction addition is $N+1$ steps each. Thus, the total delay of the MAD SRT divider is $(X+2)(N+4)-6$ steps.

Figure 8.11 8-bit MAD SRT divider

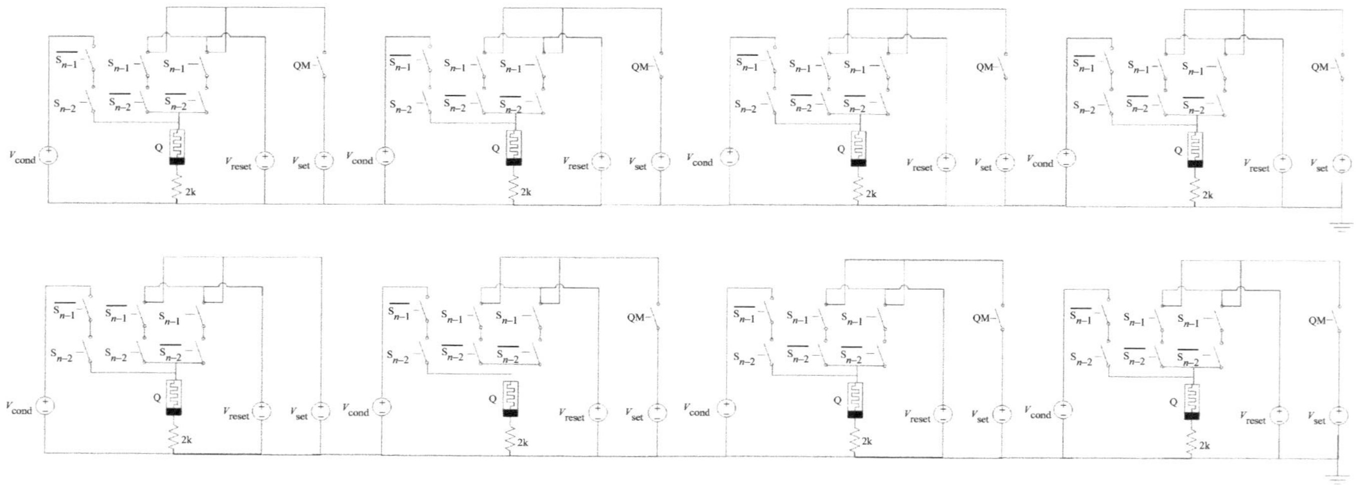

Figure 8.11 (Continued)

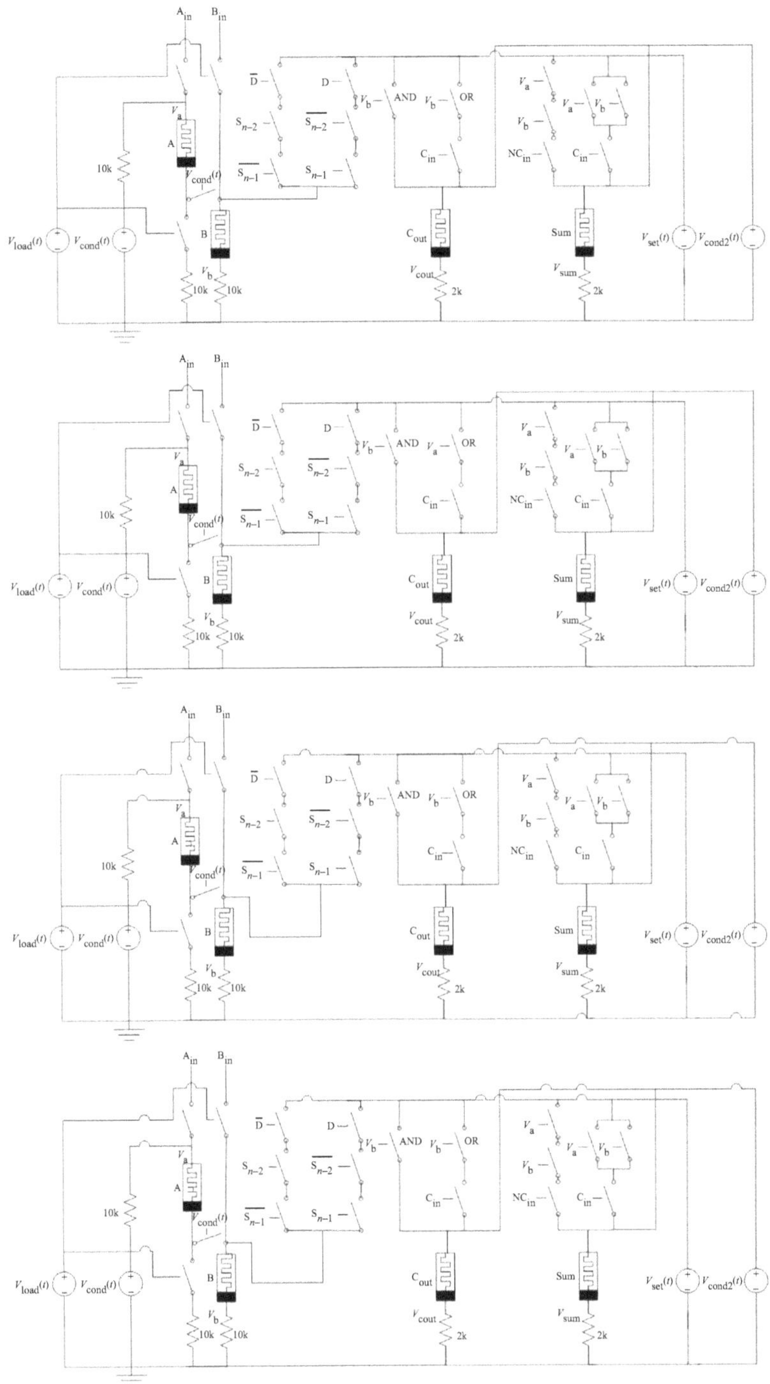

Figure 8.11 (Continued)

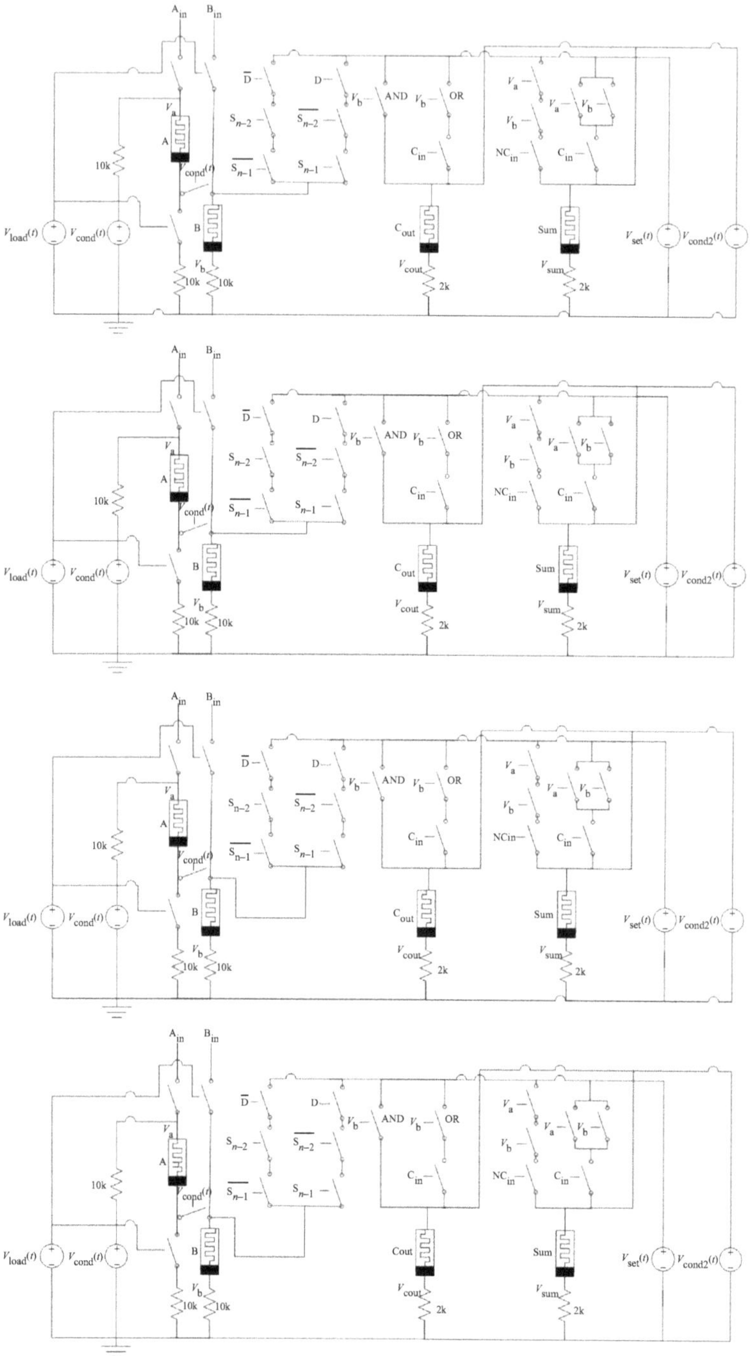

Figure 8.11 (Continued)

8.2.5 SRT divider analysis and comparison

Table 8.10 shows the delay and complexity for the proposed SRT dividers.

All of the presented memristor-based SRT dividers offer competitive alternatives to the traditional CMOS approach. Both the IMPLY and MAD approaches are able to remove fundamental components of the traditional divider due to their inherent characteristics. For example, neither design requires a shift register or multiplexer and can instead incorporate this functionality into the existing adder with little to no impact on component area or delay. The IMPLY and MAD approaches require the fewest overall components and the fewest memristors out of all presented designs. As is customary with MAD designs, it requires fewer memristors and drivers, but more switches than the IMPLY design.

The threshold-gate approach is also competitive, requiring only $6N + 14$ MOSFETs as compared to $88N + 4$ MOSFETs in the traditional CMOS approach. This is less than 7 percent of the original number of CMOS components. However, the total number of components for the threshold-gate approach is greater than the CMOS design. A trade-off must be made between the number of components and their individual form factors to decide which is more appropriate. The hybrid-CMOS approach again has the highest complexity, mostly attributed to the $2N$-bit shift register and its high use of MOSFETs.

The IMPLY design has a similar delay to the CMOS design, requiring 20 more steps per iteration, while the MAD design requires less than half the delay. Each iteration of the MAD approach is only $N + 4$ steps, nearly the same delay as a single addition. This approach is the fastest of all of the presented designs. The threshold-gate design uses the second fewest steps, needing $N + 12$ steps per iteration. Although it is not the fastest delay, it is still almost twice as fast as the CMOS design. The hybrid-CMOS approach suffers from extensive buffering on the adder outputs and shift register outputs, which fall on the critical path. This causes the design to have the greatest delay with $3N + 1$ gate delays per iteration.

Table 8.10 Delay and complexity comparisons for the SRT dividers

	CMOS	**IMPLY**	**Hybrid-CMOS**	**Threshold**	**MAD**
Delay	$(X+2)(2N+10)$	$(X+2)(2N+30)-22$	$(X+2)(3N+13)-15$	$(X+2)(N+12)-20$	$(X+2)(N+4)-6$
Area	$88N+4$ MOSFETs	$16N+1$ memristors	$120N+6$ memristors	$89N-6$ memristors	$8N$ memristors
		$16N+1$ drivers	$102N+4$ MOSFETs	$28N-3$ GOTO pairs	$5N+7$ drivers
		$16N+1$ switches		$6N+14$ MOSFETs	$35N-2$ switches

8.3 Goldschmidt dividers

Memristor-based Goldschmidt dividers are presented for the IMPLY, hybrid-CMOS, threshold, and MAD contexts. Goldschmidt dividers perform iterative multiplications of the normalized dividend N and divisor D by a series of factors until the divisor converges to 1, in which case, the dividend has converged to the quotient.

At initialization, F' is obtained from a lookup table and F_0 is calculated as $F_0 = F' - 2D_0$. Then, on each iteration, the value F is updated using the equation $F_i = 2 - D_i$, and the two multiplications execute. It is possible to perform the two multiplications in parallel or in series. This trade-off depends on whether speed or area is the greatest concern.

For all of the discussed designs in this chapter, the multiplications are performed in parallel. The lookup for F' is assumed to have already been performed since this is a simple read operation.

8.3.1 *IMPLY Goldschmidt divider*

The IMPLY approach to an N-bit Goldschmidt division requires an N-bit adder and two N-bit Dadda multipliers. The presented optimized 8-bit Dadda multiplier is used, which requires 577 memristors, 577 drivers, 577 switches, and a delay of 109 steps. In total, the two multipliers require 1,154 memristors, 1,154 drivers, and 1,154 switches and a delay of 109 steps. The N-bit adder can be optimized to improve a Goldschmidt divider. The remaining logic in the traditional CMOS design for data movement is not necessary.

At initialization, the divider must calculate the initial factor F_0 from F'. This calculation executes a left shift operation on the divisor followed by a subtraction. Each iteration, F_i, is updated by computing the 2's complement of the latest divisor and subtracting it from the value 2. Then the two multiplications execute.

In a Goldschmidt divider, the initial operation is $F_0 = F' - 2D_0$. As is known, in the context of IMPLY, the shift operation can be performed inherently by simply storing each bit i of the divisor D_0 into the B input memristor in the full adder for bit $i + 1$. This shift operation does not impact delay or area. In general, the equation $F' - 2D_0$ can be rewritten for an adder as

$$
\begin{aligned}
F_0 &= F' - 2D_0 \\
F_0 &= F' + (\text{2's complement of } 2D_0) \\
F_0 &= F' + \overline{(2D_0)} + 1
\end{aligned}
$$

The original equation, which required a shift, a 2's complement and an addition operation, now only requires a NOT and an addition operation. The 2's complement is computed during the same addition that adds F'. The "+1" calculation for the 2's complement is represented with the carry-in signal being set to "1" in the addition. The N-bit NOT operation can be performed in parallel across all the bits in one step. Thus, the calculation of F_0 requires 1 step for the shift and the NOT, and $2N + 19$ steps for the addition. Overall, the initialization stage of the adder uses $2N + 20$ steps and $7N + 1$ memristors.

After initialization, each iteration i performs a subtraction with the equation $F_i = 2 - D_i$ to obtain the new F value. Similar to the transformation for the initialization, this equation can be rewritten as

$$F_i = 2 - D_i$$
$$F_i = 2 + (\text{2's complement of } D_i)$$
$$F_i = 2 + \overline{D_i} + 1$$

However, this equation cannot be further simplified to

$$F_i = \overline{D_i} + 3$$

In an 8-bit Goldschmidt divider, the adder is set to have two bits for the integer portion of the divisor and six bits for the fractional part. Thus, in order to add 1 to the reciprocal of D_0 and add the integer value 2 to the result, the value 8'b10.000001 or 8'h81 is added to the reciprocal of D_i rather than 8'b00000011.

Each calculation of F_i is simplified to a NOT operation and an addition. One of the operands to this addition each iteration is the current divisor. The second operand is the constant 8'h81 as described.

When one of the operands to a full adder is a constant, the full adder can be simplified. Note that this is only appropriate if the adder will use the same constant for its entire lifetime. This is a rare scenario and should only be used in situations in which the expected application is known and finite.

First, assume that the input B to the full adder is a 1. The sum and carry-out equations in the full adder simplify to

$$\begin{aligned} \text{Sum} &= \text{A XOR "1" XOR } C_{in} \\ &= \overline{\text{A}} \text{ XOR } C_{in} \\ &= \text{A XOR } C_{in} \\ C_{out} &= (\text{A AND "1"}) \text{ OR } ((\text{A OR "1"}) \text{ AND } C_{in}) \\ &= \text{A OR } C_{in} \end{aligned}$$

These equations both simplify to a single Boolean operation. Table 8.11 shows an IMPLY implementation of these simplified full adder equations.

Table 8.11 Steps for an optimized IMPLY adder for B = 1

Step	Functionality	Goal
1	A IMP 0 → M0	NOT A
2	M0 IMP 0 → M1	Copy A
3	C_{in} IMP 0 → M2	NOT C_{in}
4	M2 IMP M1	C_{out} = A OR C_{in}
5	*Next bit reads C_{out} result*	
6	C_{in} IMP M0	A NAND C_{in}
7	C_{out} IMP 0 → M3	A NOR C_{in}
8	*Next bit reads C_{out} result*	
9	M0 IMP M3	Sum = A XNOR C_{in}

This adder requires nine steps and six memristors including inputs. The carry signal takes priority over the sum output and is available four cycles after execution begins. The adder maintains the original full adder carry propagation delay of two steps. An N-bit ripple-carry adder would have a delay of $2N + 7$ (as compared to $2N + 19$) and would reduce the complexity to $5N + 1$ memristors.

Similarly, if the input B to the full adder is a constant 0, the sum and carry-out equations simplify to

$$\begin{aligned} \text{Sum} &= \text{A XOR "0" XOR C}_{\text{in}} \\ &= \text{A XOR C}_{\text{in}} \\ \text{C}_{\text{out}} &= (\text{A AND "0"}) \text{ OR } ((\text{A OR "0"}) \text{ AND C}_{\text{in}}) \\ &= \text{A AND C}_{\text{in}} \end{aligned}$$

Table 8.12 shows an IMPLY implementation of these equations in a simplified full adder.

Notice that the value of C_{out}, M2 is overwritten in step 9. However, this is a safe operation because the value of C_{out} has already been used by its destination in steps 5 and 8. The value in memristor M2 is not used again and can be overwritten. This full adder requires ten steps and seven memristors. The carry-out signal is ready after four steps, and the carry propagation delay through the adder is still two steps. An N-bit ripple-carry adder would have a delay of $2N + 8$ and would use $6N + 1$ memristors.

Now, the ripple-carry adder in the iterations of the Goldschmidt divider can incorporate this simplified driver logic for the full adders. As discussed, for each iteration of the divider, the constant binary value 8'b10.000001 must be added to the inverse of the running divisor. To do this, full adders 0 and 7 of the ripple-carry adder will implement the full adder equations for a constant B operand equal to 1 and bits 1 through 6 will implement the full adder for a constant B operand equal to 0.

Note that the original full adder introduced for the initialization steps can be reused to perform the iterative additions. This adder has $7N + 1$ memristors, which is more than required for either constant-input full adder design. The modifications

Table 8.12 Steps for an optimized IMPLY adder for B = 0

Step	Functionality	Goal
1	A IMP 0 → M0	NOT A
2	M0 IMP 0 → M1	Copy A
3	C_{in} IMP M0	A NAND C_{in}
4	M0 IMP 0 → M2	C_{out} = A AND C_{in}
5	*Next bit reads C_{out} result*	
6	C_{in} IMP 0 → M3	NOT C_{in}
7	M3 IMP M1	A OR C_{in}
8	*Next bit reads C_{out} result*	
9	M1 IMP M2	(A OR C_{in}) NAND (A NAND C_{in})
10	M2 IMP 0 → M4	Sum = A XOR C_{in}

presented for the constant-input adders only impact the driver logic on the voltage operations in the adder, not the topology itself. Thus, the same adder can be reused with the updated operation steps in all of the iterations.

The use of a constant-input adder adds no complexity to the design and decreases the delay of the addition in each operation to one step for the NOT operation and $2N+8$ steps for the addition for a total of $2N+9$ steps per iteration to obtain the new F_i. A waveform for the operation $F_i = 2-D_i$ for $D = 5/8$ using the proposed adder is shown in Figure 8.12.

The bits resolve every two steps, beginning at step 10 and completing in step 24. The adder correctly resolves the value to 1 3/8, where bits 6 and 7 are the integer portion and bits 0–5 are the fraction.

As another optimization, the F_i result of the addition in each iteration can be stored directly into the multiplier inputs to remove the need for the sum memristors in each bit. This also removes the two copy steps from the overall delay that were required to copy the sum in the adder to the input operand in the multipliers. Now the complexity of the adder is $6N+1$ memristors.

Similarly, the results of the two Dadda multipliers can be directly stored into the next module that uses them rather than the local output memristors. For the Dadda multiplier, which resolves the dividend N, N can be stored back as the first

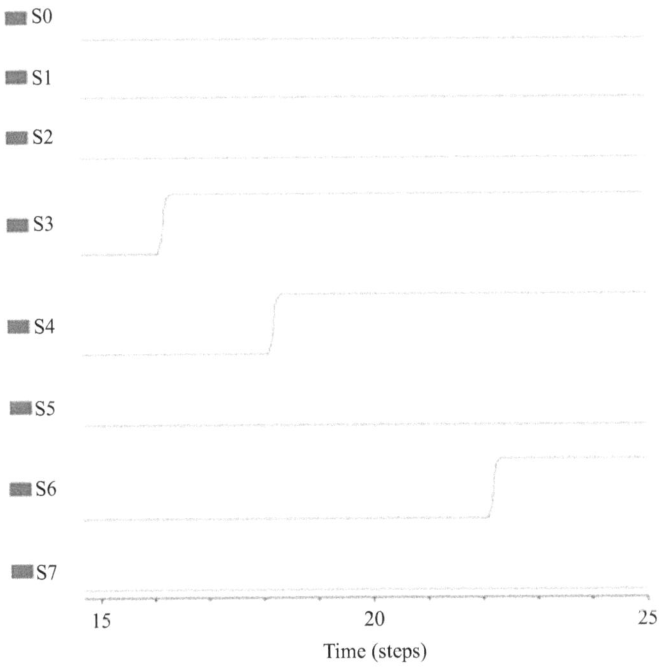

Figure 8.12 8-bit IMPLY constant-input addition for 2−D for D = 5/8

operand in the multiplier for the next iteration since this is the only place it is used. The result of the divisor D in the second Dadda multiplier can be stored directly into the A operand of the adder since it is needed to calculate the next value of F_i. Since the final adder in each 8-bit Dadda multiplier is 14 bits, this saves another two steps and 28 memristors.

When the execution of the adder to resolve F_i in each iteration was first discussed, a NOT operation on the divisor needed to be performed before beginning the addition. First, execute [(SUM IMP 0) IMP 0] per bit to perform an N-bit copy operation from the multiplier to the A inputs in the adder. Then perform (A IMP 0) to invert the divisor. Instead, a single step (SUM IMP 0) is performed with the A inputs in the adder serving as the N reset destination memristors. This executes the NOT operation and puts the necessary A input into the adder in a single step.

A full schematic for the optimized 8-bit Goldschmidt divider is shown in Figure 8.13.

Note that it is not possible to overlap the area required for the two Dadda multipliers even though both have a common input parameter, F_i. This is because the first step of the multipliers is to perform a matrix of AND operations between the bits of both input operands. This creates disjoint inputs to the remaining stages of full adders and half adders, preventing any operation overlap. However, it is possible to share the drivers across the two instantiations since they perform their operations in lockstep. In total, the complexity is 49 memristors, 49 drivers, and 49 switches for the adder and 2(577 memristors + 577 switches) + 577 drivers for the Dadda multipliers, for a total of 1,203 memristors, 1,203 switches, and 626 drivers.

Since the adder circuitry is used across iterations of the division, it is not possible to pipeline divisions in the design. It is left as an exercise to modify the current divider and select an appropriate pipelining granularity to enable divisions to be pipelined. The initialization addition takes $2N + 19$, or 35 steps for an 8-bit divider to complete. The iterative additions take $2N + 9$ steps, or 25 steps for each 8-bit addition. The multiplication takes 109 steps per iteration plus one step to invert the result back into the adder. The delay per iteration is 135 steps. The total delay for an 8-bit Goldschmidt divider that takes X iterations to converge is $135X + 10$ steps.

8.3.2 Hybrid-CMOS Goldschmidt divider

The hybrid-CMOS approach to a Goldschmidt divider is designed to minimize the use of CMOS while optimally reducing delay. The design consists of an 8-bit adder and two 8-bit Dadda multipliers with latches to store values between the units.

The adder cannot be optimized in the way it was for the IMPLY Goldschmidt divider to become a constant-input adder. In the IMPLY context, the adder could be modified by changing the programed driver logic without changing the circuitry. In the hybrid-CMOS context, the circuitry needs to change. Since the addition in the initialization uses different inputs than the additions in the iteration, there is no constant-input circuitry that would accommodate both operations. This is indicative of the key difference between the IMPLY approach and the hybrid-CMOS

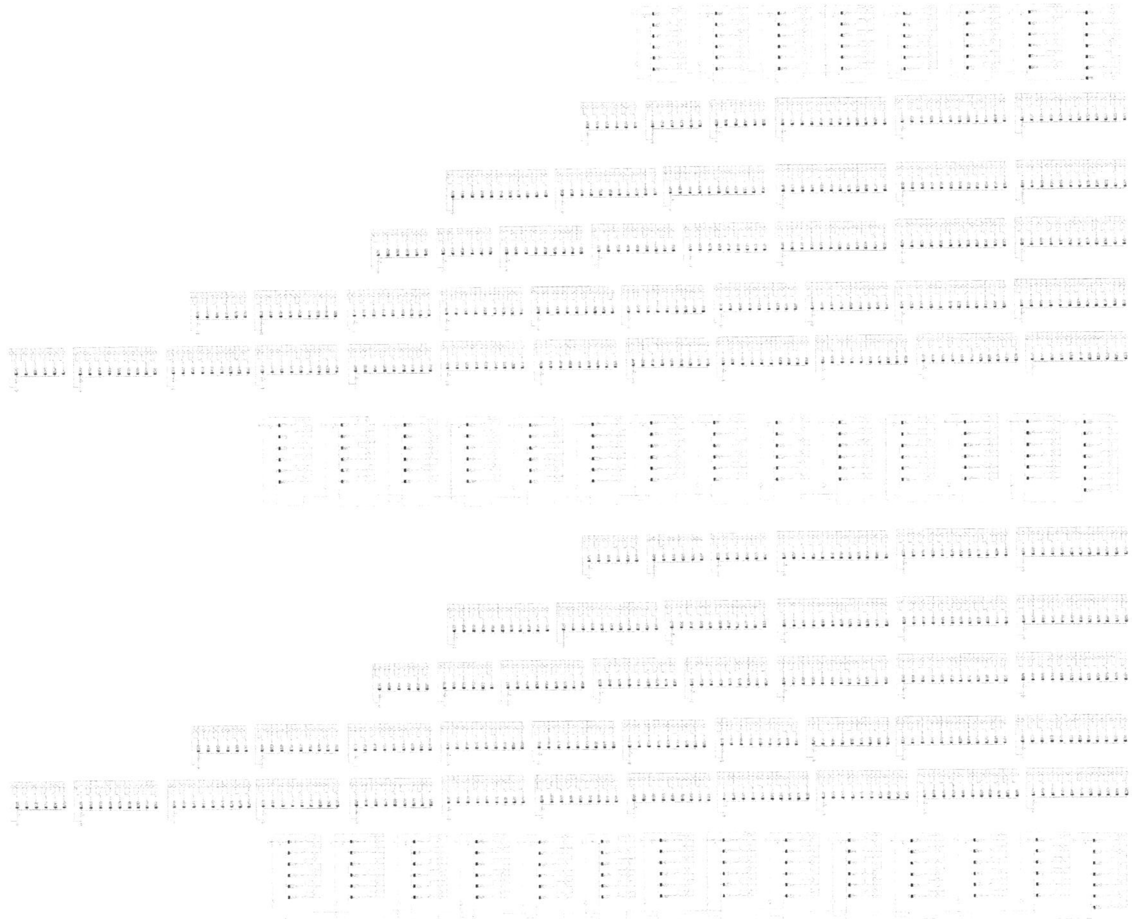

Figure 8.13 IMPLY Goldschmidt divider

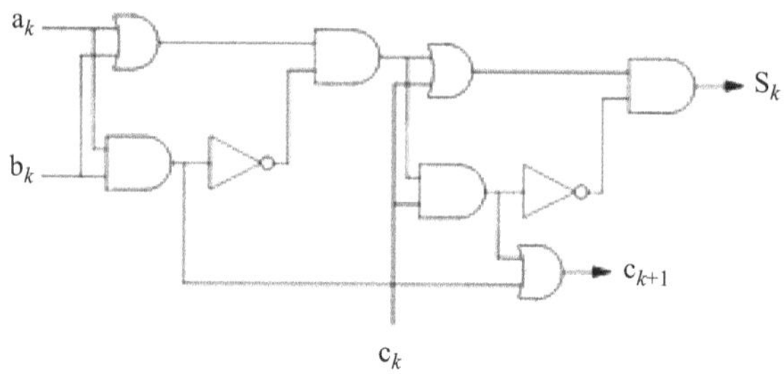

Figure 8.14 Traditional full adder gate structure

approach: The IMPLY approach performs computations with the drivers whereas the hybrid-CMOS approach performs computations with the memristors themselves.

To get around this limitation, there are two options. First, the design could use two separate adders, the first for the initial computation $F'-2D_0$ and the second for iterative $F_i=2-D_i$ operations. The second adder could be a constant-input adder. Alternatively, a single adder can be used which is not optimized but can perform all the necessary additions. It seems clear that the latter option is preferential, but the two alternatives can be characterized to be complete.

Consider the implementation of an N-bit ripple-carry adder with a known input operand. The traditional full adder uses the equations

$$\begin{aligned} \text{Sum} &= \text{A XOR B XOR C}_{\text{in}} \\ \text{C}_{\text{out}} &= (\text{A AND B})\,\text{OR}((\text{A OR B})\,\text{AND C}_{\text{in}}) \end{aligned}$$

A traditional CMOS or hybrid-CMOS full adder has the schematic shown in Figure 8.14.

As discussed for the IMPLY implementation, if the B input is known, the equations simplify to the following:

$$\begin{aligned} \text{B} = 1:\ & \text{Sum} = \text{A XNOR C}_{\text{in}} \\ & \text{C}_{\text{out}} = \text{A OR C}_{\text{in}} \\ \text{B} = 0:\ & \text{Sum} = \text{A XOR C}_{\text{in}} \\ & \text{C}_{\text{out}} = \text{A AND C}_{\text{in}} \end{aligned}$$

The circuitry for the traditional CMOS full adder can now be simplified to implement the new equations as shown in Figure 8.15.

The implementation for the full adder with B = requires one NOT gate, one AND gate, and two OR gates. The delay from the inputs to the sum is three gate delays, and the delay from carry-out to carry-out is one delay. For B = 0, the delays are the same as the B = 1 case, but the circuitry requires one NOT gate, two AND gates, and one OR gate.

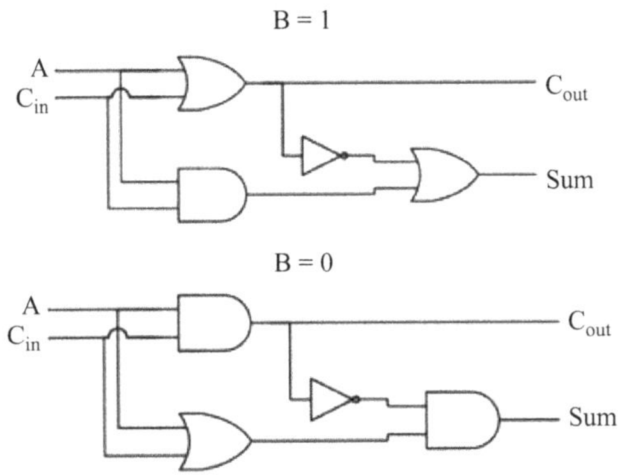

Figure 8.15 Optimized full adder for known B inputs

In the context of hybrid-CMOS designs, the AND and OR gates have the same complexity. Thus, a complete N-bit hybrid-CMOS adder would require a total of $6N$ memristors and $2N$ MOSFETs prior to signal restoration buffers being added. This can be compared to the baseline full adder, which requires over twice as many components: $14N$ memristors and $4N$ MOSFETs. The delay of the constant-input adder is $2N+2$ steps as compared to the baseline adder's delay of $2N+4$ steps (before signal buffering).

Since the complexity of the full adders has been reduced, the necessity for signal restoration buffers has also been reduced. Now, the sum outputs do not need buffers, only the carry-signals do. If the carry signals are buffered, the strength of the sum outputs cannot be significantly degraded. Note that the sum output has a maximum gate chain of three gates. However, along this path, the second gate is a NOT gate from MOSFETs. This gate restores the signal inherently. The other gate chain for the sum output is only two gates, so a signal restoration buffer is not needed.

Each signal restoration buffer requires four MOSFETs for a total of $4N$ MOSFETs dedicated to signal restoration. These buffers are along the critical path of each full adder, increasing the carry propagation from two to three gate delays. The total complexity of the optimized ripple-carry adder is $6N$ memristors and $6N$ MOSFETs, and the delay is $3N+2$ steps.

The optimized design has little improvement in terms of delay over the baseline adder. It is only two gate delays faster per iteration. Recall that if the design incorporates this constant-input adder, a second adder is needed to perform the initial addition. Thus, there is a performance vs. complexity trade-off here.

If the constant-input adder is used in the implementation, it increases the complexity by the size of the adder, $6N$ memristors and $6N$ MOSFETs. However, this reduces the delay by two gate delays per iteration. For a division that takes X iterations to converge, this is a reduction of $2X$ steps. For the purposes of this design, reducing the complexity is favored over reducing the delay. This is partially due to the fact that hybrid-CMOS designs are generally large in complexity as is.

For the hybrid-CMOS Goldschmidt divider, a single instantiation of the standard ripple-carry adder, which requires $14N$ memristors + $12N$ MOSFETs with a delay of $3N+4$, is used. No constant-input adder is incorporated. Similar to the adder in the IMPLY Goldschmidt divider, the hybrid-CMOS adder performs the initial addition and the iterative additions. The initial addition performs as usual, inverting the inputs, shifting their input, and performing $F_0 = F' + \overline{2D_0} + 1$. The iterative additions perform $F_i = \overline{2D_i} + 8\text{'h}81$ rather than the customary subtraction equations since the Goldschmidt divider has two bits for the integer and six bits for the fraction result.

In the context of an 8-bit Goldschmidt divider, the sum output of the 8-bit adder serves as input to both Dadda multipliers each iteration. Within these multipliers, each bit of the sum serves as an input to N gates in the initial $N \times N$ AND gate array in each multiplier. Thus, each bit of the sum output has a fan-out of $2N$. Since the fan-out is high, each sum output in the adder must be buffered. This is incorporated into the original complexity measurement for the ripple-carry adder.

As discussed in Chapter 7, each of the 8-bit Dadda multipliers requires 728 memristors and 530 MOSFETs with a delay of 73. When incorporated into the Goldschmidt divider, the outputs of the multipliers, divisor D_i and dividend N_i, must be stored in flip-flops at the end of each iteration. Call these two flip-flops DFF and NFF. At the end of an iteration, D_i and N_i are latched into DFF and NFF. On the next cycle, the reciprocal of the latched value D_i can feed back into the adder as the A input. At the same time, the latched values of D_i and N_i can feed as inputs to their respective Dadda multipliers. The flip-flops are necessarily to avoid interference across iterations.

Each of the flip-flops in DFF and NFF requires 16 memristors and 20 MOSFETs as shown for the hybrid-CMOS shift-and-add multiplier. However, for the Goldschmidt divider, each flip-flop buffers its outputs, adding another $8N$ MOSFETs to the design. Recall that each buffer is essentially two NOT gates. The flip-flops provide the reciprocal of D_i in parallel with D_i as part of the buffer functionality. The total complexity for the hybrid-CMOS approach is

8-bit adder:

$$= 14N \text{ memristors} + 12N \text{ MOSFETs}$$
$$= 112 \text{ memristors} + 96 \text{ MOSFETs}$$

Two 8-bit Dadda multipliers:

$$= 2\,(728 \text{ memristors} + 530 \text{ MOSFETs})$$
$$= 1{,}456 \text{ memristors} + 1{,}060 \text{ MOSFETs}$$

$2N$ flip-flops:

$$= 2N\,(16 \text{ memristors and } 24 \text{ MOSFETs})$$
$$= 256 \text{ memristors} + 384 \text{ MOSFETs}$$

Total complexity:

$$= 1{,}824 \text{ memristors} + 1{,}540 \text{ MOSFETs}$$

The delay is 28 steps for the adder, 73 for the multipliers, and 2 for the flip-flops, for a total delay of 103 steps for each iteration of the division. The total delay is $103X$ steps where X is the number of iterations performed before the divisor of the divider converges to 1.

8.3.3 Threshold-gate Goldschmidt divider

The threshold-gate Goldschmidt divider performs similar optimizations as the other designs to achieve lower latency and area. For this design, threshold gates constructed from GOTO pairs are used.

First, the N-bit adder can be optimized for known inputs. Figure 8.16 shows the implementations for the optimized threshold-based full adders for the case when input $B=0$ and $B=1$.

The complexity of each full adder is reduced by two memristors with the simplification. The B input originally served as the input to both threshold gates in the adder, using two memristors. These are both removed. The remainder of the design remains the same, with an alteration on the gate thresholds.

The delay also remains unchanged. The full adders both still require one gate delay for the carry-out signal and three for the sum signal. An N-bit adder constructed from these gates requires $5N$ memristors, $2N$ MOSFETs, and $2N$ GOTO pairs and has a delay of $N+2$ gate delays.

Threshold-gate-based designs are similar to hybrid-CMOS designs in that the computation is performed by the hardware rather than on the hardware. Because of this, the circuitry of the adder itself changes when there is a constant-input B. In the IMPLY context, the hardware did not need to change, only the driver logic did. This causes a similar design decision as the hybrid-CMOS design.

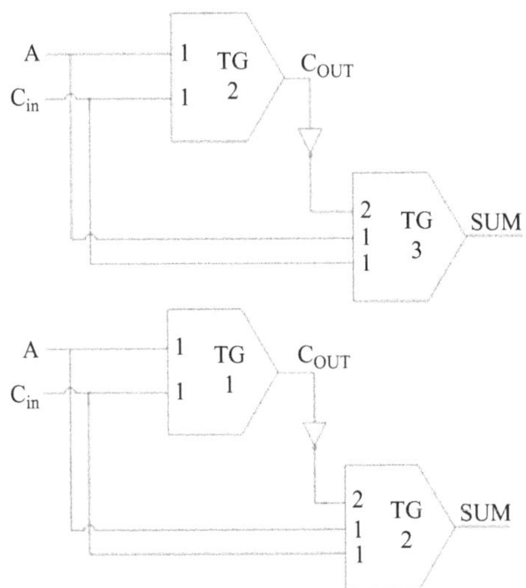

Figure 8.16 Optimized threshold full adder for $B=0$ and $B=1$

There are two choices on how to implement the adder functionality in the Goldschmidt divider. The addition in the initialization phase does not have a known input. The additions in the iterative phases do. One option for the Goldschmidt implementation uses a single standard adder for all additions. In the case where one of the inputs is known, this value is simply driven and held throughout the iterations. The second option for the Goldschmidt implementation uses two adders—one standard two-input adder and the other constant-input adder. In this case, the first adder is used for the initialization phase and the second adder is used for the iterative phases.

In general, a trade-off between delay and complexity must be made between the two options. The first option has a simpler complexity but will generally require a greater delay since there are no optimizations for the known input. The second option has a higher complexity but will generally require a smaller delay since optimizations for the iterative additions are performed. However, the trade-off is different in this case.

In the first option, there is a single standard adder. This adder requires $7N$ memristors, $2N$ MOSFETs, and $2N$ GOTO pairs. In the second option, there are two adders, one requiring $7N$ memristors, $2N$ MOSFETs, and $2N$ GOTO pairs and the other requiring $5N$ memristors, $2N$ MOSFETs, and $2N$ GOTO pairs. The second option has a total complexity of $12N$ memristors, $4N$ MOSFETs, and $4N$ GOTO pairs. This is almost double the complexity of the first option.

As shown, the delay of both the standard adder and the constant-input adder are $N+2$ gate delays. Thus, there is no benefit to the critical path when the constant-input adder is used. Since its use adds the complexity of the adder to the design, there is no convincing argument to use it in the design. Thus, the first option, the use of a single standard adder for all additions in the divider, is selected. The component area of the 8-bit adder for the Goldschmidt divider is 56 memristors, 16 MOSFETs, and 16 GOTO pairs with a delay of 10 gate delays.

The threshold-gate-based version of an 8-bit Dadda multiplier requires 615 memristors and 91 GOTO pairs and 44 MOSFETs with a delay of 16 steps. A threshold-gate-based latch can be implemented with GOTO pairs as shown for the shift-and-add multiplier. Each latch requires seven memristors, two GOTO pairs, and four MOSFETs. In an N-bit latch, the inverters can be shared across the bits for a total complexity of $7N$ memristors, $2N$ GOTO pairs, and four MOSFETs. The delay of the latch is two gate delays.

As described for the hybrid-CMOS design, two 8-bit latches are needed, one to latch the result of each multiplier. The outputs of these latches feed back into the adder and the Dadda multipliers each iteration. Thus, for an 8-bit Goldschmidt divider, the total complexity is

8-bit adder:

$$= 56\text{ memristors} + 16\text{ MOSFETs} + 16\text{ GOTO pairs}$$

Two 8-bit multipliers:

$$= 1{,}230\text{ memristors} + 88\text{ MOSFETs} + 182\text{ GOTO pairs}$$

Two 8-bit registers:

$$= 14N \text{ memristors} + 4 \text{ MOSFETs} + 4N \text{ GOTO pairs}$$
$$= 112 \text{ memristors} + 4 \text{ MOSFETs} + 32 \text{ GOTO pairs}$$

Total complexity:

$$= 1{,}398 \text{ memristors} + 108 \text{ MOSFETs} + 230 \text{ GOTO pairs}$$

The delay of each iteration is 29 steps for a total delay of $29X$ steps. A full schematic for the 8-bit Goldschmidt divider with the Kautz counter structure and the accompanying adder and latches is shown in Figure 8.17.

8.3.4 MAD-gate Goldschmidt divider

The 8-bit Goldschmidt divider based on MAD gates consists of an 8-bit adder and two 8-bit Dadda multipliers. Flip-flops and other storage blocks are not necessary for the MAD implementation because the values are inherently latched as party of the MAD operations. This is similar to the reasoning behind these modules being unnecessary in the IMPLY design.

First, focus on the 8-bit adder. Again, consider how it can be optimized for a constant input when the input B is known. An 8-bit MAD-based adder for a known input B=8'b0 is shown in Figure 8.18.

Each bit of the adder only requires three memristors now because B is not needed. Instead, B is replaced with a single resistor R_b, resistance of which is R_{off}. This is equivalent to a memristor in high-resistance, or "0," state. Again, the equations for the carry-out and sum when the B input is 0 are

$$\begin{aligned} C_{out} &= \text{A AND } C_{in} \\ \text{Sum} &= \text{A XOR } C_{in} \end{aligned}$$

The implementation of a MAD AND and XOR gate has already been shown. However, in the context of the adder, the second operand for the gates, C_{in}, is not in a resident memristor. Instead, it is a signal from the previous adder. Thus, the form of the AND and XOR operations take an altered form.

The C_{out} memristor is resolved by gating the V_{set} voltage with two switches—one driven by the A input and the other by the C_{in} value. In this way, the carry-out memristor will only be applied the V_{set} voltage (thereby setting it to logical "1") if both A and C_{in} have a high enough voltage, corresponding to a "1." This effectively implements the AND operation.

The sum memristor has two parallel paths to gate V_{set} to the memristor. The two paths correspond to the two scenarios that satisfy the XOR operation, $(\text{A AND } \overline{\text{B}})$, and $(\overline{\text{A}} \text{ AND B})$. Each of these paths is implemented in the same manner as the logic for the carry-out memristor. If either of the AND statements is true, a closed path will exist between the V_{set} voltage and the sum memristor, and the sum will be set to "1." This effectively implements the XOR operation.

The total complexity of the adder is reduced to $3N$ memristors, $7N$ switches, and $3N + 1$ drivers.

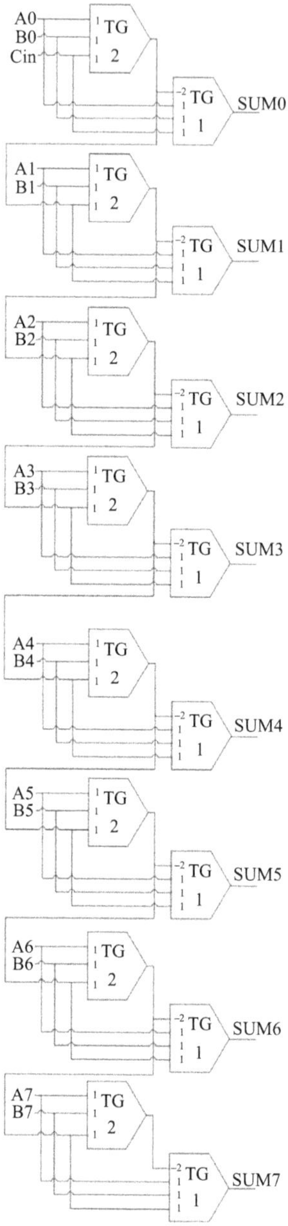

Figure 8.17 8-bit threshold-gate Goldschmidt divider using the Kautz method

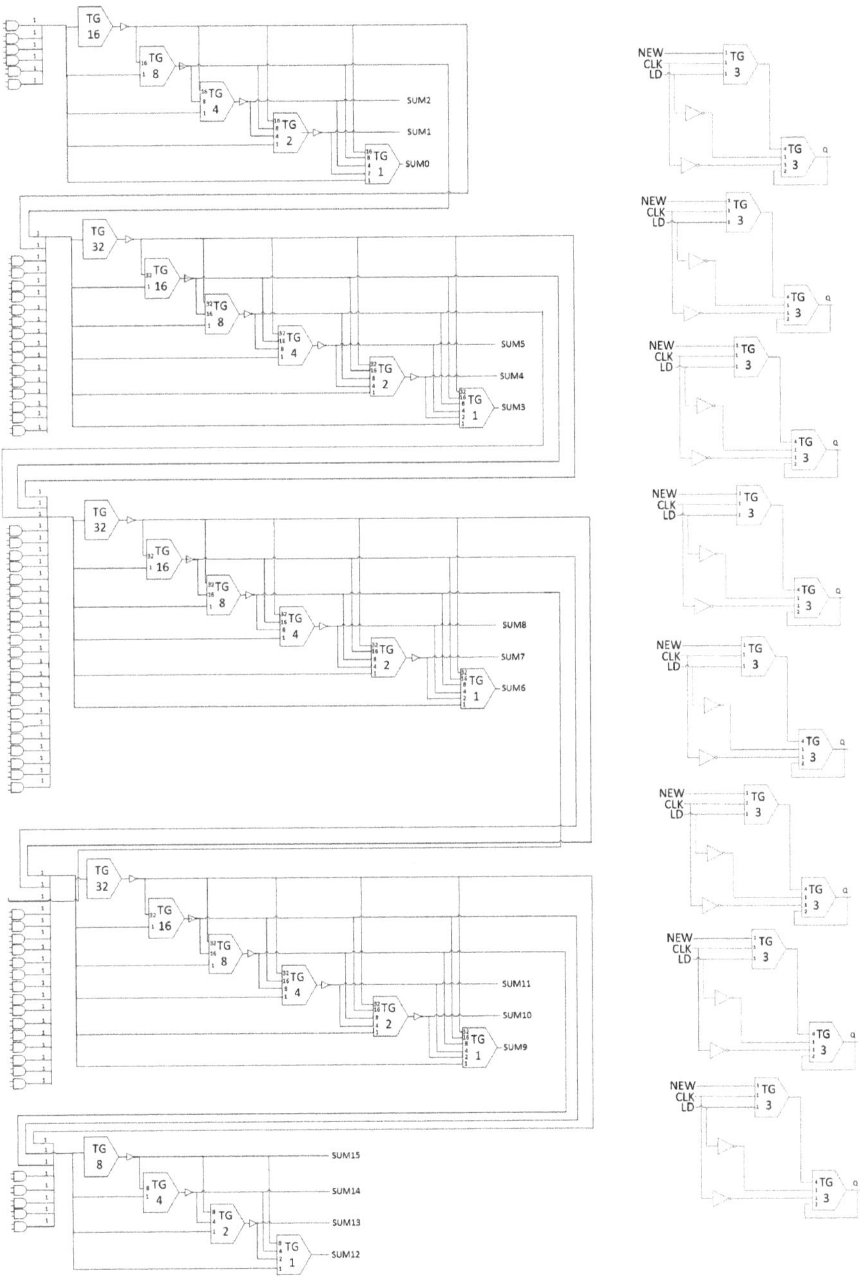

Figure 8.17 (Continued)

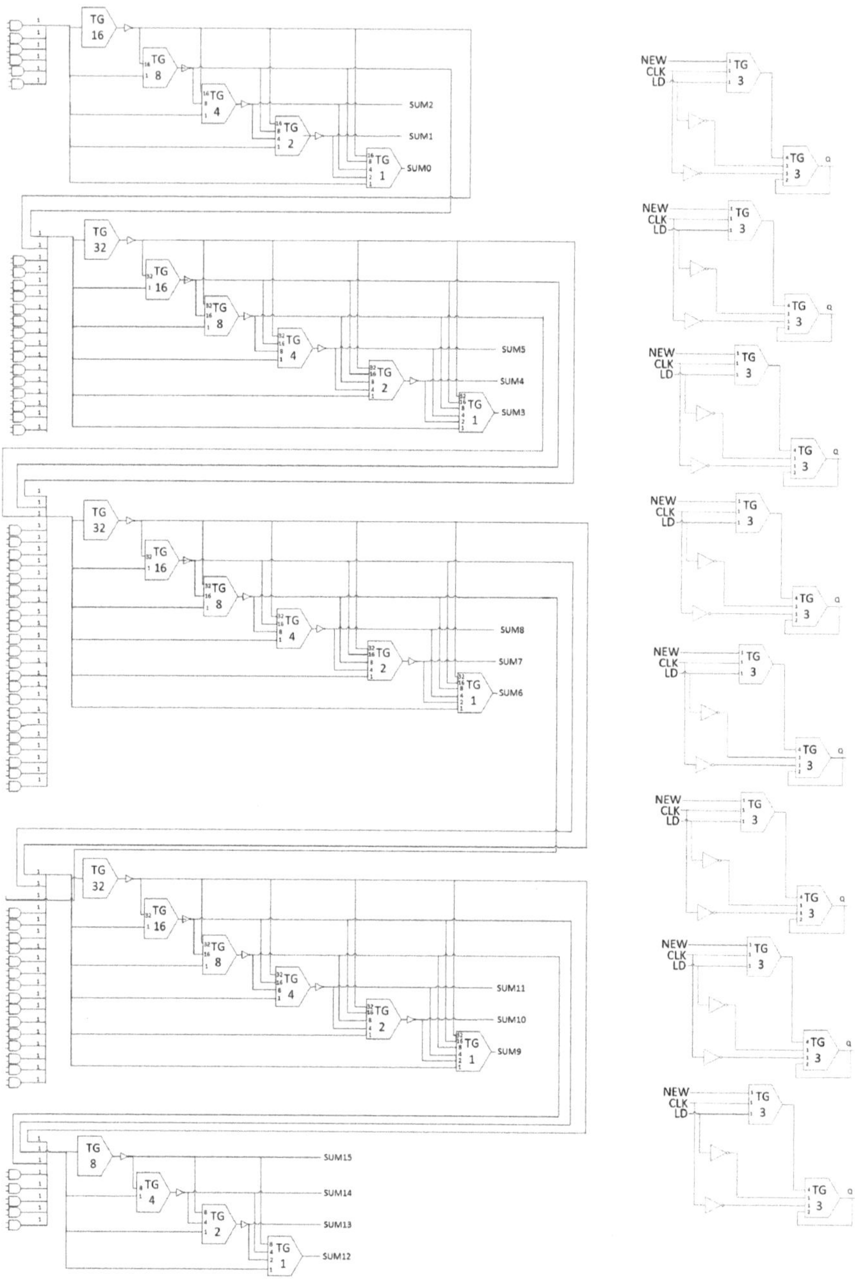

Figure 8.17 (Continued)

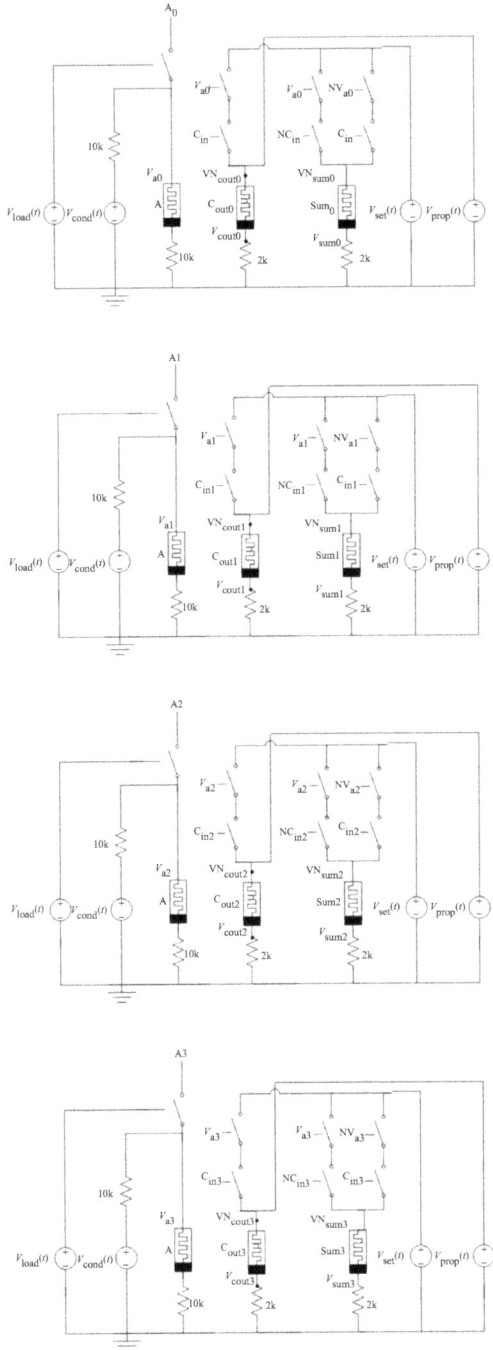

Figure 8.18 8-bit MAD optimized adder schematic for known input $B = 0$

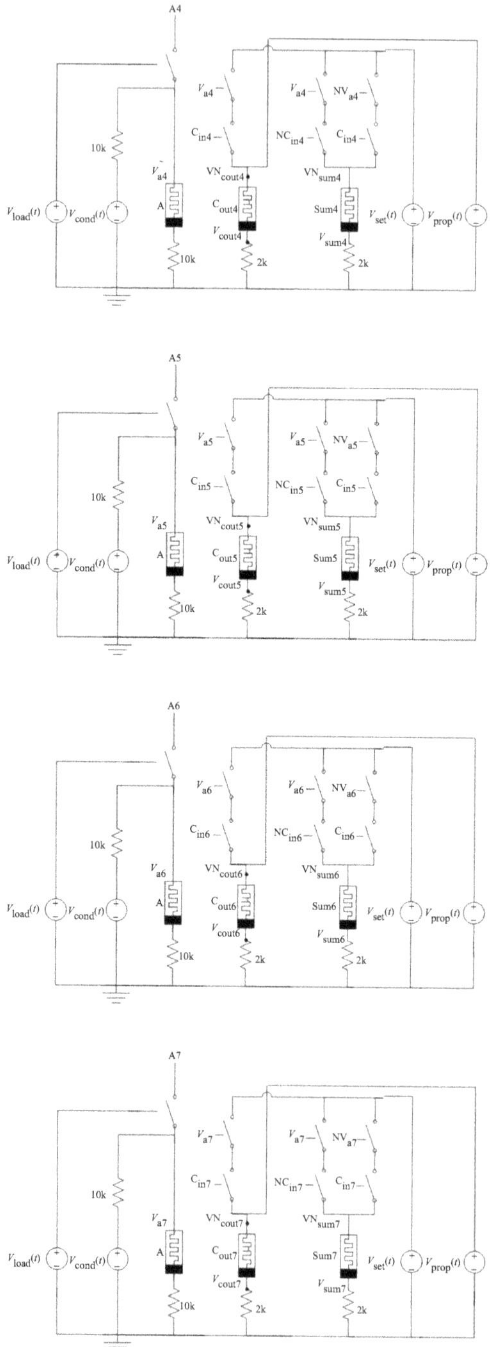

Figure 8.18 (Continued)

Similarly, the MAD-based adder for a known input B = 8'b11111111 is shown in Figure 8.19.

The circuitry looks similar to the adder for which input B is 0. The only difference is that the resistor representing B now has a resistance of R_{on} rather than R_{off}. This corresponds to the value B="1" if it were stored in a memristor. Also, the driver circuitry for the carry-out and sum memristors has changed. The equations for the carry-out and sum when the B input is 1 are

$$\begin{aligned} \mathrm{C_{out}} &= \mathrm{A\ OR\ C_{in}} \\ \mathrm{Sum} &= \mathrm{A\ XNOR\ C_{in}} \end{aligned}$$

The carry-out signal is resolved by gating the V_{set} signal with two paths. The first path corresponds to A being "1" and has a single switch driven by the voltage on the A memristor. The second path corresponds to C_{in} being "1" and has a single switch driven by C_{in}. If either is true, the path is closed, and the carry-out memristor is set to "1." This implements the OR operation.

The sum memristor still has two parallel paths, each with two switches. However, the drivers of the switches have changed slightly. Rather than implementing the two AND expressions for the XOR operation, the circuitry implements the two AND expressions for the XNOR operation. Specifically, the two paths correspond to $(\mathrm{A\ AND\ B})$ and $(\overline{\mathrm{A}}\ \mathrm{AND}\ \overline{\mathrm{B}})$. If either of the AND statements is true, a closed path will exist between the V_{set} voltage, and the sum memristor and the sum will be set to "1." This effectively implements the XNOR operation.

The total complexity remains $3N$ memristors, $7N$ switches, and $3N+1$ drivers. The total delay of both constant-input adders is maintained as N steps. An example of the functionality of this adder for the operation $2-D$ for $D=5/8$ is shown in Figure 8.20.

The sum correctly resolves after eight steps to 0×58, corresponding to 8'b01.011000.

However, notice that the circuitry of the adder must change to accommodate a known input. This is the same issue encountered for the hybrid-CMOS and threshold-gate designs. Since the delay is not improved with the constant-delay adder, there is no design trade-off to be made. If both a standard and constant-input adder were used, the total complexity would be $7N$ memristors, $21N$ switches, and $6N+2$ drivers. If a standard ripple-carry adder were used, only $4N$ memristors, $13N$ switches, and $3N+1$ drivers are needed. Thus, it is cheaper to use a single instantiation of the standard MAD ripple-carry adder for the Goldschmidt divider.

The 8-bit MAD-based Dadda multiplier requires 303 memristors, 63 drivers, and 960 switches, and a delay of 20 steps. The two instantiations of this multiplier can share drivers since they operate in lockstep, but they cannot share the remainder of the circuitry. This removes 63 drivers from the second multiplier. Also, the memristors used to store the result of the multiplications can be removed. Instead, the sum can store directly into the inputs in the adder. This is performed in the same manner as described for many previous designs. In all, this removes 14 memristors from the last stage in each multiplier in the design.

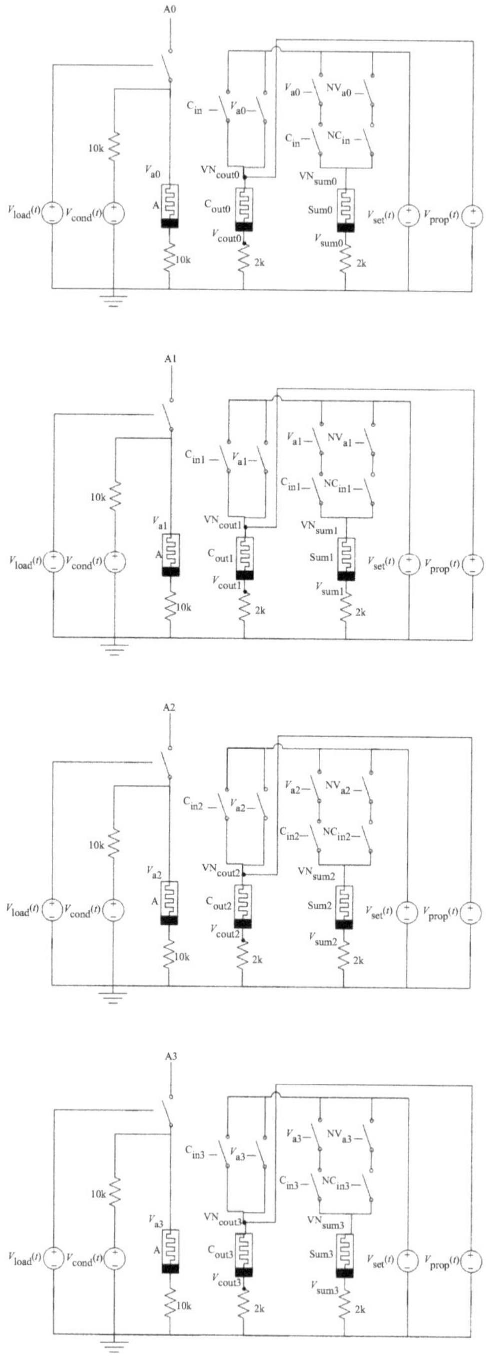

Figure 8.19 8-bit MAD optimized adder schematic for known input B = 8'hFF

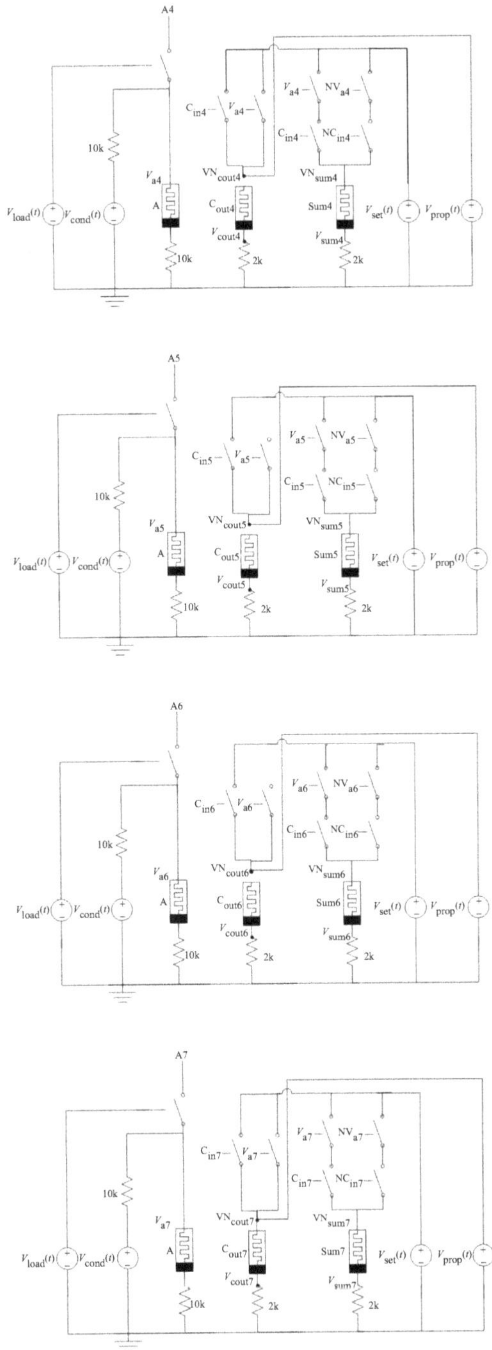

Figure 8.19 (Continued)

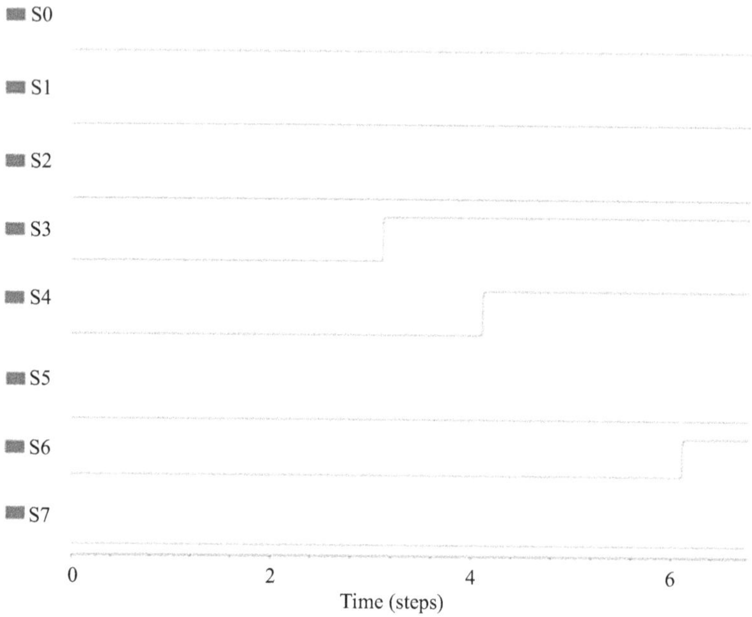

Figure 8.20 MAD optimized adder execution for 2−D for D = 5/8

The divider is complete with the addition of driver logic to propagate values from the adder to the multiplier and vice versa. Because of the driver-based nature of MAD designs, this requires no additional area.

The total complexity of the design is

One 8-bit optimized adder:

$$= 32 \text{ memristors} + 104 \text{ switches} + 25 \text{ drivers}$$

Two 8-bit Dadda multipliers:

$$= 2(303 \text{ memristors} + 960 \text{ switches}) + 63 \text{ drivers} - 28 \text{ memristors}$$

Total complexity:

$$= 610 \text{ memristors} + 2{,}024 \text{ switches} + 88 \text{ drivers}$$

The total delay is 8 steps for the adder and 20 steps for the multiplier with no additional delay for movement of data due to the design's driver optimizations. The total delay is 28 steps. Similar to the binary-restoring divider, it is not possible to pipeline divisions since the adder and multiplier are pipelined across iterations of a single division.

8.3.5 Goldschmidt divider analysis and comparison

The complexity and delay breakdown for the proposed Goldschmidt dividers are shown in Table 8.13 for a divider that performs X iterations to converge.

All of the presented values are for 8-bit Goldschmidt dividers and the threshold implementation uses GOTO pairs for the underlying gate constructs. The four memristor-based designs span a large spectrum of complexity and delay characteristics, but all exhibit some benefit over the CMOS implementation.

The threshold- and MAD-gate-based designs are significantly faster than the other proposed designs and the CMOS design. Each iteration of the division only requires 29 steps in the threshold design and 28 in the MAD design. This trend is expected since the Dadda multiplier and ripple-carry adder implementations for these two approaches also exhibit low delay.

Depending on the size of the divider, the threshold-gate design's delay could be further improved by using the Minnick method rather than the Kautz method for the Dadda multiplier. This method has a higher complexity in exchange for a faster delay.

Both the CMOS and the hybrid-CMOS approaches have a total delay about 2.5 to 3 times that of the threshold and MAD designs. The additional delay for the hybrid-CMOS over the CMOS design is due to the high delay of the Dadda multipliers. In fact, the hybrid-CMOS Dadda multiplier alone has nearly the same delay as the CMOS Goldschmidt divider. The IMPLY-based design is the slowest of the proposed designs, requiring 135 steps each iteration. Even with the optimizations involving the constant-input adder, the IMPLY design still suffers from serialization.

In terms of component area, the threshold-gate design has the fewest components. Again, this is partially due to the selection of the Kautz method for the underlying Dadda multipliers. This method optimizes for component area instead of delay. The threshold-gate design also has minimal CMOS, using only 108 MOSFETs. This is compared to the nearly 5,000 MOSFETs used for the CMOS implementation.

The IMPLY and MAD approaches have roughly equivalent component counts, with the MAD implementation slightly edging out IMPLY. Both designs require

Table 8.13 Complexity and delay comparison for the proposed 8-bit Goldschmidt dividers

	CMOS	**IMPLY**	**Hybrid-CMOS**	**Threshold**	**MAD**
Delay	$79X$	$135X + 10$	$103X$	$29X$	$28X$
Area	4,968 MOSFETs	1,203 memristors 626 drivers 1,203 switches	1,824 memristors 1,540 MOSFETs	1,398 memristors 108 MOSFETs 230 GOTO pairs	610 memristors 88 drivers 2,024 switches

fewer than the CMOS and hybrid-CMOS designs. Although the Dadda multipliers introduce a significant number of switches to both designs, the number of drivers is kept relatively low by sharing them across the two instances of the Dadda multipliers.

The hybrid-CMOS approach has the greatest component area of the memristor-based designs with 1,824 memristors and 1,540 MOSFETs. However, the amount of CMOS is reduced to less than 1/3 of the traditional CMOS design, and the overall component count is also less than the CMOS design. Overall, all of the proposed memristor-based Goldschmidt divider designs improve in either delay or performance (or both) as compared to the traditional CMOS design.

Chapter 9
Proposed future work

Given the infancy of memristor research, there are many avenues for future directions in this area. It is essential that students, researchers, and industry colleagues continue to actively explore the world of memristors and its impact on the future of modern systems.

9.1 Model and tool support

First and foremost, improved simulation models and understanding of memristors are critical to the advancement of their use. Without introducing them to the academic community and university students, memristors will not become commonplace in logic design.

One catalyst for the growth of memristors is to establish accurate, available, easy-to-use models for memristors. This would allow students, researchers, and industry members to explore memristor characteristics and applications at a more progressive rate. This would in turn lead to the rapidization of unique ideas and development in the field of memristors. It would also offer shorter design cycles for memristor designs because a starting place would exist.

Power and energy measurements should be incorporated into these models. Currently, it is not possible to collect power or energy numbers for large memristor-based designs. In a field where energy is one of the most crucial metrics for a design, it is critical that the tool support evolves to mirror this trend. Most measurements must be taken manually by summing the $V*I$ properties of each individual memristor in the design over time. This is not practical, accurate, or scalable for complex simulations.

Ideally, memristor models would be prolific and flexible enough to support various implementations of memristive devices. As seen in this text, memristive devices can be unipolar or bipolar and can have a wide variety of physical properties. There are even more physical factors left undiscussed in this text. Depending on the applications, researchers need to be able to use all variations of a memristor. Thus, models need to be able to represent many or all of them. Additionally, these models need to be lean enough that simulation times do not become egregious.

One of the major challenges to memristor simulation is the lack of programability in modern simulation tools. Most simulation tools are based on the assumption that Von Neumann architectures are being constructed and simulated.

However, in some designs of memristor-based circuits and memories, the computations are brought to the data and exist as a series of voltage pulses on the memristive devices. This is true for both IMPLY and MAD designs. This poses a key challenge to the simulation of these circuits.

Currently, the only way to represent this in modern simulation tools is via piecewise linear (PWL) signals to each individual memristor in the design. For each memristor, the designer must indicate what voltage should be applied to that memristor in each time step of the simulation. As the design grows, this can become exhaustive. Also, in many tools, there is no option to import these PWL signals from an external source, forcing the designer to embed these PWL signals into the design, limiting the portability across schematics and across design tools. Separate from these issues, there is the challenge of debugging these signal files. A simple change from applying voltage V in timestep T to timestep $T+1$ causes a change to all subsequent timesteps in the design. This is often a manual effort. Tools can be enhanced to address many of these challenges to encourage the use of memristors in modern circuit design and exploration. Tool support can shorten the design cycle, removing the manual effort from the researcher.

By the same token, tool support for signal restoration buffers in hybrid-CMOS designs would greatly enhance the quality and design time of the designs. Currently, it is a trial-and-error process to determine where to place signal restoration buffers in a hybrid-CMOS design. The faster alternative is to just conservatively place buffers at the output of every gate in the design. However, this solution is unscalable. Tools can algorithmically determine how to correctly and minimally use signal restoration buffers for a given hybrid-CMOS design. This is similar to how modern tools can perform place-and-route on behalf of the use.

Additionally, there is currently no model for GOTO pairs or CSTG threshold dates that can be used in simulation or design. Establishing one would enable researchers to simulate and measure nanoscale designs such as the threshold-gate designs presented in this text. Since designs that use these components are combinational, they do not suffer from the same challenges outlined above for the IMPLY- and MAD-based designs.

9.2 Application exploration

Second, research can focus on the application of memristors to the remainder of modern system architecture. As the complexity of the system increases, there are more opportunities that could be leveraged from memristors, especially since their logic can be pipelined much more efficiently and at a finer granularity than traditional CMOS. Research into the impact of this new way of thinking about dataflow should be further investigated.

9.2.1 Memory design

Memristors were first introduced as a viable candidate for memory structures due to their nonvolatility and use in a crossbar. It was not until later in the memristor's

Table 9.1 Possible encodings for a memristor with 2-bit storage with resistance R and a resistance range from R_{on} to R_{off}

Resistance range	Binary value
$R_3 < R < R_{off}$	00
$R_2 < R < R_3$	01
$R_1 < R < R_2$	10
$R_{on} < R < R_1$	11

lifetime that logic began to be explored as it was in this text. Although crossbars are the most abundantly researched application for memristors, this field of research is still in its mere infancy. The original idea of implementing memristors in a crossbar can be extended and explored in many orthogonal directions.

First, memristors have the potential to be used for multibit storage [63–67]. In other words, it is possible for a single memristor or memristor cell to hold more than one bit of information. In this text, it was assumed that high-resistance R_{off} represents binary "0" and low-resistance R_{on} represents binary "1." However, there is no fundamental limitation on the granularity of this representation. Instead, imagine three additional resistance values R_1, R_2, and R_3 such that $R_{on} < R_1 < R_2 < R_3 < R_{off}$. Now, a memristor with a resistance R such that $R_3 < R < R_{off}$ represents binary "00," a resistance R such that $R_2 < R < R_3$ represents binary "01," a resistance R such that $R_1 < R < R_2$ represents binary "10" and a resistance R such that $R_{on} < R < R_1$ represents binary "11." This is summarized in Table 9.1.

This memristor would be capable of 2-bit storage. The idea can be extended to n-bit storage for a memristor with a resistance range split into n subranges. This can be visualized in the step function shown in Figure 9.1.

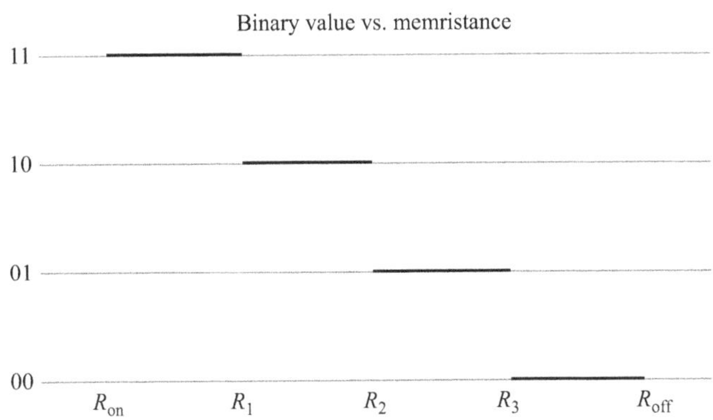

Figure 9.1 Step function for resistance values in an n-bit storage memristor device

The physical properties of the memristor determine the upper limit of n. As n increases, the sensitivity and reliability of the device become more crucial. For the largest n, $n = (R_{off} - R_{on})$, each range of resistances is only a single resistance value. The memristor needs to be able to distinguish between 100 and 101 Ω. Clearly, this is a tall task.

If a single device behaves differently over time or multiple instances of a device are not identical, these ranges will become inaccurate. The precision of the device must be able to deterministically exist for multibit storage to be feasible. Second, the detection or sensing circuitry for the device must have equal or higher precision in order to be able to differentiate the n resistance ranges and correctly read or write resistance for a given value. Both these limitations become exponentially more difficult as n increases.

9.2.2 Analog logic

The idea of multibit storage can be extended to let memristors serve as analog devices [68–71]. In this context, the memristor is able to hold any value in its resistance range, from R_{off} to R_{on}. Each distinct resistance represents a distinct value. Essentially, this is multibit storage with a maximum number of ranges n.

When memristors are used as analog devices, the logic performed on them is also analog. IMPLY and MAD operations are exchanged in favor of analog circuits such as op-amps.

Op-amp circuitry can be used to set the value of a memristor to a desired value V_{in} as shown with the circuit in Figure 9.2 [71].

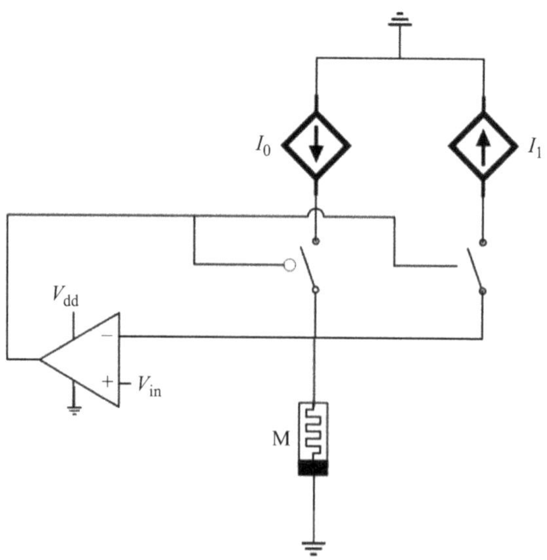

Figure 9.2 Analog set circuitry for a memristor [71]

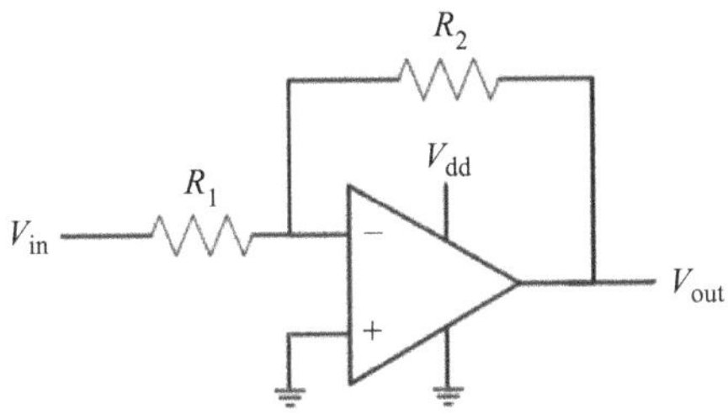

Figure 9.3 Traditional negative amplifier op-amp circuit

The desired voltage V_{in} is applied to the negative terminal of the op-amp. The voltage at the n-terminal of the memristor M is applied to the other terminal of the op-amp. If the voltage at the n-terminal of the memristor is greater than the input value, the output of the op-amp will be high, driving the second current to the memristor. The current decreases the resistance of the memristor. This in turn lowers the voltage across the M memristor and the input to the positive node of the op-amp. The op-amp circuit continues in this manner until the resistance of M is decreased such at the voltage across the memristor equals V_{in}.

If the voltage at the n-terminal of the memristor is less than the input value, the output of the op-amp will be 0 and the first current will be driven to the memristor. The current increases the resistance of the memristor and increases the voltage across it. This increases the voltage at the positive input node of the op-amp. After this iterative process completes, the memristor converges to the input value V_{in}. Effectively, this circuit performs the set operation.

Beyond the set operation, it is possible to use a similar circuitry to perform the main four arithmetic operations: addition, subtraction, multiplication, and division. Again, this is analog arithmetic. It is greatly simplified compared to digital arithmetic in terms of complexity. For example, examine division. First, consider the traditional op-amp circuit shown in Figure 9.3.

For an ideal op-amp, this circuit behaves as a negative voltage amplifier. The current through R_1 and R_2 is V_{in}/R_1 amps, and the voltage at the terminals of the op-amp is 0 V. the output voltage V_{out} is expressed as

$$V_{out} = -\frac{R_2 V_{in}}{R_1}$$

For example, if the input voltage $V_{in} = -1$ V, the output is R_2/R_1 volts. This circuitry is capable of performing R_2/R_1. However, R_1 and R_2 are resistors with static values that cannot bc changed. Thus, only a single division computation can be performed by this circuitry through its lifetime. Clearly, this is not useful in modern systems and has minimal application.

Now, replace resistors R_1 and R_2 with memristors M_1 and M_2 as shown in Figure 9.4.

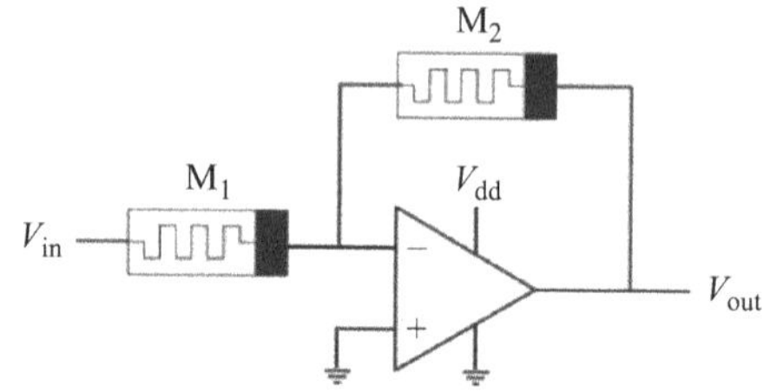

Figure 9.4 Analog memristor division circuit

Now, the equation becomes

$$V_{\text{out}} = -\frac{M_2 V_{\text{in}}}{M_1}$$

Or $V_{\text{out}} = M_2/M_1$ volts when $V_{\text{in}} = -1$ V. This circuit now essentially acts as an analog divider for the values M_2 and M_1. However, now the values of M_2 and M_1 are not static. They can be set to the desired operands for a specific division operation.

To perform a division, the dividend input is set in memristor M_2 and the divisor is set in memristor M_1. The set operation can be done using any of the approaches in this text, including the one shown above in Figure 9.2. Now, if V_{in} is applied to the circuitry, V_{out} resolves to the division M_2/M_1. This output can then be stored into a memristor too.

Similar op-amp circuits can be used to perform analog addition, subtraction, and multiplication. Many traditional op-amp circuits can be transformed into dynamic arithmetic modules by simply replacing the resistors with memristors.

One downside of analog memristor logic is that the circuits require reliability, consistency, and highly sensitive components. The approach also requires extensive conversion circuitry and CMOS for the op-amps and other components, increasing the area.

Second, distributed memory is an emerging application that is becoming more popular as the memory bottleneck becomes the limiting factor on overall performance. Distributed memory is the idea of separating the memory into small regions that are then placed in different physical locations in the system to improve the locality of the data to the computation. The motivation is similar to the motivation behind caches: move data expected to be used closer to the point of use to improve the delay of data movement.

Distributed memory concepts can be extended to systems constructed from memristors. Consider the IMPLY and MAD contexts. In these designs, the computation is done on the memory itself. The memristors serve as both logic and memory. Thus, the location of data and the location of its "use" or computation is the same thing. Thus, in a distributed memory design constructed from memristors, the memory would be allocated amongst the various logic structures in the device.

There is no need to dedicate memory purely for storage. Instead, the "memory," or memristor cells, will exist as the various components in the overall system, each storing the current elements being processed. In this way, the distributed memory system becomes somewhat of a hybrid system. The system consists of a series of distributed blocks, each of which serves as memory and logic as needed.

9.2.3 Reprogrammability

Consider the threshold gate designs discussed in this text. In all of the designs, the memristors had fixed weights in the threshold gates that remained constant for the lifetime of the design. However, it is possible to vary these weights dynamically [72].

Recall that the weight of a memristor in a threshold gate is determined by its resistance. How to change and program this resistance is already known. By driving a high-magnitude current through the memristor, the resistance can be increased or decreased and its resultant weight can be changed. By varying the weights of each memristor in the threshold gate, the equation of the gate can be changed. As a result, it is possible to reprogram a memristor-based blocks of threshold gates during execution to perform a new logic function.

This is essentially dynamic, on-the-fly reprograming of the logic. Care must be taken to ensure no computations are performed at the same time as the reprograming steps. Note that modern systems are not capable of this type of dynamic programability. FPGAs and other reprogrammable hardware need to be statically reprogramed, which often entails multiple hours of overhead. During reprograming, no useful computation can be executed. In comparison, a memristor threshold–gate block only takes N logic steps to reprogram in which N is the number of memristor weights that need to be changed.

9.2.4 Machine learning and neural networks

One key characteristic of memristors is their ability to remember their past states over time. Memristors are capable of remembering multiple previous states, not just the current and most recent states. In some cases, the current state of a memristor is enough to determine all past states of that memristor and when those states were present. This state information can in turn elucidate the past computations and values in the system.

This property has not been explored in this text. In this text, memristors were assumed to be ideal devices, behaving solely as a "0" or "1" without the use of previous state data. When previous state information of memristors is leveraged, the devices can begin to learn over time and transform their functionality as a result. This is a key property for the design of neural networks with memristors.

The fundamental idea behind machine learning is to mimic the way a human's brain works at a basic level. The human brain is composed of neurons and synapses such that the synapses connect various neurons to represent the interactions and relationships of the neurons. Over time, as a human learns, these synapses increase, strengthen, and change to represent new connections and relationships. Many complex algorithmic challenges would benefit from a computing system that behaves in a similar manner.

Consider a problem, which requires a multitude of inputs collected over time that are each used to eventually converge on a result. Until now, most solutions to these machine-learning problems were implemented in software with complex algorithms. Many times, these algorithms retain all prior inputs and information in their original form and recompute the algorithm entirely each time a new datapoint is introduced. These algorithms are compute-intensive, power-hungry algorithms that often have impractical runtimes and energy requirements. Also, sometimes the algorithms are imperfect equivalents to the systems they represent.

Instead, imagine a system that did not need to recompute using any of its earlier data when new data are added. This eliminates the need for the storage of old data and inputs across execution. In turn, this eliminates the performance and energy overhead to recompute with old data repeatedly.

Now, the system does not perform iterative computations on old data. In fact, the software is no longer responsible for the algorithm at all. Rather, the hardware itself will perform the algorithm by transforming in response to inputs across time. In this way, the state of the hardware at a given time *T* will represent a distinct state determined by all previous inputs and computations.

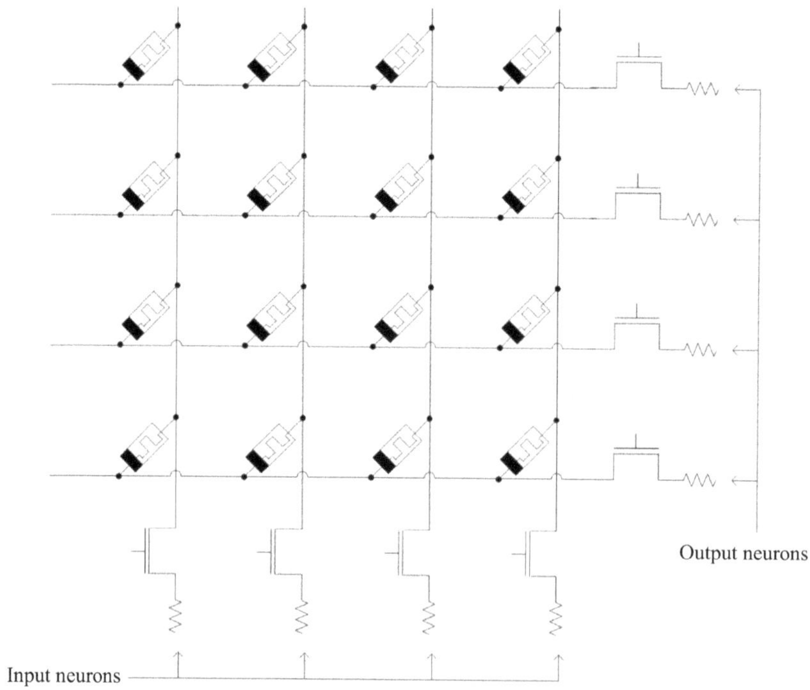

Figure 9.5 Memristors as "synapses" in a crossbar for machine learning applications

Each time a new datapoint or input is introduced, the state of the hardware changes to a new distinct state, possible only through the exact channel of inputs collected thus far. This is the same as how a human's brain has been formed and transformed incrementally over time as new inputs, experiences, and information are gained. Memristors are an ideal candidate for these types of applications since they can act as analog memory devices [73–78]. The memristors function as the "synapses" of the brain in the system, while traditional logic serves as the neurons. In this way, the memristor state, its internal resistance, is learned overtime as inputs, or currents, or supplied to it from the neurons. The memristors can be laid out in a traditional crossbar with inputs and outputs serving as "neurons." A visual representation for this can be seen in Figure 9.5 [78].

Examples of research areas that can leverage learning machines, or neuromorphic chips, are speech recognition, facial recognition, advanced modeling, and robotics.

9.2.5 Chaotic systems

Chaotic systems are based on the mathematical theory of chaos. Chaotic systems are complex dynamic systems, which are apparently random, but depend heavily on their initial conditions. Slight changes in the initial conditions of the system can cause vast differences in its behavior for the remainder of time. Examples of chaotic systems in the computing domain are random number generators, weather models and cryptography.

Since memristors are a nonlinear element, they are a popular candidate for randomness in chaotic systems [79–86]. Rather than a complex circuit to create randomness, this single element can be used. The use of memristors can greatly simplify the overall design. Recently, research has shown the ability to produce a chaotic oscillator using a single memristor, inductor, and capacitor circuit as shown in Figure 9.6 [86].

The circuit in Figure 9.6 can be expanded upon to create more complex chaotic systems that exhibit different characteristics, including periodicity and compounded chaos.

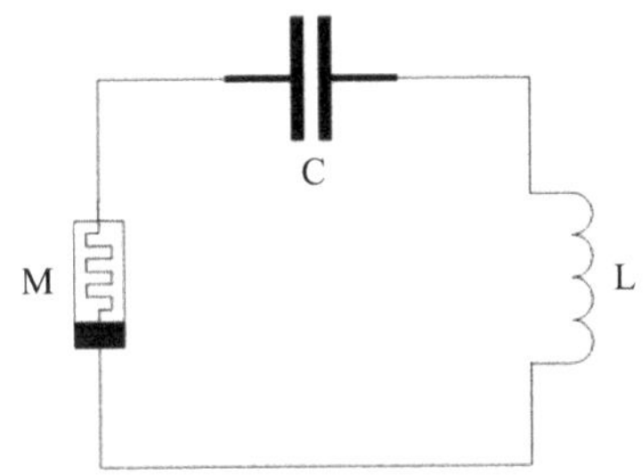

Figure 9.6 Chaotic oscillator using a memristor [86]

9.3 Memristor fabrication

A large majority of the research that has been performed on larger scale memristor designs has been done via simulation. Minimal implementations have been run and measured on real memristors. Research into real-world memristor fabrication mainly focuses on discovering new memristive devices with varying physical properties. Less focus is given to the use of these devices in complex or even simple systems such as a logic block. This slows down the community in fully understanding and accepting memristors as a design component.

Ideally, these two paths of research would grow together, providing feedback across the groups to evolve the state of memristor research as a whole. Without practical implementations of simulation designs, little is understood about the scalability, reliability, or energy implications of larger designs. Without simulations, there is little motivation or insight into the requirements and capabilities of the fabricated memristors being developed for them. As both of these fields gain more traction, the lineage of memristors will become more well defined, and the community as a whole will begin to understand and accept these devices as a viable component in future systems.

9.4 Overcoming memristor shortcomings

Last, it is important to address the concerns and criticisms with respect to memristive devices in the research community. As is natural with any new technology, memristors are in their infancy, and the technology and physical fabrication have not yet been fully perfected. Because of this, the field has not yet reached a consensus on the practicality or applicability of memristors. It is crucial to establish this universal consensus before real changes to computer architectural design can occur. Research can focus on addressing and alleviating the criticisms surrounding memristive devices.

One concern about memristors is their longevity across time. Memristors do not have the endurance of other modern technologies. Even more risky is the fact that memristors can begin to behave differently over time as the device wears. This poses an extra challenge of detecting and correcting devices when they begin to show signs of wear. It is not a simple binary between working and not working.

Second, memristors do not exhibit consistent repeatability across devices [44]. For example, the set voltage and duration for the set operation may not be identical for two memristors. Thus, after a set operation on two memristors, the exact resistance of the two may not be identical. This becomes more of a concern in designs that use memristors for multibit storage or analog logic, because the memristor must be reliable on a fine-grained scale. For applications where the memristors are used as binary values, such as the IMPLY operation or single-bit storage, this is of less concern.

Third, the use of memristors in large, dense systems poses many challenges. As shown briefly for the crossbar examples in this text, sneak paths and other forms of interference become issues when memristors are tightly packed together. This raises reliability and scalability concerns. They also currently require high voltages for programing, degrading the efficiency of their designs. Also, memristors currently exhibit improved performance and characteristics as compared to NAND flash but still fall short in comparison to DRAM. Until these design gaps are eliminated, memristors will not likely replace large memory structures.

Chapter 10

Conclusion

This text aims to provide a thorough and complete introduction to memristive devices and their use in modern arithmetic units. The memristive devices can exhibit a wide range of characteristics and properties depending on their chemical and physical makeup. Often times, the expected application for the device determines many of these characteristics. The flexibility of the design of memristive devices is just one of its strongpoints as a candidate for future system design.

Memristors are nonvolatile devices making them prime candidates for use in memory structures. Specifically, crossbars have been shown as the frontrunner in memristor-based storage. Memristors are capable of exhibiting high density and low power in these contexts. They have competitive read and write times and endurance. Perhaps most importantly, memristors are compatible with traditional CMOS designs as well as 3D architectures.

Besides their benefits in memory structures, this text outlines many of the approaches for using memristors in logic structures. IMPLY, hybrid-CMOS, threshold and MAD Boolean logic are all presented as viable memristor-based logic gates. Each of these approaches has trade-offs in terms of complexity and delay, and all offer benefits over traditional CMOS Boolean logic.

IMPLY and MAD designs stray from traditional Von Neumann architectures to systems that leverage logic-in-memory. Both these approaches use memristors as memory and storage while bringing the computations to the memristors to perform logic. These designs have the opportunity to revolutionize the view of modern systems and the flow of data and information in a system. Both these design approaches can also leverage pipelining in nearly every instance, greatly improving delay and efficiency in systems in a world where efficiency is becoming even more crucial.

The hybrid-CMOS- and threshold-gate approaches to memristor-based logic preserve the fundamentals of Von Neumann architectures. They have similar gate-level circuitry to traditional CMOS designs. This makes them more accessible and more compatible with traditional CMOS designs and their users. They also generally have simpler designs, which require less optimization and transformation.

Memristor-based Boolean logic is extended to numerous adder, multiplier, and divider designs. The arithmetic units span the spectrum of complexity and delay from the simplest and slowest designs to the fastest and most complex. The designs are implemented using each of the four discussed approaches and optimized for the given logic block and approach. The designs are analyzed in terms of delay and complexity and compared against each other, prior work and traditional CMOS designs.

Last, directions for future work and research in the memristor domain are motivated. Memristors have the potential to exist in nearly every part of the architecture ecosystem. From the smallest logic gate or storage element to the most intricate unknowns of future systems, memristors play a vital role in every level of architectural design. Just some of the possible futures areas of exploration include machine learning, chaotic system, analog logic, and simulation and tool support.

With memristors, modern architecture can be reenvisioned from the top level down, applying the insight in this work to all aspects of system design. The key takeaways from this work, non–Von Neumann memristor architectures, the low complexity and delay of memristor-based designs, and the efficiency of pipelined memristor circuitry, can be leveraged to design entirely novel systems that break the limits of traditional CMOS design. Memristors offer a solution to many of the current problems in the field of computer architecture as well as many of the problems that are currently unimagined.

References

[1] Chua L. "Memristor—The Missing Circuit Element," *IEEE Transactions on Circuit Theory*, vol. 18, pp. 507–519, 1971.

[2] Miao F., Strachan J.P., Yang J.J., *et al.* "Anatomy of a Nanoscale Conduction Channel Reveals the Mechanism of a High-Performance Memristor," *Advanced Materials*, vol. 23, pp. 5633–5640, 2011.

[3] Zhang Y., Shen Y., Wang X., Cao L. "A Novel Design for Memristor-Based Logic Switch and Crossbar Circuits," *IEEE Transactions on Circuits and Systems I: Regular Papers*, vol. 62, pp. 1402–1411, May 2015.

[4] Devolder T., Hayakawa J., Ito K., *et al.* "Single-Shot Time-Resolved Measurements of Nanosecond-Scale Spin-Transfer Induced Switching: Stochastic Versus Deterministic Aspects," *Physical Review Letters*, vol. 100, pp. 057206, 2008.

[5] Strukov D., Snider G., Stewart D., Williams R. "The Missing Memristor Found," *Nature*, vol. 453, pp. 80–83, May 2008.

[6] Lehtonen E., Laiho M. "CNN Using Memristors for Neighborhood Connections," *12th International Workshop on Cellular Nanoscale Networks and Their Applications (CNNA)*, pp. 1–4, Berkeley, CA, Feb. 2010.

[7] Kvatinsky S., Friedman E., Kolodny A., Weiser U. "TEAM: ThrEshold Adaptive Memristor Model," *IEEE Transactions on Circuits and Systems I: Regular Papers*, vol. 60, pp. 211–221, Jan. 2013.

[8] Biolek Z., Biolek D., Biolkova V. "SPICE Model of Memristor with Nonlinear Dopant Drift," *Radioengineering*, vol. 18, pp. 210–214, Jun. 2009.

[9] Pickett M., Strukov D., Borghetti J., *et al.* "Switching Dynamics in Titanium Dioxide Memristive Devices," *Journal of Applied Physics*, vol. 106, pp. 1–6, Oct. 2009.

[10] Zhang Y., Shen Y., Wang X., Guo Y. "A Novel Design for a Memristor-Based OR Gate," *IEEE Transactions on Circuits and Systems II: Express Briefs*, vol. 62, pp. 781–785, Aug. 2015.

[11] Zhu X., Yang X., Wu C., Xiao N., Wu J., Yi X. "Performing Stateful Logic on Memristor Memory," *IEEE Transactions on Circuits and Systems II: Express Briefs*, vol. 60, pp. 682–686, Oct. 2013.

[12] Kvatinsky S., Belousov D., Liman S., *et al.* "MAGIC Memristor Aided LoGIC," *IEEE Transactions on Circuits and Systems II: Express Briefs*, vol. 61, pp. 1–5, 2014.

[13] Yang Y., Mathew J., Pontarelli S., Ottavi M., Pradhan D. "Complementary Resistive Switch-Based Arithmetic Logic Implementations Using Material

Implication," *IEEE Transactions on Nanotechnology*, vol. 15, pp. 94–108, Jan. 2016.

[14] Borghetti J., Snider G., Kuekes P., Yang J. "Memristive' Switches Enable 'Stateful' Logic Operations via Material Implication," *Nature*, vol. 464, pp. 873–876, Apr. 2010.

[15] Bickerstaff K., Swartzlander, Jr. E. "Memristor-Based Arithmetic," *44th Asilomar Conference on Signals, Systems and Computers*, pp. 1173–1177, Pacific Grove, CA, Nov. 2010.

[16] Mahajan D., Musaddiq M., Swartzlander, Jr. E. "Memristor Based Adders," *48th Asilomar Conference on Signals, Systems and Computers*, pp. 1256–1260, Pacific Grove, CA, Nov. 2014.

[17] Shaltoot A., Madian A. "Memristor Based Carry Lookahead Adder Architectures," *IEEE 55th International Midwest Symposium on Circuits and Systems (MWSCAS)*, pp. 298–301, Boise, ID, Aug. 2012.

[18] Teimoory M., Amirsoleimani A., Shamsi J., Ahmadi A., Alirezaee S., Ahmadi M. "Optimized Implementation of Memristor-Based Full Adder by Material Implication Logic," *21st IEEE International Conference on Electronics, Circuits and Systems (ICECS)*, pp. 562–565, Marseille, France, Dec. 2014.

[19] Teimoory M., Amirsoleimani A., Ahmadi A., Alirezaee S., Salimpour S., Ahmadi M. "Memristor-Based Linear Feedback Shift Register Based on Material Implication Logic," *Proceedings of 22nd European Conference on Circuit Theory and Design (ECCTD'2015)*, pp. 1–4, Trondheim, Norway, Aug. 2015.

[20] Kvatinsky S., Satat G., Wald N., Friedman E., Kolodny A., Weiser U. "Memristor-Based Material Implication (IMPLY) Logic: Design Principles and Methodologies," *IEEE Transactions on Very Large Scale Integration (VLSI) Systems*, vol. 22, pp. 2054–2066, 2014.

[21] Guckert L., Swartzlander, Jr. E. "Optimized Memristor-Based Ripple Carry Adders," *50th Asilomar Conference on Signals, Systems and Computers*, pp. 1575–1579, Pacific Grove, CA, Nov. 2014.

[22] Kvatinsky S., Wald N., Satat G., Kolodny A., Weiser U., Friedman E. "MRL—Memristor Ratioed Logic," *13th International Workshop on Cellular Nanoscale Networks and Their Applications (CNNA)*, pp. 1–6, Turin, Italy, Aug. 2012.

[23] Singh T. "Hybrid Memristor-CMOS (MeMOS) Based Logic Gates and Adder Circuits," *CoRR*, pp. 1–11, Jun. 2015.

[24] Rose G., Rajendran J., Manem H., Karri R., Pino R. "Leveraging Memristive Systems in the Construction of Digital Logic Circuits," *Proceedings of the IEEE*, vol. 100, pp. 2033–2049, Jun. 2012.

[25] Collier C., Jeppesen J., Luo Y., *et al.* "Molecular-Based Electronically Switchable Tunnel Junction Devices," *Journal of the American Chemical Society*, vol. 123, pp. 632–641, Dec. 2001.

[26] Rose G., Stan, Jr. M. "A Programmable Majority Logic Array Using Molecular Scale Electronics," *IEEE Transactions on Circuits and Systems I: Regular Papers*, vol. 54, pp. 2380–2390, Nov. 2007.

[27] Goto E., Murata K., Nakazawa K., *et al.* "Esaki Diode High-Speed Logical Circuits," *IRE Transactions on Electronic Computers*, vol. EC-9, pp. 25–29, Mar. 1960.

[28] Zhou J., Tang Y., Wu J., Fang X., Zhu X., Huang D. "Design of Counters Based on Memristors," *Proceedings of the 2013 International Conference on Information System and Engineering Management (ICISEM)*, pp. 483–486, Washington, DC, Aug. 2013.

[29] Guckert L., Swartzlander, Jr. E. "MAD Gates—Memristor Logic Design Using Driver Circuitry," *IEEE Transactions on Circuits and Systems II: Express Briefs*, vol. 64, pp. 171–175, Feb. 2017.

[30] Corinto F., Ascoli A. "A Boundary Condition-Based Approach to the Modeling of Memristor Nanostructures," *IEEE Transactions on Circuits and Systems I: Regular Papers*, vol. 59, pp. 2713–2726, Nov. 2012.

[31] Deng Y., Huang P., Chen B., *et al.* "RRAM Crossbar Array with Cell Selection Device: A Device and Circuit Interaction Study," *IEEE Transactions on Electron Devices*, vol. 60, pp. 719–726, 2013.

[32] Rose G. "Overview: Memristive Devices, Circuits and Systems," *Proceedings of 2010 IEEE International Symposium on Circuits and Systems*, pp. 1955–1958, Paris, France, Jun. 2010.

[33] Chua L., Kang S. "Memristive Devices and Systems," *Proceedings of the IEEE*, vol. 64, pp. 209–223, Feb. 1976.

[34] Shin S., Kim K., Kang S. "Reconfigurable Stateful NOR Gate for Large-Scale Logic-Array Integrations," *IEEE Transactions on Circuits and Systems II: Express Briefs*, vol. 58, pp. 442–446, Jul. 2011.

[35] Balatti S., Ambrogio S., Ielmini D. "Normally-off Logic Based on Resistive Switches—Part II: Logic Circuits," *IEEE Transactions on Electron Devices*, vol. 62, pp. 1839–1847, Jun. 2015

[36] Siemon A., Menzel S., Chattopadhyay A., Waser R., Linn E. "In-Memory Adder Functionality in 1S1R Arrays," *Proceedings of 2015 IEEE International Symposium on Circuits and Systems (ISCAS)*, pp. 1338–1341, Lisbon, Portugal, Jul. 2015.

[37] Breuer T., Siemon A., Linn E., Menzel S., Waser R., Rana V. "A HfO_2-Based Complementary Switching Crossbar Adder," *Advanced Electronic Materials*, vol. 1, pp. 1500138, 2015.

[38] Linn E., Rosezin R., Tappertzhofen S., Waser R. "Beyond Von Neumann—Logic Operations in Passive Crossbar Arrays Alongside Memory Operations," *Nanotechnology*, vol. 23, pp. 305205, Jul. 2012.

[39] Zhu Q., Akin B., Sumbul E., *et al.* "A 3D-Stacked Logic-in-Memory Accelerator for Application-Specific Data Intensive Computing," *2013 IEEE International 3D Systems Integration Conference*, pp. 1–7, San Francisco, CA, Oct. 2013.

[40] Pala D., Causapruno G., Vacca M., *et al.* "Logic-In-Memory Architecture Made Real," *2015 IEEE International Symposium on Circuits and Systems (ISCAS)*, pp. 1542–1545, Lisbon, Portugal, May 2015.

[41] Zhou F., Guckert L., Chang Y.F., Swartzlander E., Lee J. "Bidirectional Voltage Biased Implication Operations Using SiO_x Based Unipolar Memristors," *Applied Physics Letters*, vol. 107, pp. 183501, Nov. 2015.

[42] Gale E. "TiO_2-Based Memristors and ReRAM: Materials, Mechanisms and Models (A Review)," *Semiconductor Science and Technology*, vol. 29, pp. 104004, Sep. 2014.

[43] Yang J., Strukov D., Stewart D. "Memristive Devices for Computing," *Nature Nanotechnology*, vol. 8, pp. 13–24, Dec. 2012.

[44] Campbell K. *The Trouble with Oxide-Based Memristors* [online]. 2015. Available from http://knowm.org/the-trouble-with-oxide-based-memristors [Accessed 11 Sep. 2017].

[45] Wang X., Chen Y., Xi H., Dimitrov D. "Spintronic Memristor through Spin-Torque-Induced Magnetization Motion," *IEEE Electron Device Letters*, vol. 30, pp. 294–297, Mar. 2009.

[46] http://eeweb.poly.edu/hli/index_files/memristors.ht [Accessed 4 Aug. 2017].

[47] Yang J., Pickett M., Li X., Ohlberg D., Stewart D., Williams R. "Memristive Switching Mechanism for Metal/Oxide/Metal Nanodevices," *Nature Nanotechnology*, vol. 3, pp. 429–433, Jul. 2008.

[48] Strukov D., Williams R. "Exponential Ionic Drift: Fast Switching and Low Volatility of Thin-Film Memristors," *Applied Physics A, Material Science and Processing*, vol. 94, pp. 515–519, Mar. 2009.

[49] Simmons J., "Generalized Formula for the Electric Tunnel Effect between Similar Electrodes Separated by a Thin Insulating Film," *Journal of Applied Physics*, vol. 34, pp. 1793–1803, Jan. 1963.

[50] Chang Y., Zhou F., Fowler B., *et al.* "Memcomputing (Memristor + Computing) in Intrinsic SiO_x-Based Resistive Switching Memory: Arithmetic Operations for Logic Applications," *IEEE Transactions on Electron Devices*, vol. 64, pp. 2977–2983, 2017.

[51] Babbage C., Campbell-Kelly M. (eds.). *Passages from the Life of a Philosopher*. New Brunswick, NJ: Rutgers University Press; 1864. pp. 59–63, 114–116.

[52] Ercegovac M., Lang T. "On-the-Fly Conversion of Redundant into Conventional Representations," *IEEE Transactions on Computing*, vol. 36, pp. 895–897, 1987.

[53] Guckert L., Swartzlander, Jr. E. "Implementing Adders with Memristors," *ETCMOS Conference*, Montreal, Canada, May 2016.

[54] Guckert L., Swartzlander, Jr. E. "Memristor-Based Threshold Gate Adders," *DAC Conference*, Austin, TX, Jun. 2016.

[55] Huisman T. *Counters and Multipliers with Threshold Logic*, M. S. Thesis, Delft University of Technology, May 1995.

[56] Kautz W. "The Realization of Symmetric Switching Functions with Linear Input Logical Elements," *IRE Transactions on Electronic Computers*, vol. EC-10, pp. 371–378, Sep. 1961.

[57] Minnick R. "Linear Input Logic," *IRE Transactions on Electronic Computers*, vol. EC-10, pp. 6–16, Mar. 1961.

[58] Guckert L., Swartzlander, Jr. E. "Carry Lookahead Adder Design Using Memristors," Chapter in *Devices, Circuits and Systems*. Boca Raton, FL: CRC Press; Oct. 2016.

[59] Celinski P., López J., Al-Sarawi S., Abbott D. "Low Depth, Low Power Carry Lookahead Adders Using Threshold Logic," *Microelectronics Journal*, vol. 33, pp. 1071–1077, Dec. 2002.

[60] Guckert L., Swartzlander, Jr. E. "Dadda Multiplier Designs Using Memristors," *ICICDT Conference*, Austin, TX, May 2017.

[61] Guckert L., Swartzlander, Jr. E. "Optimized Memristor-Based Multipliers," *IEEE Transactions on Circuits and Systems I: Regular Papers*, vol. 64, pp. 373–385, Feb. 2017.

[62] Luigi Dadda. "Some Schemes for Parallel Multipliers," *Alta Frequenza*, vol. 34, pp. 349–356, Aug. 1965.

[63] Taherinejad N., Manoj S., Jantsch A. "Memristors' Potential for Multi-Bit Storage and Pattern Learning," *2015 IEEE European Modelling Symposium (EMS)*, pp. 450–455, Madrid, Spain, 2015.

[64] Wang X., Chen Y. "Spintronic Memristor Devices and Application," *2010 Design, Automation & Test in Europe Conference & Exhibition (DATE)*, pp. 667–672, Dresden, Germany, Mar. 2010.

[65] Taherinejad N., Manoj S., Rathmair M., Jantsch A. "Fully Digital Write-in Scheme for Multi-Bit Memristive Storage," *2016 13th International Conference on Electrical Engineering, Computing Science and Automatic Control (CCE)*, pp. 1–6, Mexico City, Mexico, Sep. 2016.

[66] Fey D. "Using the Multi-Bit Feature of Memristors for Register Files in Signed-Digit Arithmetic Units," *Semiconductor Science and Technology*, vol. 29, pp. 104008, Sep. 2014.

[67] Kvatinsky S., Nacson Y., Etsion Y., *et al.* "Memristive Multistate Pipeline Register," *2014 14th International Workshop on Cellular Nanoscale Networks and their Applications (CNNA)*, pp. 1–2, Notre Dame, IN, Jul. 2014.

[68] Laiho M., Lehtonen E. "Arithmetic Operations within Memristor-Based Analog Memory," *2010 12th International Workshop on Cellular Nanoscale Networks and their Applications (CNNA 2010)*, pp. 1–4, Berkeley, CA, 2010.

[69] Shin S., Kim K., Kang S. "Memristor-based Fine Resolution Programmable Resistance and its Applications," *International Conference on Communications, Circuits and Systems (ICCCAS)*, pp. 948–951, *Milpitas, CA*, Jul. 2009.

[70] Shin S., Kim K., Kang S. "Memristor Applications for Programmable Analog ICs," *IEEE Transactions on Nanotechnology*, vol. 10, pp. 266–274, Jan. 2010.

[71] Merrikh-Bayat F., Bagheri Shouraki S. "Memristor-based Circuits for Performing Basic Arithmetic Operations," *Procedia Computer Science*, vol. 3, 2010.

[72] Gao L., Alibart F., Strukov D. "Programmable CMOS/Memristor Threshold Logic," *IEEE Transactions on Nanotechnology*, vol. 12, pp. 115–119, Mar. 2013.

[73] Jo S., Chang T., Ebong I., Bhadviya B., Mazumder P., Lu W. "Nanoscale Memristor Device as Synapse in Neuromorphic Systems," *Nano Letters*, vol. 10, pp. 1297–1301, Mar. 2010.

[74] Pershin Y., Di Ventra M. "Experimental Demonstration of Associative Memory with Memristive Neural Networks," *Neural Networks*, vol. 23, pp. 881–886, Sep. 2010.

[75] Pérez-Carrasco J., Zamarreño-Ramos C., Serrano-Gotarredona T., Linares-Barranco B. "On Neuromorphic Spiking Architectures for Asynchronous STDP Memristive Systems," *Proceedings of 2010 IEEE International Symposium on Circuits and Systems (ISCAS)*, pp. 1659–1662, Paris, France, May 2010.

[76] Snider G. "Spike-Timing-Dependent Learning in Memristive Nanodevices," *IEEE International Symposium on Nanoscale Architectures (NANOARCH)*, pp. 85–92, Anaheim, CA, 2008.

[77] Zomaya A. *Handbook of Nature-Inspired and Innovative Computing: Integrating Classical Models with Emerging Technologies*. New York, NY: Springer; 2006.

[78] Saighi S., Mayr C., Serrano-Gotarredona T., *et al.* "Plasticity in Memristive Devices for Spiking Neural Networks," *Frontiers in Neuroscience*, vol. 9, pp. 51, 2015.

[79] Itoh M., Chua L. "Memristor Oscillators," *International Journal of Bifurcation and Chaos*, vol. 18, pp. 3183–3206, Nov. 2008.

[80] Muthuswamy B. "Implementing Memristor Based Chaotic Circuits," *International Journal of Bifurcation and Chaos*, vol. 20, pp. 1335–1350, May 2010.

[81] Muthuswamy B., Przybylski A., Feilbach C., Mossbrucker J. "Two Element Chaotic and Hyperchaotic Circuits," *Proceedings of the 4th International Interdisciplinary Chaos Symposium*, pp. 1–10, Antalya, Turkey, 2012.

[82] Bo-Cheng B., Zhong L., Jian-Ping X. "Transient Chaos in Smooth Memristor Oscillator," *Chinese Physics B*, vol. 19, pp. 030510, Mar. 2010.

[83] Bao B., Liu Z., Xu J. "Steady Periodic Memristor Oscillator with Transient Chaotic Behaviors," *Electronic Letters*, vol. 46, pp. 237–238, Feb. 2010.

[84] Kyprianidis I., Volos C., Stouboulos I. "Experimental Synchronization of Two Unidirectionally Coupled Double Scroll Circuit," *7th International Conference of the Balkan Physical Union*, vol. 1203, pp. 403–408, Jan. 2010.

[85] Buscarino A., Fortuna L., Frasca M., Gambuzza L. "Memristive Chaotic Circuits Based on Cellular Nonlinear Networks," *International Journal of Bifurcation and Chaos*, vol. 22, pp. 023136, Mar. 2012.

[86] Muthuswamy B., Chua L. "Simplest Chaotic Circuit," *International Journal of Bifurcation and Chaos*, vol. 20, pp. 1567–1580, May 2010.

Index

www.ingramcontent.com/pod-product-compliance
Lightning Source LLC
LaVergne TN
LVHW070827220926
848486LV00027B/40